THE FLAVONOIDS

Extraction and Applications

AAP Advances in Nutraceuticals

THE FLAVONOIDS

Extraction and Applications

Edited by

Deepika Saini, PhD
Rajesh K. Kesharwani, PhD
Raj K. Keservani, PhD, MPharm

APPLE ACADEMIC PRESS

First edition published 2024

Apple Academic Press Inc.
1265 Goldenrod Circle, NE,
Palm Bay, FL 32905 USA

760 Laurentian Drive, Unit 19,
Burlington, ON L7N 0A4, CANADA

CRC Press
2385 NW Executive Center Drive,
Suite 320, Boca Raton FL 33431

4 Park Square, Milton Park,
Abingdon, Oxon, OX14 4RN UK

Library and Archives Canada Cataloguing in Publication

Title: The flavonoids : extraction and applications / edited by Deepika Saini, PhD, Rajesh K. Kesharwani, PhD, Raj K. Keservani, PhD, MPharm.
Other titles: Flavonoids (Apple Academic Press)
Names: Saini, Deepika, editor. | Kesharwani, Rajesh Kumar, 1978- editor. | Keservani, Raj K., 1981-editor.
Series: AAP advances in nutraceuticals.
Description: First edition. | Series statement: AAP advances in nutraceuticals | Includes bibliographical references and index.
Identifiers: Canadiana (print) 20230513816 | Canadiana (ebook) 20230513867 | ISBN 9781774913772 (hardcover) | ISBN 9781774913789 (softcover) | ISBN 9781003399964 (ebook)
Subjects: LCSH: Flavonoids. | LCSH: Flavonoids—Health aspects. | LCSH: Flavonoids—Therapeutic use.
Classification: LCC QK898.F5 F53 2024 | DDC 615.3/21—dc23

Library of Congress Cataloging-in-Publication Data

CIP data on file with US Library of Congress

ISBN: 978-1-77491-377-2 (hbk)
ISBN: 978-1-77491-378-9 (pbk)
ISBN: 978-1-00339-996-4 (ebk)

ABOUT THE BOOK SERIES: AAP ADVANCES IN NUTRACEUTICALS

SERIES EDITORS:

Raj K. Keservani, PhD, MPharm
Associate Professor, Faculty of B. Pharmacy, CSM Group of Institutions, Prayagraj, Uttar Pradesh, India

Rajesh K. Kesharwani, PhD
Associate Professor, Department of Computer Application, Nehru Gram Bharati (Deemed to be University), Prayagraj, India

Anil K. Sharma, PhD, MPharm
Assistant Profesoor, Department of Pharmacy, School of Medical and Allied Sciences, GD Goenka University, Gurugram, India

In the modern era, mankind has witnessed a paradigm shift with respect to fundamental eating behavior. The lack of physical workouts and busy schedules at offices and in households have promoted the consumption of junk foods, which eventually results in numerous diseases and disorders since the nutritional content from fast food is inadequate. Plumpness and obesity have become a global health threat. The leading causes of death from most developing nations are noncommunicable diseases such as cardiovascular diseases, cancer, arthritis, osteoporosis, and liver toxicity. Patients suffering from such lifestyle ailments are a bit apprehensive towards prolonged use of costly modern therapeutics, which encourages instead the use of alternative approaches for management of such diseases and disorders.

The emerging sector of the nutraceutical industry encompasses products derived from nature, dietary supplements, and functional foods. Nutraceuticals are used for treatment and prevention of a broad range of diseases, such as the common cold, arthritis, sleep-related disorders, cancers, cardiovascular complications, metabolic disorders, and others.

The research on nutraceuticals is increasing day by day, considering the beneficial effects of food or food supplements in the management of diverse diseases. The issue of paramount concern is standardization and establishment of clinical efficacy of nutraceuticals, which is in fact a challenge for researchers around the globe.

This book series aims at realizing the significance of the variety of nutraceuticals for human well-being. The books in the series will also emphasize the role of dieticians and nutritionists for the prescription of judicious eating habits. The key food components such as carbohydrates, proteins, and lipids as well as micronutrients (vitamins, minerals) are demonstrated to maintain good health, obviating the need for medicines. Thus, nutraceuticals that are indeed derived from food ingredients are believed to be potential alternative therapeutics.

Asymptomatic diseases/disorders necessitate proper diagnosis. Stressed circumstances have been reported to cause weight loss. Workplace stress is a key variable in the genesis and propagation of metabolic disorders. The nuances of the underlying mechanisms are still to be deciphered. Poor lifestyle trends have been attributed along with the consumption of junk foods in several such instances. Sugar-rich carbonated beverages, the use of contraband drugs, and the consumption of liquor in excess are assigned as risk factors. This book series will cover advances and applications in addition to providing the basics of nutraceuticals.

The books in this series are valuable resources for industry professionals to design and develop quality products for end users. In addition, the book series will be fortified by ideas of innovation, concept building, manufacturing aspects, quality control, and regulatory status of nutraceuticals.

Ostensibly, business has seen a rise in the nutraceutical consumption in recent years. Pure, safe, efficacious products are the need of the hour globally. There are a number of books on nutraceuticals in the market, yet these are only a few books presenting certain aspects of nutraceuticals therein. This book series offers a holistic view of this promising strategy, which is picking up pace with time. This book series is a unique endeavor to bring manufacturing, research and development, and marketing strategies under a single umbrella.

CURRENT BOOKS IN THE SERIES

- **Micronutrients and Macronutrients as Nutraceuticals**
  Editors: Prakash Chandra Gupta, PhD,
  Sayan Bhattacharyya, MBBS, MD, Nisha Sharma, PhD, MPharm,
  Rajesh K. Kesharwani, PhD, and Raj K. Keservani, PhD, MPharm
- **Formulations, Regulations, and Challenges of Nutraceuticals**
  Editors: Tingirikari Jagan Mohan Rao, PhD, Rajesh K. Kesharwani, PhD,
  Raj. K. Keservani, PhD, MPharm, and Anil K. Sharma, PhD

- **Food Supplements and Dietary Fibers in Health and Disease**
  Editors: Bhushan R. Rane, PhD, MPharm,
  Raj K. Keservani, M Pharm, Durgesh Singh, DPhil,
  Nayan A. Gujarathi, PhD, and Ashish S. Jain, PhD
- **Nutraceuticals in Cancer Prevention, Management, and Treatment**
  Editors: Raj K. Keservani, MPharm, Bui Thanh Tung, PhD,
  Sippy Singh, DPhil, and Rajesh K. Kesharwani, PhD
- **Advances in Flavonoids for Human Health and Prevention of Diseases**
  Editors: Nisha Sharma, PhD, Deepika Saini, PhD,
  Rajesh K. Kesharwani, PhD, Prakash Chandra Gupta, and
  Raj. K. Keservani, PhD, MPharm
- **The Flavonoids: Extraction and Applications**
  Editors: Deepika Saini, PhD, Rajesh K. Kesharwani, PhD,
  and Raj K. Keservani, PhD, MPharm
- **Immune-Boosting Nutraceuticals for Better Human Health: Novel Applications**
  Editors: Urmila Jarouliya, PhD, Raj K. Keservani, MPharm,
  Rajesh K. Kesharwani, PhD, Virendra K. Patel, PhD, and Adi D. Bharti
- **Applications of Functional Foods in Disease Prevention**
  Editors: Raj K. Keservani, MPharm, and Eknath D. Ahire, MSPharm
- **Plant Metabolites and Vegetables as Nutraceuticals**
  Editors: Raj K. Keservani, MPharm, Bui Thanh Tung, PhD,
  Rajesh K. Kesharwani, PhD, MPharm, and Eknath D. Ahire
- **Preventive and Therapeutic Role of Vitamins as Nutraceuticals**
  Editors: Khemchand R. Surana, Eknath D. Ahire,
  Raj K. Keservani, PhD, MPharm, and Rajesh K. Kesharwani, PhD
- **Nutraceuticals and Bone Health**
  Editors: Deepak Sharma, PhD, Madan Mohan Gupta, PhD,
  Anil K. Sharma, PhD, MPharm, Raj K. Keservani, PhD, MPharm,
  and Rajesh K. Kesharwani, PhD
- **Flavonoids as Nutraceuticals**
  Editors: Rajesh K. Kesharwani, PhD, MTech, Deepika Saini, PhD,
  Raj K. Keservani, PhD, MPharm, and Anil Kumar Sharma, PhD, MPharm

- **Herbals as Nutraceuticals: Their Role in Healthcare**
  Editors: Raj K. Keservani, MPharm,
  Rajeshwar Kamal Kant Arya, PhD, MPharm, and
  Rajesh K. Kesharwani, PhD, MTech
- **Nutraceutical Fruits: Overview and Disease Prevention**
  Editor: Raj K. Keservani, PhD, MPharm
- **Nutrigenomics and Nutraceuticals**
  Editors: Raj K. Keservani, PhD, MPharm, Eknath D. Ahire,
  Shubham J. Khainar, PhD, Sanjay J. Kshirsagar, PhD, and
  Rajesh K. Kesharwani, PhD
- **The Nature of Nutraceuticals: History, Properties, Sources, and Nanotechnology**
  Editors: Rajesh K. Kesharwani, PhD, Prashant Kumar, PhD, and
  Raj K. Keservani, PhD, MPharm
- **Nutraceuticals for the Treatment and Prevention of Sexual Disorders**
  Editors: Raj K. Keservani, PhD, Sharangouda J. Patil, PhD, and
  Ivan Aranha, PhD
- **Antioxidants as Nutraceuticals**
  Editors: Nayan A. Gujaratih, PhD, Raj K. Keservani, PhD
  Rajesh K. Kesharwani, PhD, Bhushan R. Rane, PhD, and
  Yogeeta Sameer Goyal, PhD
- **Nutraceuticals in Cardiac Health Management**
  Editors: Raj K. Keservani, PhD, Rajesh K. Kesharwani, PhD,
  Praveen Kumar Jain, PhD, Bhushan R. Rane, PhD, and
  Nayan A. Gujarathi, PhD
- **Nutraceuticals in Obesity Management and Control**
  Editors: Raj K. Keservani, PhD, and Alka Lohani, PhD
- **Nutraceuticals in Arthritis and Psoriasis**
  Editors: Meenakshi Jaiswal, PhD, Raj K. Keservani, PhD,
  Rajesh K. Kesharwani, PhD, and Swati G. Talele, PhD

ABOUT THE EDITORS

Deepika Saini, PhD
Assistant Professor, Department of Zoology, Chamanlal Mahavidyalaya, Haridwar, Uttarakhand, India

Deepika Saini, PhD, is currently working as an Assistant Professor in the Department of Zoology, Chamanlal Mahavidyalaya, Haridwar, Uttarakhand. She has a teaching experience of about five years. She has also worked as an adjunct faculty member in the Natural Science Department of the University of Maryland Eastern Shore, Salisbury (MD), USA. She has published in about 14 national and international publications. She has presented more than 14 papers in national and international seminars/conferences and has also attended many workshops all around India. She has also been awarded the Best Teacher Award 2020 by the Society of Research in Biological Studies. Dr. Saini is also providing her services as a reviewer and editorial board member to more than eight international and national journals including *The American Journal of Life Sciences, The International Journal of Zoological Studies, The International Journal of Biological Studies,* etc. She has successfully edited four books. During the pandemic year 2020, she organized five webinars with reputed institutes like the Zoological Survey of India and the Botanical Survey of India. She has organized two national conferences funded by the Uttarakhand Council of Science and Technology, Dehradun, and the Uttarakhand Council for Biotechnology, Dehradun.

Rajesh K. Kesharwani, PhD
Associate Professor, Department of Computer Application, Nehru Gram Bharati (deemed to be university), Prayagraj, Uttar Pradesh, India

Rajesh K. Kesharwani, PhD, MTech, is currently working as an Associate Professor in the Department of Computer Application at Nehru Gram Bharati (deemed to be university), Prayagraj, India. He has more than 11 years of research and 9 years of teaching experience in various institutes in India, imparting bioinformatics and biotechnology education. He has received several awards, including the NASI-Swarna Jayanti Puruskar by The National Academy of Sciences of India. He has supervised one PhD, and more than 20 UG and PG students for their research work, has authored over 49 peer-reviewed articles, 20 book chapters, and 14 edited books with international publishers (e.g., Springer Nature, IGI Global, Taylor & Francis, etc.). He has been a member of many scientific communities as well as a reviewer for many international journals. He has presented many papers at various national and international conferences. Dr. Kesharwani received his PhD from the Indian Institute of Information Technology, Allahabad, and worked at NIT Warangal for two semesters. He has been a recipient of the Ministry of Human Resource Development (India) Fellowship and a Senior Research Fellowship from the Indian Council of Medical Research, India. His research fields of interest are medical informatics, protein structure and function prediction, computer-aided drug designing, structural biology, drug delivery, cancer biology, nano-biotechnology, and biomedical sciences.

Raj K. Keservani, PhD, MPharm
Associate Professor, Faculty of B. Pharmacy, CSM Group of Institutions, Allahabad, Uttar Pradesh, India

Raj K. Keservani, PhD, MPharm, is working as an Associate Professor in the Faculty of B. Pharmacy at CSM Group of Institutions, Allahabad, India. He has more than 12 years of academic (teaching) experience in pharmaceutical education from various Indian institutes. He has supervised more than 25 UG students for their research work, has published 31 peer-reviewed papers in the field of pharmaceutical sciences in national and international journals, one patent, 43 book chapters, three co-authored books, and 19 edited books. He is also active as a reviewer for several international scientific journals. Dr. Keservani graduated with a pharmacy degree from the Department of Pharmacy, Kumaun University, Nainital (Uttarakhand), India. He received his Master of Pharmacy (MPharm) (specialization in pharmaceutics) from the School of Pharmaceutical Sciences, Rajiv Gandhi Proudyogiki Vishwavidyalaya, Bhopal, India, and recently was awarded his PhD. Dr. Keservani is a life member of the Society of Pharmaceutical Education and Research (SPER). His research interests include nutraceutical and functional foods, novel drug delivery systems (NDDS), transdermal drug delivery/drug delivery, health science, cancer biology, and neurobiology.

CONTENTS

CONTRIBUTORS

Chithra A.
Bioprospecting Research Laboratory, Department of Biotechnology, Bannari Amman Institute of Technology, Sathyamangalam, Erode, India

Eknath D. Ahire
Department of Pharmaceutics, MET's Institute of Pharmacy, Affiliated to Savitribai Phule Pune University, BKC, Adgaon, Nashik, Maharashtra, India

Sana Ansari
Department of Biotechnology, Invertis University, Bareilly, Uttar Pradesh, India

Tahira Ansari
Department of Zoology, DDU Gorakhpur University, Gorakhpur, Uttar Pradesh, India

Chinmoyee Bhattacharyya
Department of Life Sciences, Dibrugarh University, Dibrugarh, Assam, India

Pankaj Bhatt
Department of Pharmaceutical Sciences, Gurukul Kangri (Deemed to be University), Haridwar, Uttarakhand, India
Department of Pharmaceutics, KIET Groups of Institutions (KIET School of Pharmacy), Muradnagar, Ghaziabad, Uttar Pradesh, India

Chandra Biswas
Laboratory of Cancer Biology, National Centre for Cell Sciences, Pune, Maharashtra, India

Nikalje G. C.
Seva Sadan's R. K. Talreja College of Arts, Science and Commerce, Affiliated to University of Mumbai, Ulhasnagar, India

Anupriya Chatterjee
Health Sciences, University of the People, USA

Nilesh Gorde
Konkan Gyanpeeth Rahul Dharkar College of Pharmacy & Research Institute, Dahivali, Parade Vengaon Road, Maharashtra, India

Mhatre K. J.
Seva Sadan's R. K. Talreja College of Arts, Science and Commerce, Affiliated to University of Mumbai, Ulhasnagar, India

Khanderao R. Jadhav
Department of Pharmaceutics, RGS College of Pharmacy, Anjaneri, Nashik, India

Soniya Joshi
Department of Agriculture, Invertis University, Bareilly, Uttar Pradesh, India

Darshana K.
Bioprospecting Research Laboratory, Department of Biotechnology, Bannari Amman Institute of Technology, Sathyamangalam, Erode, India

Moni Philip Jacob K.
Department of Biotechnology, Bannari Amman Institute of Technology, Sathyamangalam, Erode, India

Mohan Kale
Konkan Gyanpeeth Rahul Dharkar College of Pharmacy & Research Institute, Dahivali, Maharashtra, India

Kapinder
Department of Zoology, DDU Gorakhpur University, Gorakhpur, Uttar Pradesh, India

Ajay Kharche
Konkan Gyanpeeth Rahul Dharkar College of Pharmacy & Research Institute, Dahivali, Parade Vengaon Road, Maharashtra, India

Shubham J. Khairnar
Department of Pharmaceutics, MET's Institute of Pharmacy, Affiliated to Savitribai Phule Pune University, BKC, Adgaon, Nashik, Maharashtra, India

Sanjay J. Kshirsagar
Department of Pharmaceutics, MET's Institute of Pharmacy, Affiliated to Savitribai Phule Pune University, BKC, Adgaon, Nashik, Maharashtra, India

Anand Kumar
Department of Chemistry, SGRR (PG) College, Dehradun, Uttarakhand, India

Nitesh Kumar
Amity Institute of Advanced Research and Studies (Materials and Devices) Amity University, Noida, India

Tarun Kumar
Department of Pharmaceutical Sciences, Gurukul Kangri (Deemed to be University), Haridwar, Uttarakhand, India

Vipin Kumar
Department of Pharmaceutical Sciences, Gurukul Kangri (Deemed to be University), Haridwar, Uttarakhand, India
Department of Pharmaceutics, KIET Groups of Institutions (KIET School of Pharmacy), Muradnagar, Ghaziabad, Uttar Pradesh, India

Sunil K. Mahajan
Department of Pharmaceutical chemistry, MGV's Pharmacy College, Panchvati, Nashik, India

Mayank Kumar Malik
Department of Pharmaceutical Sciences, Gurukul Kangri (Deemed to be University), Haridwar, Uttarakhand, India

Shrawan Kumar Mangawa
Department of Chemistry, Motilal Nehru College, University of Delhi, South Campus, Delhi, India

Vineet K. Maurya
Department of Botany and Microbiology, H.N.B. Garhwal University, Srinagar Garhwal, India

Akansha Mehra
Amity Institute of Advanced Research and Studies (Materials and Devices) Amity University, Noida, India

Naima Mirza
Department of Biotechnology, Invertis University, Bareilly, Uttar Pradesh, India

Abhirup Mitra
Department of Agriculture, Invertis University, Bareilly, Uttar Pradesh, India

Manisha Arora Pandit
Department of Zoology, Kalindi College, University of Delhi, Delhi, India

Dhananjay M. Patil
Department of Pharmaceutics, Shreeshakti Shaikshanik Sanstha, Divine College of Pharmacy, Satana, Nashik, India

Dharanyshri R.
Bioprospecting Research Laboratory, Department of Biotechnology, Bannari Amman Institute of Technology, Sathyamangalam, Erode, India

Girishree R.
Bioprospecting Research Laboratory, Department of Biotechnology, Bannari Amman Institute of Technology, Sathyamangalam, Erode, India

Patil J. R.
Seva Sadan's R. K. Talreja College of Arts, Science and Commerce, Affiliated to University of Mumbai, Ulhasnagar, India

Pankaj Kumar Rai
Department of Biotechnology, Invertis University, Bareilly, Uttar Pradesh, India

Deepika Saini
Department of Zoology, Chamanlal Mahavidyalaya, Haridwar, India

Rohit Saluja
Department of Biochemistry, AIMS Bibinagar Hyderabad, India

Babita Saroha
Department of Biotechnology, UIET, MDU, Rohtak, Haryana, India

Richa Saxena
Department of Biotechnology, Invertis University, Bareilly, Uttar Pradesh, India

Rajaseetharama Sekar
Bioprospecting Research Laboratory, Department of Biotechnology, Bannari Amman Institute of Technology, Sathyamangalam, Erode, India

Richa Sharma
Department of Microbiology, Mahatma Gandhi University of Medical Science & Technology, Jaipur, India

Shailja Singh
Department of Chemistry, Bhaskaracharya College of Applied Sciences, University of Delhi, New Delhi, India

Nishtha Srivastava
Department of Applied Science, Invertis University, Bareilly, Uttar Pradesh, India

Khemchand R. Surana
Department of Pharmaceutical Chemistry, Divine College of Pharmacy, Nampur Road, Satana, Nashik, Maharashtra, India

Tarkeshwar
Department of Zoology, Kalindi College, University of Delhi, Delhi, India

Tanwi Trushna
ICMR-National Institute for Research in Environmental Health (NIREH), India

Sandeep O. Waghulde
Konkan Gyanpeeth Rahul Dharkar College of Pharmacy & Research Institute, Dahivali, Parade Vengaon Road, Maharashtra, India

ABBREVIATIONS

AD	Alzheimer's disease
ADHD	attention deficit hyperactivity disorder
AEDs	antiepileptic drugs
ALS	amyotrophic lateral sclerosis
ANS	anthocyanin synthase
APP	amyloid precursor protein
ARE	antioxidant response element
ASD	autism spectrum disorders
Aβ	amyloid beta
BMI	body mass index
CAT	catalase
CDKs	cyclin-dependent kinases
cGMP	cyclic guanosine monophosphate
CHS	chalcone synthase
CHI	chalcone flavanone isomerase
CI	confidence interval
CNP	cyclic nucleotide phosphodiesterase
CNS	central nervous system
CO	carbon monoxide
COX	cyclooxygenase
CPDs	cyclobutane pyrimidine dimers
CRP	C-reactive protein
CVD	cardiovascular disease
DAMPs	damage-associated molecular patterns
DCs	dendritic cells
DD	depressive disorders
DFR	dihydroflavonol 4-reductase
DVT	deep vein thrombosis
EAE	enzyme-assisted extraction
EAUMSE	enzyme-assisted ultrasonic microwave
EC	epicatechin
ECG	epicatechin-3-gallate
EDHF	endothelium-derived hyperpolarizing factor
EGC	epigallocatechin

EGCG	epigallocatechin-3-gallate
eNOS	endothelial nitric oxide synthase
ER	endoplasmic reticular
ER	estrogen receptors
EV-A71	Enterovirus A71
FLS	flavonol synthase
FLW	food loss and waste
GC	guanylyl cyclase
GLA	galacturonic acid
GP	glycoprotein
GRAS	generally recognized as safe
HCA	hydroxycitric acid
HD	Huntington's disease
HES	hesperidin
HHP	high hydrostatic pressure
HHPE	high hydrostatic pressure extraction
HIV-1	human immunodeficiency virus
HO-1	heme oxygenase-1
HSV-1	herpes simplex virus type 1
ID	intellectual disability
IDF	International Diabetes Federation
IFR	isoflavone reductase
IFS	isoflavonoid synthase
IHD	ischemic heart disease
iNos	inducible nitric oxide synthase enzymes
I2'H	isoflavone 2′ hydroxylase
LDH	lactase dehydrogenase
LIF	leukemia inhibitory factor
LPDM	lipoprotein-deficient medium
LPH	lactase phloridzin hydrolase
LOX	lipoxygenase
LPS	lipopolysaccharide
LTs	leukotrienes
MAE	mechanochemical-assisted extraction
MAE	microwave-assisted extraction
MAPK	mitogen-activated protein kinase
MDA	malondialdehyde
MG	methylglyoxal
MIC	minimum repressive concentration

MMP-1	matrix metalloproteinase-1
MRI	magnetic resonance imaging
MS	multiple sclerosis
MSPD	matrix solid-phase dispersion
MTP	microsomal triacylglyceride transfer protein
MUAE	microwave and ultrasound-assisted extraction
NBC	niacin, bound chromium
NGF	nerve growth factor
NHEKs	normal human epidermal keratinocytes
NO	nitric oxide
NOS3	nitric oxide synthase 3
NOX	NADH oxidase
NPCE	negative pressure cavitation extraction
OH	own hydroxyl
PAD	peripheral arterial disease
PAH	polycyclic aromatic hydrocarbon
PD	Parkinson's disease
PDEs	phosphodiesterase
PEF	pulsed electric field
Pf-3	parainfluenza virus type 3
PG	prostaglandin
PHWE	pressurized hot water extraction
PKA	protein kinase A
PKC	protein kinase C
PKG	protein kinase G
PLA2	phospholipase A2
PLE	pressurized liquid extraction
PM	particulate matter
PNS	peripheral nervous system
PPARS	peroxisome proliferator-activated receptors
PPET1	preproendothelin-1
RA	rheumatoid arthritis
RAAS	renin–angiotensin–aldosterone system
RCT	reverse cholesterol transfer
RE	reseda extract
RHD	rheumatic heart disease
RNS	reactive nitrogen species
ROS	reactive oxygen species
RSV	respiratory syncytial virus

SAR	structure–activity relationship
SCZ	schizophrenia
SDGs	sustainable development goals
SEM	scanning electron microscopy
SF	supercritical fluid
SFE	supercritical fluid extraction
STAT-6	signal transducer and activator of transcription 6
SVR	systemic vascular resistance
Th2	T helper 2
TFC	total flavonoid content
TIA	transient ischemic attack
TIMP-1	tissue inhibitor matrix metalloproteinase-1
TLR	toll-like receptors
TNFR	tumor necrosis factor receptor
TQ	total quercetin
UAE	ultrasound-assisted extraction
USDA	United States Department of Agriculture
UV	ultraviolet
VMHG	vacuum microwave hydro dispersion gravity extraction
VR	vestitone reductase
XO	xanthine oxidase

PREFACE

Flavonoids, a subclass of plant polyphenols and product of secondary metabolism of plants, have been shown to possess numerous health-promoting benefits in a wide range of investigations. Chemically, flavonoids are phenolic in nature which confers antioxidant quality on them. Approximately, 6000 varieties of this group have already been discovered. They play a variety of biological roles in plants, animals, and bacteria.

They have been ascribed to have positive effect on human and animal health and are employed for disease therapy and chemoprevention. They can be a part of one's diet by consumption of edible plants. For several decades now, they have been extensively consumed for their high bioactive components and health benefits. In the current book, we summarize the chemistry, classification of flavonoids; and highlight their utility in immune-enhancing applications including those in the development of treatment strategies for diseases.

This book *The Flavonoids: Extraction and Applications* comprises 13 chapters divided into four sections that describe the recent advances of polyphenol (flavonoids) derivatives for human health and in management and prevention of diseases.

PART I: OVERVIEW AND RECENT ADVANCES

Chapter 1, *Flavonoids as Potential Immunity Boosters: An Overview, written* by Shailja Singh and Shrawan Kumar Mangawa summarizes the classification of flavonoids and explores their potential as immunity boosting compounds.

Anand Kumar and Babita Saroha report on the role of flavonoids in various diseases and their recent applications in Chapter 2, *Recent Advances of Flavonoids and Their Applications.*

PART II: EXTRACTION AND BIOCHEMISTRY OF FLAVONOIDS

Chapter 3, *Methods of Flavonoids Extraction*, written by Moni Philip Jacob K. focuses on the main parameters involved in their extraction such as temperature, solvent used, sample quantity, time for extraction, etc.

Chapter 4, *Extraction Methods of Polyphenol Derivative (Flavonoid)*, written by Tarkeshwar et al., briefly describes different extraction methods of flavonoids.

Akansha Mehra et al., describe recent development in the research of flavonoids including their extraction, detection, characterization, and scrutinizing their applicability for health benefits in Chapter 5, *Biochemistry and Applications of Flavonoids*.

PART III: APPLICATIONS OF FLAVONOIDS IN DISEASE MANAGEMENT

Richa Saxena et al. report the potential health benefits of flavonoids from various sources and also provide insights on flavonoids as a natural bioactive compound for the development of potent pharmacological remedy in the future in Chapter 6, *Potential Pharmacological Health Benefits of Flavonoids.*

Chapter 7, *Citrus Flavonoids: Recent Advances and Future Perspective in Preventing Cardiovascular Diseases*, written by Pankaj Bhatt et al., covers the phytochemical elements of citrus fruits, as well as their biological characteristics in the prevention of cardiovascular illnesses.

Chapter 8, *Role of Flavonoids in Improving Mental Health*, is written by Tanwi Trushna et al. and discusses the properties of flavonoids' components in the improvement of brain health.

Chapter 9, *Role of Herbs in Metabolic Syndrome*, written by Sandeep O. Waghulde et al. investigates the fitness benefits of ethnomedicinal flora in defense against the improvement of fitness effects associated with metabolic syndrome.

Chapter 10, *Role of Flavonoids in Vasodilation*, written by Nikalje G. C. et al. explores the effects of flavonoids on human endothelial function particularly vasodilation and cardiovascular health.

Chapter 11, *Antimicrobial and Anti-inflammatory Action of Flavonoids*, written by Khemchand Surana and his associates aims to enhance the understanding of the biological and molecular roles of plant flavonoids as antimicrobial and anti-inflammatory.

Chapter 12, *Role of Flavonoid in Dermatology and Cosmetic Preparation*, written by Eknath D. Ahire et al. discusses the role of flavonoids in dermatology and cosmetic preparations.

PART IV: FLAVONOIDS IN AGRICULTURE

Chandra Biswas and Chinmoyee Bhattacharyya discuss the role played by flavonoids in plants and its potential use in agriculture in Chapter 13, *Flavonoids: A Potential for Sustainable Agriculture*.

PART I
Overview and Recent Advances

CHAPTER 1

FLAVONOIDS AS POTENTIAL IMMUNITY BOOSTERS: AN OVERVIEW

SHAILJA SINGH[1] and SHRAWAN KUMAR MANGAWA[2]

[1]*Department of Chemistry, Bhaskaracharya College of Applied Sciences, University of Delhi, New Delhi, India*

[2]*Department of Chemistry, Motilal Nehru College, University of Delhi, South Campus, New Delhi, India*

ABSTRACT

Immune dysfunction is one of the root causes of human diseases, including cancer. Therefore, regulation of factors influencing immune responses is a practical approach to counter the illness. Numerous synthetic drugs play an essential role in enhancing immune response but can also exert adverse side effects. The search for naturally occurring products with immunity promoting properties is gaining research interest worldwide. Plant-based compounds possess several immunomodulatory effects and increase immunity through different mode of actions, including influence on immune organs, cellular immunity, nonspecific immunity, etc. Among these naturally occurring compounds, flavonoids are one of the most studied groups of plant-based phenolics. They display an extensive range of activities depending on structures and are also responsible for particular colors and tastes. This chapter summarizes the classification of flavonoids and explores their potential as immunity booster compounds.

The Flavonoids: Extraction and Applications. Deepika Saini, Rajesh K. Kesharwani, & Raj K. Keservani (Eds.)

1.1 INTRODUCTION

Hippocrates, the father of Western medicine, believed the value of food and its potential positive impact on public health. Hence, the saying "Let food be thy medicine and medicine be thy food." He encouraged people to believe and respect nature's healing forces. Over the past few decades, people are acknowledging the power of eating right which resulted in coinage of terms nutraceuticals (substance which is either food or part of food and possesses health benefits), pharma, and functional food (foods that have health benefits further than basic nutrition). These terms are gaining acceptance among researchers and consumers are also inclined toward natural diet-based regimes (Bourgeon et al., 2007; Butt et al., 2009). These naturally occurring products could have multiple properties like antioxidant defenses, cell proliferation, and gene expression. These compounds also act as immune boosters, thus improving the quality of life (Ares et al., 2009; Atta et al., 2009).

Immune system is known as one of the most complex and sophisticated biological systems in the body. It is a multifaceted network of organs, tissues, cells, proteins, and chemicals which detects and responds to various pathogens, cancer cells, and other potential dangerous agents (Carr and Maggini, 2017). It prevents from infection by engulfing, modulating, and moderating suspicious foreign cells. The earlier published reports demonstrate that host immunity is a well-coordinated system of innate nonspecific (innate) and adaptive specific (adaptive) immunity (Orlowsky and Kraus, 2015; Nicholson, 2016). Innate immunity is already present in the body, which is nonspecific, rapid, limited and lower in potency, and fast acting. On the other hand, adaptive immunity is generated slowly when the body is exposed to foreign pathogens, which is specific, slow, and highly potent with a slower response. Dysfunction of immune system renders an organism sensitive to pathogens, which can develop allergies, autoimmune diseases, inflammatory diseases, and even cancer.

Since immunity is fundamental to health and plays an essential role in preventing diseases, researchers are constantly exploring the immunomodulatory agents to enhance the immune system. Plant-based compounds already proved to have beneficial effects in diabetes, cancer, cardiovascular disease (Keservani et al., 2016a), hyperlipidemia, and hyperuricemia and are considered immunity boosters (Andrea et al., 2017; Gong et al., 2019; Jiang et al., 2021; Mehmood et al., 2019). Plant-based functional foods which have positive effects on immunity system response include alkaloids, polysaccharides, saponins, flavonoids, and among all; flavonoids are the most active class of nutraceutical (DeFelice, 1992).

Flavonoids, a subclass of plant polyphenols and product of secondary metabolism of plants, have been shown to possess numerous health-promoting benefits in a wide range of investigations (Keservani and Sharma, 2014; Liu et al., 2018; Ding et al., 2018). Chemically, flavonoids are phenolic in nature which confers antioxidant activity on them. Approximately, 6000 varieties of this group have already been discovered. Flavonoids play a variety of biological activities in plants, animals, and bacteria. They have been ascribed positive effects on human and animal health and employed for disease therapy and chemoprevention. They can be the part of one's diet by consumption of edible plants. From past several decades, they have been extensively consumed because of their high bioactive components and health benefits (Mohamad et al., 2020; Shafabakhsh et al., 2019; Davoodvandi et al., 2019). In this chapter, we summarize the chemistry and classification of flavonoids, and highlight their effectiveness as immunity booster molecules including those which could be used directly or modified in the treatment of diseases.

1.2 CHEMISTRY OF FLAVONOIDS

Flavonoids are naturally occurring polyphenolic compounds with 15 carbon skeleton of diphenylpropanes consisting of two aromatic rings linked through three carbons (Middleton, 1984). It consists of two benzene rings A and C, connected by a pyran ring B as shown in Figure 1.1. One of the benzene rings (A) is fused with pyran ring while the other benzene ring (C) is attached as a substituent to the pyran ring. Various derivatives of flavonoids with unique physicochemical properties and biological activities can be synthesized (Panche et al., 2016).

1.3 CLASSIFICATION OF FLAVONOIDS

Flavonoids are further divided into subgroups depending on: position at which C ring is attached to heterocyclic ring, degree of unsaturation, and oxidation of the C ring. These flavonoids have a dominant role in different biological activities. These subgroups are discussed below and summarized in Table 1.1. Structures of different flavonoids and their natural sources are shown in Figure 1.2.

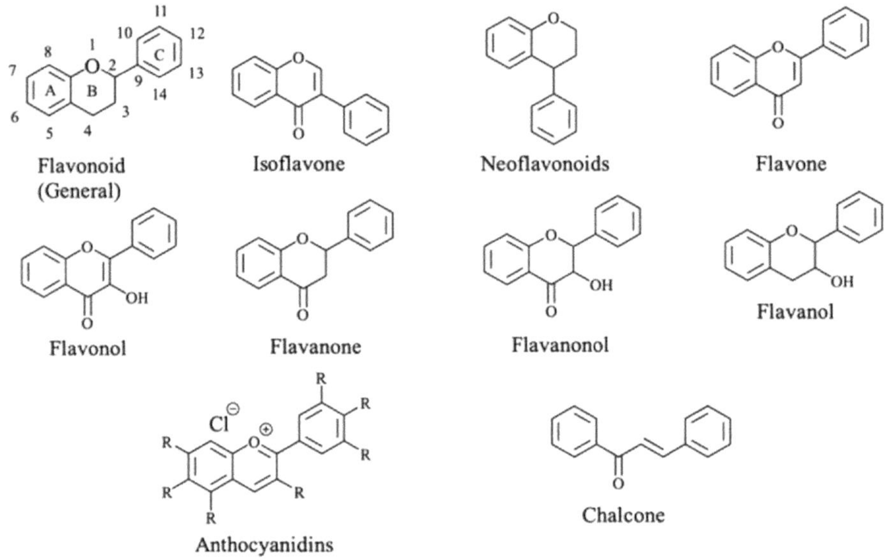

FIGURE 1.1 Chemical structures of flavonoid subgroups.

1.3.1 *ISOFLAVONOIDS*

The isoflavonoids are a distinct group of flavonoids, predominately found in soybeans and other leguminous plants. Despite their limited distribution in plant kingdom isoflavonoids are remarkably diverse as far as their structure is concerned. In this subgroup benzene ring is attached to the position 3 of the pyran ring. The important dietary isoflavones include genistein and daidzein, which are found in four chemical structures respectively aglycones, 7-O-glycosides, 6′-O-acetylglycosides, and 6′-O-malonylglycosides (Clavel et al., 2005). They actively participate in the development of phytoalexins during plant microbe interaction (Aoki et al., 2000; Dixon and Ferreira, 2002).

Genistein was first isolated from the plant Dyer's *Genistatinctoria L.,* and is predominantly found in the Fabaceae family. Genistein suppresses antigen-specific immune response in vivo and lymphocyte proliferation response in vitro. Due to its unique effect on immune system, genistein has been used for the treatment of diverse and it also inhibits allergic inflammatory response. It is also well reported that ingestion of genistein increases lymphocyte proliferation and release of lactase dehydrogenase (LDH) (Ghaemi et al., 2012). Moreover, the treatment with genistein caused a rise in interferon gamma which promotes macrophase activation and facilitates antiviral and antibacterial activities. John Wei et al. in 2012 tested the effect of isoflavones on Th1 immune response in mice and reported that the presence of isoflavones could suppress the antigen uptake and maturation.

TABLE 1.1 Substitution Patterns of Series of Flavonoids.

Group	3	5	6	7	8	3′	4′	5′	C2=C3
Flavones									
Apigenin	H	OH	H	OH	H	H	OH	H	+
Diosmin	H	OH	H	OH	H	OH	OH	H	+
Luteolin	H	OH	H	OH	H	OH	OH	H	+
Flavonol									
Quercetin	OH	OH	H	OH	H	OH	OH	H	+
Kaempferol	OH	OH	H	OH	H	H	OH	H	+
Fisetin	OH	H	H	OH	H	OH	OH	H	+
Myricetin	OH	OH	H	OH	H	OH	OH	OH	+
Flavanone									
Naringenin	H	OH	H	OH	H	H	OH	H	–
Eriodictyol	H	OH	OH	OH	H	OH	OH	H	–
Pinocembrin	H	OH	H	OH	H	H	H	H	–
Liquiritigenin	H	H	H	OH	H	H	OH	H	–
Flavanonol									
Taxifolin	H	OH	H	OH	H	OH	OH	H	–
Isoflavone									
Genistein	–	OH	H	OH	H	H	OH	H	+
Tectorigenin	–	OH	OCH_3	OH	H	H	OH	H	+
Daidzein	–	H	H	OH	H	H	OH	H	+
Formononetin	–	H	H	OH	H	H	OCH_3	H	+
Flavan-3-ols									
(+) Catechin	βOH	OH	H	OH	H	OH	OH	H	–
(-) Epicatechin	ἀOH	OH	H	OH	H	OH	OH	H	–
(-) Epigallocatechin	ἀOH	OH	H	OH	H	OH	OH	OH	–
Flavylium salts									
Cyanidin	OH	OH	H	OH	H	OH	OH	H	–
Pelargonidine	OH	OH	H	OH	H	H	OH	H	–

Source: Reprinted from Tapas et al. (2008). Open Access.

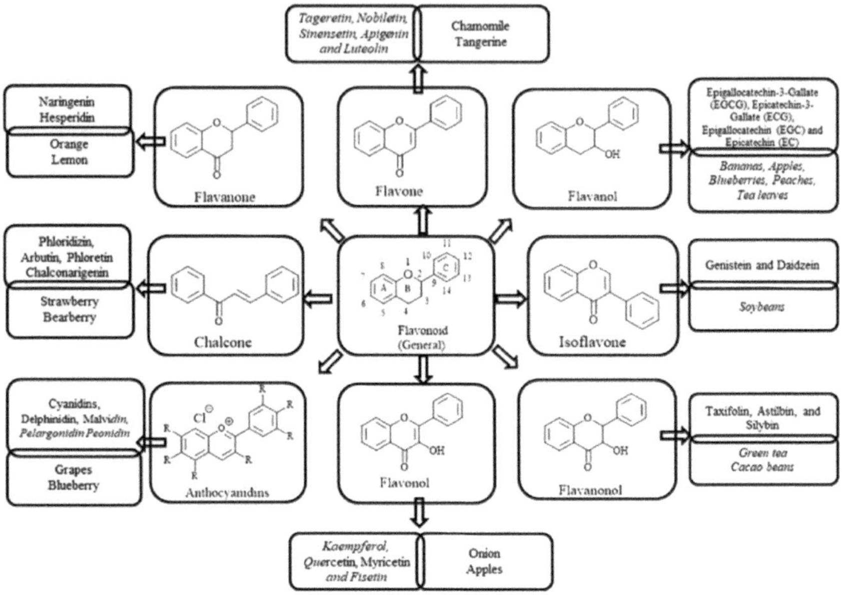

FIGURE 1.2 Structure of different flavonoids and their natural sources.

1.3.2 NEOFLAVONOIDS

The neoflavonoids are structurally and biogenetically related to isoflavonoids and comprise 4-acylcoumarins (4-aryl-2H-1-benzopyran-zones), 3, 4-dihydro-4-arylcoumarins, and neoflavenes. In this class the benzene ring is attached to the position 4 of pyran ring. Neoflavonoids are well known for their antidiabetic activity (Raut et al., 2016).The first neo flavonoids isolated in 1951 were calophyllodide from caloplylluminophyllum seeds (Linuma et al., 1987; Nishimura et al., 2000; Garazd et al., 2003).

1.3.3 FLAVONES

Flavones are one of the important groups of flavonoid containing a double bond on the pyran ring between carbons 2 and 3 and carbonyl group at position 4. They are commonly found as glucosides in leaves, flowers, and even fruits. Most flavones of vegetable and fruits have a hydroxyl group at 5 position of a ring. Red peppers, mint, celery, parsley, chamomile, and *GinkobaBiloba* are rich in flavonoids. The skins of citrus fruits are a good source of polymethoxylated flavonoids such as tageretin, nobiletin, and sinensetin. Apigenin and luteolin also belong to this subclass (Panche et al., 2016; Martínez et al., 2019).

Apigenin (4′,5,7-trihydroxy flavone) is a common dietary flavonoid found in parsley, celery, onions, oranges, maize, rice, and tea It shows strong therapeutic effects against several diseases. Recent research also reports the effect of apigenin on immunity or change in immune response. Cardenas et al. (2016) reported that apigenin significantly modulated nuclear factor kappa B (NF-kB) activity in lungs. This study showed the ability of apigenin to exert an immune-regulatory activity in an organ-specific manner.

Luteolin or 3′,4′,5,7-tetrahydroxyflavone and its glycosylated analog lutelin-7-glycoside are a member of flavone subclass of flavonoids and are usually present in aromatic plants. It is well distributed in vegetables like cucumber, peppermint, thyme, carrot, oregano, pepper, celery, and lettuce. Although its glycosylated form is more common, but reports suggest that luteolin is absorbed in the aglycone form only. Apart from antioxidant and anticarcinogenic effects, features like anti-inflammatory and antiallergic reports are also published. Studies highlight that it has beneficial effects on modulation of immune responses, the mechanism is not very clear yet (López-Lázaro, 2009; Lapidot et al., 2002).

Tangeretin (4′,5,6,7,8-pentamethoxyflavone) is a citrus peel methoxy flavone and has different favorable biological activities such as anti-oxidant, anti-inflammatory, hepato protective, and neuroprotective effects. Ashrafizadeh et al. in 2020 reported that citrus-derived polymethoxy flavone has the potential to prevent neurodegenerative and neurological disorders because of their anti-inflammatory and antioxidant properties.

1.3.4 FLAVONOLS

Flavonols are the alcoholic derivatives of flavones. They have hydroxyl group at position-3 of pyran ring, and thus they are also called 3-hydroxyflavones. The most studied flavonols are kaempferol, quercetin, myricetin, and fisetin (Guven et al., 2019).

Rutin (3,3′,4′,5,7-pentahydroxyflavone-3-rhamnoglycoside) is abundant in different types of plants. It has antibacterial, antifungal, antiallergic, and several other biological activities. Rutin can also modulate signal transduction pathways and can also improve thyroid functions (Al-Dhabi et al., 2015). Biological activities of rutin are very much similar to its aglycone form quercetin.

Quercetin (3,3′,4′,5,7-pentahydroxy flavone) is the most studied and one of the most abundant flavonols in nature. It is aglycone form of a number of other flavonoid glycosides like rutin and quercitrin which are present in a variety of vegetables, fruits, nuts, seeds, flower, bark, and leaves (Rogerio et al., 2010; Patil et al., 2009). Several studies indicate the potential role of quercetin in modulating immune system (Chirumbolo, 2010).

Kaempferol (3,4′,5,7-tetrahydroxyflavone), also known as kaempferol3 or kaempferide,naturally occurs in tea, vegetables like green cabbage, broccoli, spinach, kale, parsley, and fruits like raspberry and black berry (Calderon-Montano et al., 2011). It blocks the synthesis of pro-inflammatory chemokine and inhibits cyclooxygenase (COX), lipoxygenase (LOX), and inducible nitric oxide synthase enzymes (iNos) (Yahfoufi et al., 2018). It is also known for altering important elements in signal transduction pathways of apoptosis and angiogenesis, inflammation, and metastasis (Chen and Chen, 2013).

Myricetin (3,3′,4′,5,5′,7-hexahydroxy flavone) is a natural flavonol which is commonly consumed with dietary food like fruits, red wine, berries, and vegetables. Various studies show that they have iron-chelating, antioxidant, anticancer, and anti-inflammatory activities. Myricetin modulates immune response by changing the level of cytokines. It shows anti-inflammatory by inhibiting inducible nitric oxide synthase enzymes (iNos), cyclooxygenase-z, and production of proinflammatory cytokines.

Fisetin (3, 3′, 4′, 7-tetrahydroxy flavone) is present in several types of fruits and vegetables. Fisetin shows a spectrum of biological activities including antioxidant, antiangiogenic, anticancer, neuroprotective (Keservani et al., 2016b), anti-inflammatory, neurotrophic, and may more. Song et al. (2013) reported the immunosuppressive nature of fisetin against T-cell activation both in vitro and in vivo.

1.3.5 FLAVANOLS

Flavanols or flavan-3-ol is the class of flavonoids lacking carbonyl group at position 4. They are highly diversified and multi-substituted subclass. The heterocyclic ring in this type of compound is saturated and substituted into carbons 2 and 3. Unlike other flavonoids there in no double bond between positions 2 and 3. They are abundantly found in fruits like blueberries, bananas, peaches apples, and tea leaves like black, green or oolong tea are a great source of catechins (Ayaz et al., 2019). Green tea is unfermented leaves of *Camellia sinensis* containing polyphenolic flavonoids like epigallocatechin-3-gallate (EGCG), epicatechin-3-gallate (ECG), epigallocatechin (EGC), and epicatechin (EC), also named catechins.

1.3.6 ANTHOCYANIDINS

Anthocyananins are responsible for color in plant flower and fruits. The color of the anthrocynins is pH dependents and due to acylation or methylation at hydroxyl groups present at A and B rings. The most commonly studied antocynins are cyanidins, delphinidin, malvidin, pelargonidin, and peonidin (Mattioli et al., 2020). They predominantly occur in the outer layer of fruits, like cranberries, red grapes, strawberries, blackberries, etc. Anthocyanines are a glycone form of antocyanidines when the addition of carbohydrate occurs.

1.3.7 CHALCONES

Chalcones are referred to as "open chain flavonoids" due to the absence of ring C, which is the integral part of the basic skeleton of the flavonoids (Ghani, 2020). The open chain structure has a carbonyl group conjugate to a double bond which makes it an α,β-unsaturated carbonyl compound. The most studied chalcones include phloridizin, arbutin, phloretin, and chalconarigenin. They are abundantly found in tomato, peas, strawberries, barberries, and some wheat products. This class of flavonoids grabbed considerable attention because of its significant biological activities (Díaz-Tielas et al., 2016).

1.3.8 FLAVANONOLS

Flavanonols are 3-hydroxy flavanones and also called dihydroflavonols. They have saturated pyran ring with hydroxyl group and carbonyl group at positions 3 and 4, respectively. Silymarin is considered the first member of flavanonols and silybin is the main active analog. Flavonoids astilbin and taxifolin are the important member of this class and commonly found in grapes (Panche et al., 2016; Tapas et al., 2008).

1.3.9 FLAVANONES

Flavanones or dihydroflavones are saturated flavones. They are different from flavones as they do not have unsaturation between C-2 and C-3 (Grayer and Veitch, 2006). Bitter taste of juice and peel of citrus fruits is attributed to the flavanones naringenin and hesperidin. Citrus fruits are rich in flavanones and consist of numerous pharmacological properties like anti-inflammatory, antioxidant, and act as immunity boosters (Barreca et al., 2017).

1.4 CONCLUSIONS

Naturally occurring plant-based functional foods have already received worldwide attention because of a plethora of immunity-strengthening properties. Unlike current expensive drugs which cause severe side-effects, plant-based compounds are stable, safe and exhibit a long-lasting efficacy. With the discovery of phenol-based compounds as immunity booster, a wide range of natural products are identified and numerous are under investigation. Still, the application of such active compounds has certain limitations and herein we propose the following recommendations: (1) certain plant-based foods have been reported to exhibit enhancing immunity but their active ingredient has not been completely explored; hence, further investigations should be carried out, (2) active ingredients derived from plant-based functional foods should undergo in clinical trials, (3) additional in vivo studies should be carried out to explore the mode of action of plant-based functional food with immunity-enhancing capabilities. The extensive researches of such plant-based components are essential to recognize their potential and to provide a theoretical explanation for the synthesis of library of unusual immunity-boosting drugs.

CONFLICTS OF INTEREST

The authors declare no conflict of interest.

KEYWORDS

- **flavonoids**
- **immunity boosters**
- **plant phenolics**
- **kaempferol and myricetin**

REFERENCES

Al-Dhabi, N. A.; Arasu, M. V.; Park, C. H.; Park, S. U. An Up-To-Date Review of Rutin and Its Biological and Pharmacological Activities. *EXCLI J.* **2015,** *14*, 1459–1463.

Andrea, K.; Patrik, S.; Peter, K.; Pavol, Z.; Sona, U.; Martin, K. et al. Are Plant-Based Functional Foods Better Choice Against Cancer Than Single Phytochemicals? A Critical Review of Current Breast Cancer Research. *Biomed. Pharmacother.* **2017,** *96*, 1465–1477.

Aoki, T.; Akashi, T. and Ayabe, S. Flavonoids of Leguminous Plants: Structure, Biological Activity, and Biosynthesis. *J. Plant Res.* **2000,** *113*, 475–488.

Ares, G.; Gim´enez, A.; G´ambaro, A. Consumer Perceived Healthiness and Willingness to Try Functional Milk Desserts; Influence of Ingredient, Ingredient Name and Health Claim. *Food Qual.Prefer.* **2009,** *20*, 50–56.

Ashrafizadeh, M.; Ahmadi, Z.; Mohammadinejad, R.; Ghasemipour Afshar, E. Tangeretin: A Mechanistic Review of Its Pharmacological and Therapeutic Effects. *J. Basic Clin. Physiol. Pharmacol.* **2020,** *31* (4), 20190191.

Atta, A. H.; Elkoly, T. A.; Mouneir, S. M.; Kamel, G.; Alwabel, N. A. Sultan, M. T. et al. Nutritional Profile of Indigenous Cultivar of Black Cumin Seeds and Antioxidant Potential of Its Fixed and Essential Oil. *Pak. J. Bot.* **2009,** *41*, 32–40.

Ayaz, M.; Sadiq, A.; Junaid, M.; Ullah, F.;Ovais, M.; Ullah, I.; Ahmed, J.; Shahid, M. Flavonoids as Prospective Neuroprotectants and Their Therapeutic Propensity in Aging Associated Neurological Disorders. *Front. Aging Neurosci.* **2019,** *11*, 155.

Barreca, D.; Gattuso, G.; Bellocco, E.; Calderaro, A.; Trombetta, D.; Smeriglio, A.; Laganà, G.; Daglia, M.; Meneghini, S.; Nabavi, S. M. Flavanones: Citrus Phytochemical with Health-Promoting Properties. *BioFactors* **2017,** *43*, 495–506.

Bourgeon, S.; Raclot, T.; Le Maho, Y.; Ricquier, D.; Criscuolo, F. Innate Immunity Assessed by Plasma No Measurement Is Not Suppressed During the Incubation Fast in Eiders. *Dev. Comp. Immunol.* **2007,** *31*, 720–728.

Butt, M. S.; Sultan, M. T.; Butt, M. S.; Iqbal, J. Garlic: Nature's Protection Against Physiological Threats. *Crit. Rev. Food Sci. Nutr.* **2009,** *49*, 538–555.

Cardenas, H.; Arango, D.; Nicholas, C.; Duarte, S.; Nuovo, G. J.; He, W. Dietary Apigenin Exerts Immune-Regulatory Activity In Vivo by Reducing NF-κB Activity, Halting Leukocyte Infiltration and Restoring Normal Metabolic Function. *Int. J. Mol. Sci.* **2016,** *17*, 323.

Carr, A.; Maggini, S. Vitamin C and Immune Function. *Nutrients* **2017,** *9* (11), 1211.

Calderon-Montano, J. M.; Burgos-Morón, E.; Pérez-Guerrero, C.; López- Lázaro, M. A Review on the Dietary Flavonoid Kaempferol. *Mini Rev. Med. Chem.* **2011,** *11*, 298–344.

Clavel, T.; Fallani, M.; Lepage, P. et al. Isoflavones and Functional Foods Alter the Dominant Intestinal Microbiota in Postmenopausal Women. *J. Nutr.* **2005,** *135*, 2786–2792.

Chen, A. Y.; and Chen, Y. C. A Review of the Dietary Flavonoid, Kaempferol on Human Health and Cancer Chemoprevention. *Food Chem.* **2013,** *138* (4), 2099–2107.

Chirumbolo, S. The Role of Quercetin, Flavonols and Flavones in Modulating Inflammatory Cell Function. *Inflamm Allergy Drug Targets.* **2010,** *9*, 263–285.

Davoodvandi, A.; Sahebnasagh, R.; Mardanshah, O.; Asemi, Z.; Nejati, M.; Shahrzad, M. K. et al. Medicinal Plants as Natural Polarizers of Macrophages: Phytochemicals and Pharmacological Effects. *Curr. Pharm. Des.* **2019,** *25* (30), 3225–3238.

DeFelice, S. L. Nutraceuticals: Opportunities in an Emerging Market. *Scrip Mag*, **1992,** 9.

Díaz-Tielas, C.; Graña, E.; Reigosa, M. J.; Sánchez-Moreiras, A. M. Biological Activities and Novel Applications of Chalcones. *Planta Daninha, Viçosa-MG* **2016,** *34* (3), 607–616.

Ding, S. J.; Jiang, H. M.; Fang, J. Regulation of Immune Function by Polyphenols. *J. Immunol. Res.* **2018,** 1–8.

Dixon, R.; Ferreira, D. Molecules of Interest: Genistein. *Phytochemistry* **2002,** *60*, 205–211.

Garazd, M.; Garazd, Y.; Khilya, V. Neoflavones. Natural Distribution and Spectral and Biological Properties. Chem. *Nat. Comp.* **2003,** *39*, 54–121.

Ghani, S. Chapter Three-Polyphenols. *Alpha-Glucosidase Inhibitors*; Elsevier, 2020; pp 61–100.

Ghaemi, A.; Soleimanjahi, H.; Razeghi, S.; Gorji, A.; Tabaraei, A.; Moradi, A.; Alizadeh, A.; Vakili, M. A. Genistein Induces a Protective Immunomodulatory Effect in a Mouse Model of Cervical Cancer. *Iran J. Immunol.* **2012,** *9* (2), 119–127.

Gong, X.; Ji, M. Y.; Xu, J. P.; Zhang, C. H.; Li, M. H. Hypoglycemic Effects of Bioactive Ingredients from Medicine Food Homology and Medicinal Health Food Species Used in China. *Crit. Rev. Food Sci. Nutr.* **2019,** *60*, 2303–2326.

Grayer, R. J.; Veitch, N. C. Flavanones and Dihydroflavonols. *Flavonoids Chem. Biochem. App.* **2006,** 917–1002.

Guven, H.; Arici, A.; Simsek, O. Flavonoids in Our Foods: A Short Review. *J. Basic Clin. Health. Sci.* **2019,** *3*, 96–106.

Jiang, L.; Zhang, G.; Li, Y.; Shi, G.; Li, M.; Potential Application of Plant-Based Functional Foods in the Development of Immune Boosters. *Front. Pharmacol.* **2021,** *12*, 637782.

Keservani, R. K.; Sharma, A. K. Flavonoids: Emerging Trends and Potential Health Benefits. *J. Chin. Pharma. Sci.* **2014,** *23* (12), 815.

Keservani, R. K.; Sharma, A. K.; Kesharwani, R. K. Nutraceutical and Functional Foods for Cardiovascular Health. In *Food Process Engineering*; Apple Academic Press: CRC Press, 2016a; pp. 257–278, 291–312. ISBN: 9781771884020.

Keservani, R. K.; Sharma, A. K.; Kesharwani, R. K. Medicinal Effect of Nutraceutical Fruits for the Cognition and Brain Health. *Scientifica* **2016b,** Article ID 3109254, 10 pp. DOI: 10.1155/2016/3109254.

Lapidot, T.; Walker, M. D.; Kanner, J. Antioxidant and Prooxidant Effects of Phenolics on Pancreatic β-Cells In Vitro.*J. Agric. Food Chem.* **2002,** *50*, 7220–7225.

Linuma, M.; Tanaka, T.; Hamada, K. et al. Revised Structure of Neoflavone in Coutarea-hexandra. *Phytochemistry* **1987,** *26*, 3096–3097.

Liu, M.; Li, S. S.; Wang, X. X.; Zhu, Y. F.; Zhang, J. J.; Liu, H. et al. Characterization, Anti-Oxidation and Anti-Inflammation of Polysaccharides by *Hypsizygus marmoreus* Against LPS-Induced Toxicity on Lung. *Int. J. Biol. Macromol.* **2018,** *111*, 121–128.

López-Lázaro, M. Distribution and biological activities of the flavonoid luteolin. *Mini Rev. Med. Chem.* **2009,** *9* (1), 31–59.

Martínez, G. Mijares, M. R.; Sanctis, J. B. De. Effects of Flavonoids and Its Derivatives on Immune Cell Responses. *Recent Pat. Inflamm. Allergy Drug Discov.* **2019,** *13*, 84–104.

Mattioli, R.; Francioso, A.; Mosca, L.; Silva, P. Anthocyanins: A Comprehensive Review of Their Chemical Properties and Health Effects on Cardiovascular and Neurodegenerative Diseases. *Molecules* **2020,** *25* (17), 3809.

Mehmood, A.; Zhao, L.; Wang, C. T.; Nadeem, M.; Raza, A.; Ali, N. et al. Management of Hyperuricemia Through Dietary Polyphenols as a Natural Medicament: A Comprehensive Review. *Crit. Rev. Food Sci. Nutr.* **2019,** *59* (9), 1433–1455.

Middleton, E. The Flavonoids. *Trends Pharmacol. Sci.* **1984,** *5*, 335–338.

Mohamad, N. E.; Romli, M. F.; Alitheen, N. B.; Abu, N.; Yeap, S. K.; Lim, K. L. et al. Apoptosis and Metastasis Inhibitory Potential of Pineapple Vinegaragainst Mouse Mammary Gland Cells In Vitro and In Vivo. *Nutr. Metab. (Lond).* **2020,** *16*, 49.

Nicholson, L. B. The Immune System. *Essays Biochem.* **2016,** *60* (3), 275–301.
Nishimura, S.; Taki, M.; Takaishi, S. et al. Structures of 4-arylcoumarin (Neoflavone) Dimers Isolated from Pistaciachinensis BUNGE and Their Estrogen-Like Activity. *Chem. Pharm. Bull. (Tokyo).* **2000,** *48*, 505–508.
Orlowsky, E. W.; Kraus, V. B. The Role of Innate Immunity in Osteoarthritis: When Our First Line of Defense Goes on the Offensive. *J. Rheumatol.* **2015,** *42* (3), 363–371.
Panche, A.; Diwan, A.; Chandra, S. Flavonoids: An Overview. *J. Nutr. Sci.* **2016,** *5*, e47.
Patil, B. S.; Jayaprakasha, G.; Chidambara Murthy, K.; Vikram, A. Bioactivecompounds: Historical Perspectives, Opportunities, and Challenges. *J. Agric. Food Chem.* **2009,** *57*, 8142–8160.
Park, H-H.; Lee, S.; Oh, J-M.; Lee, M-S.; Yoon, K-H.; Park, BH. et al. Anti-Inflammatory Activity of Fisetin in Human Mast Cells (HMC-1). *Pharm Res.* **2007,** *55*, 31–37.
Raut, N. A.; Dhore, P. W.; Saoji, S. D.; Kokare, D. M. Chapter 9-Selected Bioactive Natural Products for Diabetes Mellitus. *Stud. Nat. Products Chem.* **2016,** *48*, 287–322.
Rogerio, A. P.; Dora, C. L.; Andrade, E. L.; Chaves, J. S., Silva, L. F.; Lemos-Senna, E. et al. Anti-Inflammatory Effect of Quercetin-Loaded Microemulsion in the Airways Allergic Inflammatory Model in Mice. *Pharmacol. Res.* **2010,** *61*, 288–297.
Semwal, D.; Semwal, R.; Combrinck, S.; Viljoen, A. Myricetin: A Dietary Molecule with Diverse Biological Activities. *Nutrients.* **2016,** *8*, 90.
Shafabakhsh, R.; Pourhanifeh, M. H.; Mirzaei, H. R.; Sahebkar, A.; Asemi, Z.; Mirzaei, H. Targeting Regulatory T Cells by Curcumin: A Potential for Cancer Immunotherapy. *Pharmacol. Res.* **2019,** *147*, 104353.
Song, B.; Guan, S.; Lu, J.; Chen, Z.; Huang, G.; Li, G.; Xiong, Y.; Zhang, S.; Yue, Z.; Deng, X. Suppressive Effects of Fisetin on Mice T Lymphocytes In Vitro and In Vivo. *J. Surg. Res.* **2013,** *185* (1), 399–409.
Tapas, A. R.; Sakarkar, D. M.; Kakde, R. B. Flavonoids as Nutraceuticals: A Review. *Trop. J. Pharm. Res.* **2008,** *7* (3), 1089–1099.
Wei, J.; Bhatt, S.; Chang, L. M.; Sampson, H. A.; Masilamani, M. Isoflavones, Genistein and Daidzein, Regulate Mucosal Immune Response by Suppressing Dendritic Cell Function. *Plos One* **2012,** *7* (10), e47979.
Yahfoufi, N.; Alsadi, N.; Jambi, M.; Matar, C. The Immunomodulatory and Anti-Inflammatory Role of Polyphenols. *Nutrients* **2018,** *10* (11), 1618.

CHAPTER 2

RECENT ADVANCES OF FLAVONOIDS AND THEIR APPLICATIONS

BABITA SAROHA[1] and ANAND KUMAR[2]

[1]*Department of Biotechnology, UIET, MDU, Rohtak, Haryana, India*

[2]*Department of Chemistry, SGRR (PG) College, Dehradun, Uttarakhand, India*

ABSTRACT

Flavonoids are the secondary metabolites with various metabolic functions in plants such as combating oxidative stress, defence against UV-B radiation, pathogen infections, nodulation, pollen fertility, and regulate cell growth. Bioavailablity, metabolism, and biological function depends upon the configuration, hydroxyl group's number, and functional group substitution. They are present in fruit, flowers, vegetables, grains roots, stem, bark, tea, and wine. Antioxidative, antibacterial, anticancerous, antiviral, hepatoprotective, anti-inflammatory, and cardioprotective activities are associated with flavonoids. They have the potential to reduce free radicals and protect against oxidative stress due to the presence of hydroxyl group. Neuroprotective effects have also been observed along with coronary vasodilation improvement and prevention of LDLs oxidation. This chapter focuses on flavonoids structural features as well as their beneficial roles in human health.

2.1 INTRODUCTION

Plants can produce number of metabolites; in the plant kingdom over 1,000,000 metabolites have been identified (Afendi et al., 2012). Plant metabolites are

The Flavonoids: Extraction and Applications. Deepika Saini, Rajesh K. Kesharwani, & Raj K. Keservani (Eds.)

categorized in two groups: primary metabolites concerned with nutrition and essential metabolic processes whereas secondary metabolites are the ones which play a significant role in plant and its environment interaction (Ruiz-Cruz et al., 2017). Flavonoids are one of the secondary metabolites which impart color, fragrance, and flavor in plants, seeds, and fruits (Dias et al., 2021; Panche et al., 2016). Flavonoids have a large family of approximately 10000 hydroxylated polyphenolic compounds (Khajuria et al., 2019; Ahmed et al., 2016; Kumar and Pandey, 2013). Until now, 12,000 alkaloids, 30,000 terpenoids, and approximately 10,000 flavonoids have been obtained (Martens et al., 2010; Dixon and Pasinetti, 2010). IUPAC recommendation, 2017 suggests word flavonoids is applicable to derivatives of a phenyl-substituted propylbenzene with C-15, C-16, and flavonolignans (Rauter et al., 2018). The term flavonoid is used for only those compounds having carbon skeleton C6–C3–C6 in a restricted sense. In plants, animals, and bacteria flavonoids play multiple biological roles. Flower color and its aroma are related to flavonoids which will be produced in particular sites; in fruits flavonids attract pollinator insects, regulate cell growth, and protect plants against abiotic and biotic stresses (Kumar and Pandey, 2013). Flavonoids act as allopathic compounds, signaling molecules, UV filters, phytoalexins (Dias et al., 2020; Saini et al., 2017; Panche et al., 2016; Ferdinando et al., 2012). They play a characteristic role in frost hardiness, drought resistance, heat and freezing tolerance also (Samanta et al., 2011). Besides their relevance in plants, they are an indispensible component in nutraceutical, pharmaceutical, medicinal, and cosmetic applications. Several studies stated that flavonoids possess a remarkable spectrum of bio-pharmacological properties like antimalarial, cytotoxic, cardioprotective, antiviral, hepatoprotective, antioxidant, antitrypanosomal anticancer, neuroprotective, antibacterial, anti-inflammatory, antithrombotic, antileishmanial, antiallergic, and antiamebial properties (Amri and Hossain, 2018; Tungmunnithum et al., 2018; Meng et al., 2018; Andreu et al., 2018; Działo et al., 2016; Chen et al., 2015; Kumar and Pandey, 2013). Biochemical and antioxidant effects are also observed related to diseases like Alzhemier's and atheroscleosis, cancer, etc. (Panche et al., 2015, 2016).

2.2 BASIC STRUCTURE OF FLAVONOIDS

Polyphenolic molecules with 15 carbon atoms, soluble in water (Olagaray and Bradford, 2019) are present in the cytosol and stored in cell vacoule (De Villiers et al., 2016; Zuiter and Zarqa, 2014). They are represented as 2-phenyl-benzo-γ-pyrone. In flavonoids carbon atoms are arranged in A and

B benzene rings which connect with oxygen-containing pyrene ring (C). All falvonoids have a similar structure, that is, carbon skeleton based on flavan system (C6-C3-C6) (Isoda et al., 2014; Ferreyra et al., 2012; Symonowicz and Kolanek, 2012). The skeleton of ring is represented in Figure 2.1.

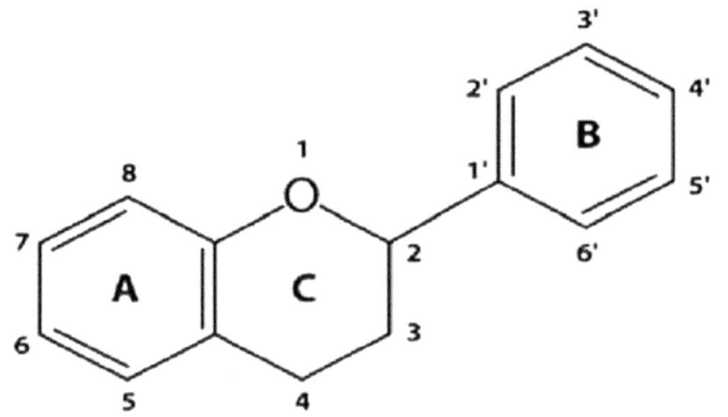

FIGURE 2.1 Basic structure of flavonoids.

2.3 BIOSYNTHESIS OF FLAVONOIDS

Flavonoids biosynthesis occurs through phenylpropanoid pathway. For initial substrates one *p*-coumaroyl CoA and three molecules of malonyl CO-A are required.*p*-coumaroyl Coumaroyl CoA is synthesized from phenylalanine obtained from the shikimate pathway which changes phosphoenolpyruvate and erythrose-4-phosphate into phenylalanine, tyrosine, and tryptophan by three enzymatic reactions (Nabavi et al., 2020) (Figure 2.2). Formation of intermediate polyketide takes place after successional condensation and decarobxylation reactions which after cyclizations and aromitizations form ring A and *p*-coumaroyl-CoA synthesize ring B and ring C. All the three aromatic rings undergo condensation and result in the formation of chalcone which after isomerization catalyzed cyclization and formed flavanone. Flavanone after hydroxylation, methylation, or glycosylation forms various flavonoids (Figure 1.3).

2.3.1 *FLAVONES*

Flavones are 2 aryl-4*H*-chromen-4-ones. They can be found in various parts of the plants. Flavones are formed by the formation of a double bond between

C2 and C3 after the removal of two hydrogen atoms (Olagaray and Bradford, 2019). Substitution pattern and distribution divides flavones into subclasses like *O*-methylated, *C*-methylated, and isoprenylated, from other flavonoids (Harborne and Baxter, 1999). Both O- and C-glycosides flavones, aglycones the apigenin and luteolin are mostly absorbed by intestine.

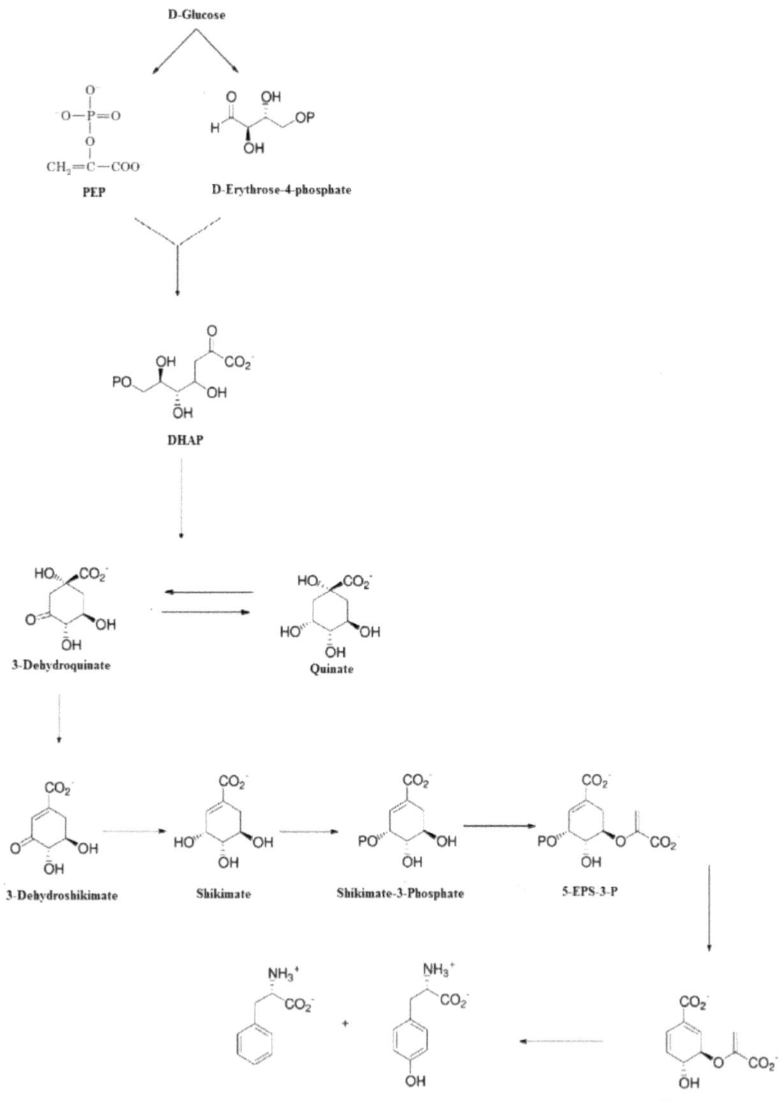

FIGURE 2.2 Shikimate pathway: PEP: Phosphoenolpyruvate, DHAP: 3-Deoxy-D-arabino-heptulosonic acid 7-phosphate, 5-EPS-3-P : 5- Enolpyruvylshikimate-3-phosphate.

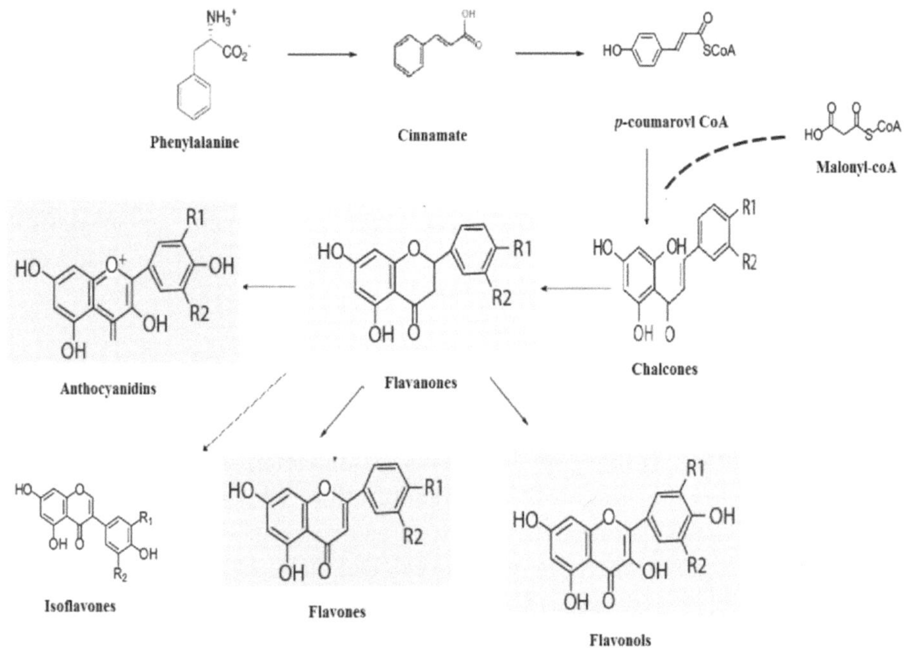

FIGURE 2.3 Phenylpropanoid pathway.

2.3.2 *FLAVONOLS*

Flavanols are also called catechins, with typical flavonoid structure having ring A and ring B with varying hydroxylation patterns and ring C with asymmetrical stereochemistry of carbon (C2 and C3) (Pons et al., 2016; Oliveira et al., 2013). Catechins are found as free catechins and esterified catechins (Vuong et al., 2010); their size, monomers, or polymeric forms (condensed tannins) make them the most complex flavonoids (Cassidy and Kay, 2013). Hydroxyl group at position 3 in ring C make them different from flavones which could also be glycosylated. Flavonols are commonly found in fruits and vegetables. Flavonols are extremely different in their methylation and hydroxylation patterns and, in view of the different glycosylation patterns.

2.3.3 *FLAVANONES*

Flavanones, also named as dihydroflavones (2-aryl-3-hydroxychroman-4-one), have the saturated C ring; in case of flavones there is a double bond between C2 and C3 but in flavonones the bond is saturated which make them different from each other. There are several hydroxyl groups which can be glycosylated or methylated due to multi-hyrdoxylated nature of flavanones. They are commonly present citrus fruits like oranges, grapes, and lemon. Hesperidin and naringenin are the most well-known and studied flavanones (Shah and Smith, 2020). Naringin, naringenin, and naringenin 7-*O*-neohesperidoside flavanone glycosides are present in grapefruits. eriocitrin, eriodictyol, and eriodictyol 7-*O*-rutinoside could be observed in lemons (Terahara, 2015; Agrawal, 2011).

2.3.4 ISOFLAVONOIDS

By rearrangement isoflavonoids (3-aryl-4*H*-chromen-4-ones) are obtained from flavanones that initiate 2, 3-aryl migration, later on dehydrogenation takes place. These flavonoids are proved to be useful signaling and nodule induction in legume–rhizobia symbioses (Del Valle et al., 2020). Estrogen hormonal activity and reduced inflammation is related to these metabolites (Pinto et al., 2020; Vitale et al., 2013). Genistein and diazein are the common isoflavonoids.

2.3.5 ANTHOCYANIDINS

Colors of many fruits and vegetables depend upon anthocyanidins. Anthocyanin or anthocyanin glycoside pigments produce red or blue color in apple, grapes, and berries (Erlund, 2004). Color changes due to the structure of the compound which depends on the hydroxylation and methylation reactions at particular positions on ring A and ring B (Panche et al., 2016). No ketone group at position 4 of ring C are present like other flavonoids except flavanols. High amount of anthocyanin is founds in berries, plums, grapes, and sweet potato (Panche et al., 2016).

2.3.6 CHALCONES

Chalcone is the precursor for mainly all the flavonoids. Chalcone has a missing C ring which makes it different from other flavonoids it has opened ring (Tsao, 2010). Apple, hops or beer, berries, tomato, and few wheat contain chalcones. Chalco-naringenin is the widely studied chalcone (Panche et al., 2016).

2.4 FLAVONOIDS IN FOOD

Flavonoids are synthesized in all parts of the plant. They provide color, fragrance, and taste to the fruits, flowers, and seeds, which make them attractants for insects, birds, and mammals, which aid in pollen or seed transmission. Table 2.1 shows various classes of flavonoids and some of their food sources.

TABLE 2.1 Classes of Flavonoids and Their Food Sources.

Class	Flavonoid	Source	References
Flavanol	Catechins, gallactocatechins, epicatechin, epigallocatechin gallate	Apples, hops, green tea, beer, kiwi, cocoa products, wine, fruit, juice, tea	Panche et al. (2016), Agrawal (2011), Bae et al. (2020), Loez et al. (2001), Bondonno et al. (2012), Toh et al. (2013), Harnly et al. (2006)
Flavanones	Hesperidin, Naringenin Neohesperidin. eriodictyol	Citrus fruits, curcumin, oranges, grapefruits, peppermint, lemon	Terahara (2015), Agarwal (2011), Benavente-Garcia et al. (1997), Tripoli et al. (2007)
Anthocyamins	Anthocyanidin, cyanidin, delphinidin, malvidin, pelargonidin, petunidin, peonidin	Fruits and flowers, blueberries, bananas, strawberries, red wine, cranberries, blackcurrants	Khoo et al. (2017), Terahara (2015), Agarwal (2011), Yao et al. (2014), Basu et al. (2010)
Flavones	Apigenin, chrysin, luteolin, Diosmetin, luteolin	Herbs, cereals, fruits, parsley, thyme, bee pollen, passiflora plant, vegetables, flowers	Kozlowska et al. (2017), Panche et al. (2016), Terahara (2015), Saric et al. (2009)
Flavonols	Isorhamnetin, kaempferol, quercetin, myricetin, rutin, pinocembrin	Onions, cherries, apples, broccoli, kale, tomatoes, berries, black tea, red wine, Tartary buckwheat, edible part of plants, leeks, bee pollen	Basu et al. (2016), Bondonno et al. (2012), Larson et al. (2012)

2.5 FUNCTIONS AND APPLICATIONS OF FLAVONOIDS

Various pharmacological activities have been reported in medicinal plants, mainly secondary metabolites, that is, flavonoids and phenolic compounds which are produced by plants for self-defense or to enhance growth during unfavorable conditions. Scientific evidence has shown prevention of some chronic diseases by consuming diet rich in flavonoids (Jucá et al., 2020; Fraga et al., 2019; Rodríguez-García et al., 2019).

2.5.1 ANTIOXIDANT ACTIVITY

Secondary metabolites like phenolics and flavonoids are considered the largest known phytochemical molecules with free-radical scavenger properties from plants (Andreu et al., 2018; Zahoor et al., 2018; Wang et al., 2016; Mishra et al., 2013). They provide protection against stress by acting as scavengers of free radicals like reactive oxygen species (ROS) also work as iron chelating by Fenton reaction (Williams et al., 2004), inhibits oxidase enzymes; therefore, it inhibits the formation of ROS and organic hydroperoxides (Kaleem and Ahmad, 2018; Kumar and Pandey, 2013; Agarwal, 2011; Escamilla et al., 2009). Free-radical scavenging donates hydrogen and an electron from the hydroxyl group on the B ring to hydroxyl, peroxyl, and peroxynitirite free radicals which stabilizes them due to the presence of the hydroxyl group radicals with high reactivity becomes inactive in flavonoids as per Eq. (1) where R* denotes a free radical and oxygen free radical is denoted by O* (Beecher, 2003; Heim et al. 2002; Nijveldt et al., 2001).

$$\text{F-OH} + \text{R}\bullet \rightarrow \text{F-O}\bullet + \text{RH}$$

As reported by Kerry and Abbey (1997) flavonoids can resist LDL oxidation in in vitro studies by scavenging free radicals.

2.5.2 ANTICANCER EFFECT OF FLAVONOIDS

Healthy and nutritious diet is vital in the prevention of cancer. Flavonoids present in fruits and vegetables have been stated as chemopreventive agents. Flavonoids help in the Inactivation of carcinogen, cancerous cell proliferation inhibition, DNA repair processes and reduction in oxidative stress, progression stage, apoptosis induction and development of cytotoxic activity toward carcinogen cell (Terahara, 2015; Kozlowska and Szostak-Wegierek, 2014; Sokolov et al., 2013). Inhibition of proliferation of tumor cell by reduced ROS and xanthine oxidase, cyclooxygenase-2 and 5-lipoxoygenase enzymes inhibition are required in the promotion and development of tumor (Ginwala et al., 2019).

How flavonoids act on cancer:

1. They downregulate mutant p53 protein
2. Cell cycle arrest at checkpoints
3. Inhibition of tyrosine, heat shock protein

4. Regulation of estrogen binding capacity
5. Ras protein expression inhibition

Several studies reported carcinogens cells reduction by induction flavonoids. Isorhamnetin and acacetin are capable of inhibiting human breast cancer proliferation (Maleki et al., 2019). In human osteosarcoma and breast, stomach, lungs canrinogen cells Kaempferol shows antiproliferative activity and apoptosis (Kozłowska et al., 2017). Genistein also shows benefit in various types of cancer like breast, prostate, and ovarian (Kaushik et al., 2019). At G2/M phase in breast cancer cell cycle arrest by Genistein and daidzein also ROS-dependent apoptosis follows (Kaushik et al., 2019; Jin et al., 2010). Superoxide dismutase, catalase, and glutathione activity increases by naringenin which reduce ROS production in cancer cells (Zaidun et al., 2018). Naringenin and hesperidin both suppress metastasis and MG-63 osteocarcinoma cells proliferation (Pourakbari et al., 2020). For hepatocellular cancer hesperidin is found to be more effective as antitumor and hepatoprotective (Pourakbari et al., 2020). Flavonoids present in green tea like epigallocatechin-3-gallate prove to be concerned in cell cycle and death in prostate cancer (Kozłowska et al., 2017). In-vitro studies of various cancer quercetin benefit in cell cycle arrest and growth inhibition on different malignant cancer cell lines (Kumar and Pandey, 2013). In ovarian cancer apigenin and luteolin shows beneficial effects (A2780, OVCAR-3, and SKOV-3) by varying signal of ROS and apoptosis induction (Tavsan and Kayali, 2019). Cyanidin prevent proliferation and stimulate apoptosis in epithelial colorectal adenocarcinoma cells (Caco-2) of humans. Thus, diet with flavonoids rich in free-radical scavengers might reduce the cancer-stimulating action of some other radicals.

2.5.3 FLAVONOIDS ON THE CARDIOVASCULAR SYSTEM

Several scientists reported cardio protection role of flavonoids including some actions, like coronary vasodilatation improving, decreasing the platelets' ability to clot blood, LDL oxidation prevention inhibition of inflammation propagation, preventing apoptosis, antinecrotic, reducing oxidative stress, and playing a vital role in the impact on capillary blood vessel (Rolnik et al., 2020; Arct and Pytkowska, 2008; Soobrattee et al., 2005; Benavente-García et al., 1997). Rice-Evans et al. (1996), in a group of Dutch population, studied the intake of flavonoids and found that taking they can decrease cardiovascular diseases up to 2.4 times as compared to those with low intake. Quercetin, naringenin, and hesperetin show vasodilator properties, blood pressure and relaxation of vascular smooth muscles is reduced by naringenin (Jucá et al., 2020; Ciumarnean et al., 2020). Isoflavones have a protective role against inflammatory vascular diseases, cardio protective properties have been shown by quercetin (Maleki et al., 2019). Baicalin shows cardiac dysfunction and inhibits apoptosis in heart tissue (Zhang et al. 2017), inhibition of platelet function by Chrysin, and antihypertensive properties by genistein (Ciumarnean et al., 2020). Anthocyanin diminishes the myocardial infarction risk in humans, improves systolic blood pressure, and reduces triglycerides level as well as total and LDL cholesterol (Cassidy, 2018). Adding up, systolic blood pressure and LDL blood concentration are reduced by quercetin (Fraga et al., 2019). Several studies with acacetin show positive effects in human arrhythmia control (Jucá et al., 2020).

2.5.4 FLAVONOIDS ON THE NERVOUS SYSTEM

Beneficial effects have been shown by food enriched in flavonoids, in maintaining cognitive functions in humans to promote memory improvement, neurons protection, neuronal regeneration stimulation, and oxidative neuronal damages prevention (Hollman and Katan, 1999). Several scientists reported neurobiological actions of these compounds in two ways: the first way is by the regulation of neuronal signal cellular cascades which occur by neurotoxic products and can harm formation of new neurons, many neuronal functions, and connectivity to brain. Second, flavonoids may improve blood flow to the brain and sensory systems and seem to exert good effects on peripheral and central nervous systems (Kozlowska et al., 2014; Sokolov et al., 2013).

KEYWORDS

- **flavonoid**
- **bioavilability**
- **oxidative stress**
- **metabolism**
- **neuroprotective**

REFERENCES

Afendi, F. M.; Okada, T.; Yamazaki, M.; Hirai-Morita, A.; Nakamura, Y.; Nakamura, K. et al. Knapsack Family Databases: Integrated Metabolite Plant Species Databases for Multifaceted Plant Research. *Plant Cell Physiol.* **2012,** *53*, e1.

Agrawal, A. D. Pharmacological Activity of Flavonoids: A Review. *Int. J. Pharma. Sci. Nano* **2011,** *4*, 1394, 1398.

Ahmed, S. I.; Hayat, M. Q.; Tahir, M.; Mansoor, Q.; Ismail, M.; Keck, K.; Bates, R. B. Pharmacologically Active Flavonoids from the Anticancer, Antioxidant and Antimicrobial Extracts of *Cassia angustifolia* Vahl. *BMC Complement. Altern. Med.* **2016,** *16*, 460.

Amri, F. S.; Hossain, M. A. Comparison of total phenols, Flavonoids and Antioxidant Potential of Local and Imported Ripe Bananas. *EJBAS.* **2018,** *5*, 245–251.

Andreu, L.; Nuncio-Jáuregui, N.; Carbonell-Barrachina, Á. A.; Legua, P.; Hernández, F. Antioxidant Properties and Chemical Characterization of Spanish Opuntia ficus-indica Mill. cladodes and Fruits. *J. Sci. Food Agric.***2018,** *98*, 1566–1573.

Arct, J.; Pytkowska, K. Flavonoids as Components of Biologically Active Cosmeceuticals. *Clin. Dermatol.* **2008,** *26* (4), 347–357.

Bae, J.; Kim, N.; Shin, Y.; Kim, S. Y.; Kim, Y. J, Activity of Catechins and Their Applications. *Biomed. Dermatol.* **2020,** *4*, 8.

Basu, A.; Das, A. S.; Majumder, M.; Mukhopadhyay, R. Anti-Atherogenic Roles of Dietary Flavonoids Chrysin, Quercetin and Luteolin. *J. Cardiovasc. Pharmacol.* **2016,** *68* (1), 89–96.

Basu, A.; Rhone, M.; Lyons, T. J. Berries: Emerging Impact on Cardiovascular Health. *Nutr. Rev.* **2010,** *68* (3), 168–177.

Beecher, G. R. Overview of Dietary Flavonoids: Nomenclature, Occurrence and Intake. *J. Nutr.* **2003,** *133* (10), 2448S–3254S.

Benavente-García, O.; Castillo, J.; Marin, F. R.; Ortuño, A.; Del Río, J. A. Uses and Properties of Citrus Flavonoids. *J. Agric. Food. Chem.* **1997,** *45* (12), 4505–4515.

Bondonno, C. P.; Yang, X.; Croft, K. D. Flavonoid-Rich Apples and Nitrate Rich Spinach Augmnt Nitric Oxide Status and Improve Endothelial Function in Healthy Men and Women a Randomized Controlled Trial. *Free Radic. Biol.Med.* **2012,** *52* (1), 98–102.

Cassidy, A. Berry Anthocyanin Intake and Cardiovascular Health. *Mol. Asp. Med.* **2018,** *61*, 76–82.

Ciumarnean, L.; Milaciu, M. V.; Runcan, O.; Vesa, S. C.; Rachis, A. L.; Negrean, V.; Perné, M. G.; Donca, V. I.; Alexescu, T. G. M.; Para, I. et al. The Effects of Flavonoids in Cardiovascular Diseases. *Molecules.* **2020,** *25*, 4320.

Cassidy, A.; Kay, C. Phytochemicals: Classification and occurrence. In *Encyclopedia of Human Nutrition*; Caballero, B., Ed.; Elsevier: Amsterdam, The Netherlands, 2013; pp 39, 46.

Chen, X.; Dang, T. T. T.; Facchini, P. J. Noscapine Comes of Age. *Phytochemical* **2015,** *111*, 7–13.

De Villiers, A.; Venter, P.; Pasch, H. Recent Advances and Trends in the Liquid-Chromatography-Mass Spectrometry Analysis of Flavonoids. *J. Chromatogr. A* **2016,** *1430*, 16–78.

Del Valle, I.; Webster, T. M.; Cheng, H. Y.; Thies, J. E.; Kessler, A.; Miller, M. K.; Ball, Z. T.; MacKenzie, K. R.; Masiello, C. A.; Silberg, J. J. et al. Soil Organic Matter Attenuates the Efficacy of Flavonoid-Based Plant-Microbe Communication. *Sci. Adv.* **2020,** *6*, eaax8254.

Dias, M. C.; Pinto, D. C. G. A.; Freitas, H.; Santos, C.; Silva, A. M. S. The Antioxidant System in *Olea europaea* to Enhanced UV–B Radiation Also Depends on Flavonoids and Secoiridoids. *Phytochem.* **2020,** *170*, 112199.

Dias, M. C.; Pinto, D. C. G. A.; Silva, A. M. S. Plant Flavonoids: Chemical Characteristics and Biological Activity. *Molecules* **2021,** *26*, 5377.

Dixon, R.; Pasinetti, G. Flavonoids and Isoflavonoids: From Plant Biology to Agriculture and Neuroscience. *Plant Physiol.* **2010,** *154*, 453–457.

Działo, M.; Mierziak, J.; Korzun, U.; Preisner, M.; Szopa, J.; Kulma, A. The Potential of Plant Phenolics in Prevention and Therapy of Skin Disorders. *Int. J. Mol. Sci.* **2016,** *17*, 160.

Erlund, I. Review of the Flavonoids Quercetin, Hesperetin, and Naringenin: Dietary Sources, Bioactivities, Bioavailability, and Epidemiology. *Nutr. Res.* **2004,** *24*, 851–874.

Escamilla, C. O.; Cuevas, E. Y.; Guevara, J. Flavonoides y sus acciones Antioxidantes. Revista de la Facultad de Medicina. **2009,** *52* (002), 73–75.

Ferdinando, M. D.; Brunetti, C.; Fini, A.; Tattini, M. Flavonoids as Antioxidants in Plants Under Abiotic Stresses. In *Abiotic Stress Responses in Plants, Metabolism, Productivity and Sustainability*; Ahmad, P.; Prasad, M. N. V., Eds.; Springer: Berlin/Heidelberg, Germany, 2012.

Ferreyra, F.; Rius, S. P.; Casati, P. Flavonoids: Biosynthesis, Biological Functions, and Biotechnological Applications. *Front. Plant Sci.* **2012,** *3*, 1–15.

Fraga, C. G.; Croft, K. D.; Kennedy, D. O.; Tomás-Barberán, F. A. The Effects of Polyphenols and Other Bioactives on Human Health. *Food Funct.* **2019,** *10*, 514–528.

Ginwala, R.; Bhavsar, R.; Chigbu, D. G. I.; Jain, P.; Khan, Z. K. Potential Role of Flavonoids in Treating Chronic Inflammatory Diseases with a Special Focus on the Anti-Inflammatory Activity of Apigenin. *Antioxidants* **2019,** *8*, 35.

Harborne, J. B.; Baxter, H. *The Handbook of Natural Flavonoids*, Vol. 1; John Wiley & Sons: Chichester, UK, 1999.

Harnly, J. M.; Dohert, F.; Beecher, G. R.; Holden, J. M.; Hayowitz, D. B.; Bhagwat, S.; Gebhardt, S. Flavonoid content of U. S. fruits, vegetables, and nuts. *J.Agric. Food Chem.* **2006,** *54* (26), 9966–9977.

Heim, K. E.; Tagliaferro, A. R.; Bobilya, D. J. Flavonoid Antioxidants: Chemistry, Metabolism and Structure Activity Relationships. *J. Nutr. Biochem.* **2002,** *13* (10), *572*, 584.

Ho, C. T.; Lee, C. Y.; Huang, M. T. *Phenolic Compounds in Food and Their Effects on Health. I. Analysis, Occurrence & Chemistry*. American Chemical Society: Austin, 1992; p 506.

Hollman, P. H.; Katan, M. B. Dietary Flavonoids: Intake, Health Effects and Bioavailability. *Food Chem. Toxicol.* **1999,** *37* (9), 937–942.

Isoda, H.; Motojima, H.; Onaga, S. et al. Analysis of the Erythroid Differentiation Effect of Flavonoid Apigenin on K562 Human Chronic Leukemia Cells. *Chem. Biol. Interact.* **2014,** *220*, 269–277.

Jin, S.; Zhang, Q. Y.; Kang, X. M.; Wang, J. X.; Zhao, W. H. Daidzein Induces MCF-7 Breast Cancer Cell Apoptosis via the Mitochondrial Pathway. *Ann. Oncol*, **2010,** *21*, 263–268.

Jucá, M. M.; Filho, F. M. S. C. de Almeida, J. C.; Mesquita, D. S.; Barriga, J. R. M.; Dias, K. C. F.; Barbosa, T. M.; Vasconcelos, L. C.; Leal, L. K. A. M.; Ribeiro, J. R. et al. Flavonoids: Biological Activities and Therapeutic Potential. *Nat. Prod. Res.* **2020,** *5*, 692–705.

Kaleem, M.; Ahmad, A. Flavonoids as Nutraceuticals. In *Therapeutic, Probiotic, and Unconventional Foods*; Grumezescu, M. A.; Holban, A. M., Eds.; Academic Press: Cambridge, 2018; pp 137, 155.

Kaushik, S.; Shyam, H.; Agarwal, S.; Sharma, R.; Nag, T. C.; Dwivedi, A. K.; Balapure, A. K. Genistein Potentiates Centchroman Induced Antineoplasticity in Breast Cancer via PI3K/Akt Deactivation and ROS Dependent Induction of Apoptosis. *Life Sci.* **2019,** *239*, 117073.

Kerry, N.; Abbey, M. Red Wine and Fractionated Phenolic Compounds Prepared from Red Wine Inhibit Low Density Lipoprotein Oxidation In Vitro. *Atherosclerosis* **1997,** *135*, 93–102.

Keservani, R. K.; Kesharwani, R. K.; Vyas, N.; Jain, S.; Raghuvanshi, R.; Sharma, A. K. Nutraceutical and Functional Food as Future Food: A Review. *Der Pharmacia Lettre* **2010a,** *2* (1), 106–116.

Keservani, R. K.; Kesharwani, R. K.; Sharma, A. K.; Vyas, N.; Chadoker, A. Nutritional Supplements: An Overview. *Int. J. Curr. Pharma. Rev. Res.* **2010b,** *1* (1), 59–75.

Keservani, R. K.; Sharma, A. K. Flavonoids: Emerging Trends and Potential Health Benefits. *J. Chin. Pharma. Sci.* **2014,** *23* (12), 815.

Keservani, R. K.; Sharma, A. K.; Kesharwani, R. K. Nutraceutical and Functional Foods for Cardiovascular Health. In *Food Process Engineering*; Apple Academic Press, CRC Press, 2016a; pp 257–278. ISBN: 9781771884020.

Keservani, R. K.; Sharma, A. K.; Kesharwani, R. K. Medicinal Effect of Nutraceutical Fruits for the Cognition and Brain Health. *Scientifica* **2016b,** Article ID 3109254, 10 pp. DOI: 10.1155/2016/3109254.

Keservani, R. K.; Sharma, A. K.; & Kesharwani, R. K. An Overview and Therapeutic Applications of Nutraceutical and Functional Foods. *Recent Adv. Drug Deliv. Technol.* **2017,** 160–201. ISBN:9781522507543.

Keservani, Raj K.; Kesharwani, Rajesh K.; Sharma, Anil K.; Ahmad, Mohd Fasih. Stress, Parkinson's, Alzemer's disease, Role of Dietary Supplements. In: *Nutraceutical and Functional Foods in Human Life and Disease Prevention*; Bagchi, D.; Preuss, H. G.; Swaroop, A., Eds.; CRC, Press, Taylor and Francis: Boca Raton, 2015; Chapter 14, pp 241–253. ISBN: 9781482237214.

Keservani, R. K.; Sharma, A. K.; Kesharwani, R. K., Eds. *Nutraceuticals and Dietary Supplements, Applications in Health Improvement and Disease Management*; CRC Press, 2020. ISBN: 9781771888738.

Khajuria, R.; Singh, S.; Bahl, A. General Introduction and Sources of Flavonoids. In: *Current Aspects of Flavonoids: Their Role in Cancer Treatment*; Singh Tuli H., Eds.; Springer: Singapore, 2019.

Khoo, H. E.; Azlan, A.; Tang, S. T.; Lim, S. M. Anthocyanidins and Anthocyanins: Colored Pigments as Food, Pharmaceutical Ingredients, and the Potential Health Benefits. *Food Nutr. Res.* **2017,** *61*, 1361779.

Koes, R. E.; Quattrocchio, F. The Flavonoid Biosynthetic Pathway in Plants: Function and Evolution. *Bio. Essays* **1994,** *16* (2), 123–132.

Kozlowska, A.; Szostak-Wegierek, D. Flavonoids-Food Sources and Health Benefits. *Roczniki Panstwowego Zakladu Higieny* **2014,** *65* (2).

Kozłowska, A.; Szostak-Wegierek, D. Flavonoids—Food Sources, Health Benefits, and Mechanisms Involved. In *Bioactive Molecules in Food*; Reference Series in Phytochemistry; Mérillon, J. M., Ramawat, K., Eds.; Springer: Cham, Switzerland, 2017.

Kumar, S.; Pandey, A. K. Chemistry and Biological Activities of Flavonoids: An Overview. *Sci. World J.* **2013,** 162750.

Larson, A. J.; Symons, J. D.; Jalili, T. Therapeutic Potential of Quercetin to Decrease Blood Pressure: Review of Efficacy and Mechanisms. *Adv. Nutr.: Int. Rev. J.* **2012,** *3* (1), 39–46.

Liang, T.; Guan, R.; Wang, Z.; She, H.; Xia, Q.; Liu, M. Comparison of Anticancer Activity and Antioxidant Activity Between Cyanidin-3-O-Glucoside Liposomes and Cyanidin-3-O-Glucoside in Caco-2 Cells In Vitro. *RSC Adv.* **2017,** *7*, 37359–37368.

Lopez, M.; Martinez, F.; Del Valle, C.; Orte, C.; Miro, M. Analysis of Phenolic Constituents of Biological Interest in Red Wines by High-Performance Liquid Chromatography. *J. Chromatogr. A*, **2001,** *922* (1–2), 359–363.

Maleki, S. J.; Crespo, J. F.; Cabanillas, B. Anti-Inflammatory Effects of Flavonoids. *Food Chem.* **2019,** *299*, 125124

Martens, S.; Preuss, A.; Matern, U. Multifunctional Flavonoid Dioxygenases: Flavonols and Anthocyanin Biosynthesis in *Arabidopsis thaliana* L. *Phytochem.* **2010,** *71*, 1040–1049.

Meng, X. H.; Liu, C.; Fan, R.; Zhu, L. F.; Yang, S. X.; Zhu, H. T.; Wang, D.; Yang, C. R.; Zhang, Y. J. Antioxidative Flavan-3-ol Dimers from the Leaves of *Camellia fangchengensis*. *J. Agric. Food Chem.* **2018,** *66*, 247–254.

Mishra, A.; Sharma, A. K.; Kumar, S.; Saxena, A. K.; Pandey, A. K. Bauhinia Variegata Leaf Extracts Exhibit Considerable Antibacterial, Antioxidant, and Anticancer Activities. *Biomed. Res. Int.* **2013,** 915436.

Nabavi, S. M.; Samec, D.; Tomczyk, M.; Milella, L.; Russo, D.; Habtemariam. S.; Suntar, I.; Rastrelli L.; Daglia, M.; Xiao, J. et al. Flavonoid Biosynthetic Pathways in Plants: Versatile Targets for Metabolic Engineering. *Biotechnol. Adv.* **2020,** *38*, 107316.

Nijveldt, R.; Van Nood, E. L. S.; Van Hoorn, D. E.; Boelens, P. G.; Van Norren, K.; Van Leeuwen, P. A. Flavonoids: A Review of Probable Mechanisms of Action and Potential Applications. *Am. J. Clin. Nutr.* **2001,** *74* (4), 418–425.

Olagaray, K. E.; Bradford, B. J. Plant Flavonoids to Improve Productivity of Ruminants—A Review. *Anim. Feed Sci. Tech.* **2019,** *251*, 21–36.

Oliveira, J.; Mateus, N.; de Freitas, V. Flavanols: Catechins and Proanthocyanidins. In *Natural Products*; Ramawat, K. G.; Mérillon, J. M.; Eds.; Springer: Berlin/Heidelberg, Germany, 2013; pp 1753–1801.

Panche, A.; Chandra, S.; Diwan, A. et al. Alzheimer's and Current Therapeutics: A Review. *Asian. J. Pharm. Clin. Res.* **2015,** *8*, 14–19.

Panche, A. N.; Diwan, A. D.; Chandra, S. R. Flavonoids: An overview. *J. Nutr. Sci.* **2016,** *5*, e47.

Pinto, D. C. G. A.; Simões, M. A. M.; Silva, A. M. S.*Genista tridentata* L.: A Rich Source of Flavonoids with Anti-Inflammatory Activity. *Medicines* **2020,** *7*, 31.

Pons, Z.; Margalef, M.; Bravo, F. I.; Arola-Arnal, A.; Muguerza, B. Grape Seed Flavanols Decrease Blood Pressure *via* Sirt-1 and Confer a Vasoprotective Pattern in Rats. *J. Funct. Foods* **2016,** *24*, 164–172.

Pourakbari, R.; Taher, S. M.; Mosayyebi, B.; Ayoubi-Joshaghani, M. H.; Ahmadi, H.; Aghebati-Maleki, L. Implications for Glycosylated Compounds and Their Anti-Cancer Effects. *Int. J. Biol. Macromol.* **2020,** *163*, 1323–1332.

Rauter, A. P.; Ennis, M.; Hellwich, K. H.; Herold, B. J.; Horton, D.; Moss, G. P. et al. Nomenclature of Flavonoids (IUPAC recommendations 2017). *Pure Appl. Chem.* **2018,** *90*, 1429–1486.

Rice-Evans, C. A.; Miller, N.; Paganga, G. Structure-Antioxidant Activity Relationships of Flavonoids and Phenolic Acids. *Free Rad. Biol. Med.* **1996,** *20* (7), 933–956.

Rodríguez-García, C.; Sánchez-Quesada, C.; Gaforio, J. J. Dietary Flavonoids as Cancer Chemopreventive Agents: An Updated Review of Human Studies. *Antioxidants* **2019,** *8*, 137.

Rolnik, A.; Zuchowski, J.; Stochmal, A.; Olas, B. Quercetin and Kaempferol Derivatives Isolated from Aerial Parts of *Lens culinaris* Medik as Modulators of Blood Platelet Functions. *Ind. Crop. Prod.* **2020,** *152*, 112536.

Ruiz-Cruz, S.; Chaparro-Hernández, S.; Hernández-Ruiz, K. L.; Cira-Chávez, L. A.; Estrada-Alvarado, M. I.; Ortega, L. E. G.; Ornelas-Paz, L. L.; Mata, M. A. L. Flavonoids: Important

Biocompounds in Food. In *Flavonoids—From Biosynthesis to Human Health*; Justino, G. C., Ed.; IntechOpen: London, 2017.
Saini, N.; Gahlawat, S. K.; Lather, V. Flavonoids: A Nutraceutical and Its Role as Anti-Inflammatory and Anticancer Agent. In *Plant Biotechnology, Recent Advancements and Developments*; Gahlawat, S., Salar, R., Siwach, P., Duhan, J., Kumar, S., Kaur, P., Eds.; Springer: Singapore, 2017.
Samanta, A.; Das, G.; Das, S. K. Roles of Flavonoids in Plants. *Int. J. Pharm. Sci. Technol.* **2011,** *6*, 12–35.
Santos, E.; Maia, B.; Ferriani, A.; Teixeira, S. Flavonoids: Classification, Biosynthesis and Chemical Ecology. **2017,** 10.5772/67861.
Saric, A.; Balog, T.; Sobocanec, S.; Kusic, B.; Sverko, V.; Rusak, G.; Likic, S.; Bubalo, D.; Pinto, B.; Reali, D.; Marotti, T. Antioxidant Effects of Flavonoid from Croatin *Cystus incanus* L. Rich Bee Pollen. *Food Chem. Toxicol.* **2009,** *47* (3), 547–557.
Shah, A. L.; Smith, D. Flavonoids in Agriculture: Chemistry and Roles in, Biotic and Abiotic Stress Responses, and Microbial Associations. *Agronomy* **2020,** *10*, 1209.
Sharma, V.; Ramawat, K. G. Isoflavonoids. In *Natural Products*; Ramawat, K., Mérillon, J. M., Eds.; Springer: Berlin/Heidelberg, Germany, 2013.
Sokolov, A. N.; Pavlova, M. A.; Klosterhalfen, S.; Enck, P. Chocolate and the Brain: Neurobiological Impact of Cocoa Flavnols on Cognition and Behavior. *Neurosci. Biobehavi. Rev.* **2013,** *37* (10), 2445–2453.
Soobrattee, M. A.; Neergheen, V. S.; Luximon-Ramma, A.; Aruoma, O. I.; Bahorun, T. Phenolics as Potential Antioxidant Therapeutic Agents: Mechanism and Actions. *Mutation Res./Fundamental Mol. Mech. Mutagen.* **2005,** *579* (1), 200–213.
Symonowicz, M.; Kolanek, M. Flavonoids and Their Properties to Form Chelate Complexes. *Biotechnol. Food. Sci.* **2012,** *76* (1), 35–41.
Tavsan, Z.; Kayali, H. A. Flavonoids Showed Anticancer Effects on the Ovarian Cancer Cells: Involvement of Reactive Oxygen Species, Apoptosis, Cell Cycle and Invasion. *Biomed. Pharmacother.* **2019,** *116*, 109004.
Terahara, N. Flavonoids in Foods: A Review. *Nat. Prod. Com.* **2015,** *10*, 521–528.
Toh, J. Y.; Tan, V. M. H.; Lim, P. C. Y.; Lim, S. T.; Chong, M. F. F. Flavonoids from Fruit and Vegetables: A Focus on Cardiovascular Risk Factors. *Curr. Atherosclerosis Rep.* **2013,** *15* (12), 1–7.
Tripoli, E.; La Guardia, M.; Giammanco, S.; Di Majo, D.; Giammanco, M. Citrus Flavonoids: Molecular Structure, Biological Activity and Nutritional Properties: A Review. *Food Chem.* **2007,** *104* (2), 466–479.
Tsao, R. Chemistry and Biochemistry of Dietary Polyphenols. *Nutrition* **2010,** *2*, 1231–1246.
Tungmunnithum, D.; Thongboonyou, A.; Pholboon, A.; Yangsabai, A. Flavonoids and Other Phenolic Compounds from Medicinal Plants for Pharmaceutical and Medical Aspects: An Overview *Med.* **2018,** *5*, 93.
Vitale, D. C.; Piazza, C.; Melilli, B.; Drago, F.; Salomone, S. Isoflavones: Estrogenic Activity, Biological Effect and Bioavailability. *Eur. J. Drug Metab. Pharmacokinet.* **2013,** *38*, 15–25.
Vuong, Q. V.; Golding, J. B.; Nguyen, M.; Roach, P. D. Extraction and Isolation of Catechins from Tea. *J. Sep. Sci.* **2010,** *33*, 3415–3428.
Wang, J.; Cao, X.; Ferchaud, V.; Qi, Y.; Jiang, H.; Tang, F.; Yue, Y.; Chin, K. L. Variations in Chemical Fingerprints and Major Flavonoid Contents from the Leaves of Thirty-One Accessions of *Hibiscus sabdariffa* L. *Biomed. Chromatogr.* **2016,** *30*, 880–887.

Williams, R.; Spencer, J.; Rice- Evans, C. Flavonoids: Antioxidants or Signalling Molecules? Freeradic. *Biol. Med.* **2004,** *36*, 838–849.

Yao, L. H.; Jiang, Y. M.; Shi, J.; Tomás-Barberán, F. A.; Datta, N.; Singanusong, R.; Chen, S. S. Flavonoids in Food and Their Health Benefits. *Plant Foods Human Nutr.* **2014,** *59,* 113–122.

Zahoor, M.; Shafiq, S.; Ullah, H.; Sadiq, A.; Ullah, F. Isolation of Quercetin and Mandelic Acid from *Aesculus indica* Fruit and Their Biological Activities. *BMC Biochem.* **2018,** *19*, 5.

Zaidun, N. H.; Thent, Z. C.; Latiff, A. A. Combating Oxidative Stress Disorders with Citrus Flavonoid: Naringenin. *Life Sci.* **2018,** *208*, 111–122.

Zhang, Z.; Liao, P.; Zhu, M.; Li, W.; Hu, D.; Chen, L.; Guan, S. Baicalin Attenuates Cardiac Dysfunction and Myocardial Remodeling in a Chronic Pressure-Overload Mice Model. *Cell. Physiol. Biochem.* **2017,** *41*, 849–864.

Zuiter, A. S.; Zarqa, J. Proanthocyanidin: Chemistry and Biology: From Phenolic Compounds to Proanthocyanidins. In *Reference Module in Chemistry, Molecular Sciences and Chemical Engineering*; Reedijk, J., Ed.; Elsevier: Amsterdam, The Netherlands, 2014; pp 1, 29.

PART II

Extraction and Biochemistry of Flavonoids

CHAPTER 3

METHODS OF FLAVONOIDS EXTRACTION

CHITHRA A.[1], DARSHANA K.[1], GIRISHREE R.[1], DHARANYSHRI R.[1], RAJASEETHARAMA SEKAR[1], and MONI PHILIP JACOB K.[2]

[1]*Bioprospecting Research Laboratory, Department of Biotechnology, Bannari Amman Institute of Technology, Sathyamangalam, Erode, India*

[2]*Department of Biotechnology, Bannari Amman Institute of Technology, Sathyamangalam, Erode, India*

ABSTRACT

Flavonoids are polyphenols found in plants. Flavonoids have several medicinal properties such as anticancer, antioxidant, antibacterial, anti-inflammatory properties, etc. It finds much application in food, cosmetics, and drug sectors. Hence, strategies employing efficient extraction from plant sources are critical. Flavonoids may be grouped as flavonols, flavanones, flavones, isoflavones, chalcones, and anthocyanidins based on their core structure. Depending on the functional groups and the type, the extraction of different flavonoids may vary significantly. This chapter focuses on the various strategies used in the extraction from traditional to modern methods. Conventional techniques use solvents such as ethanol to extract from plant sources. Nowadays, advanced techniques are used to reduce the consumption of energy and solvent with increased selectivity and efficiency. The various methods utilized for the extraction are ultrasound, pressurized liquid extraction, mechano-chemical, high hydrostatic pressure, negative pressure cavitation, supercritical fluid, microwave, infrared, and intensification of vaporization by decompression to

The Flavonoids: Extraction and Applications. Deepika Saini, Rajesh K. Kesharwani, & Raj K. Keservani (Eds.)

the vacuum, pulsed electric field, enzyme-assisted extraction, and high-voltage electrical discharges. There are no unified operation conditions to achieve high yields and purity. This chapter also focuses on the main parameters involved in the extraction, such as temperature, solvent, sample quantity, time for extraction, etc.

3.1 INTRODUCTION

Humans have made the most from plants' inherent pharmacological and nutritional properties. Plant organs like stems, leaves, roots, fruits, and their related coproducts were used. Such properties are accredited to the phytochemicals present in the various plant tissues. One such phytochemical is flavonoids. Flavonoids are synthesized by the plants as secondary metabolites and are the most abundant plant pigment found, along with carotenoids and chlorophyll. Flavonoids help the plants protect themselves from UV radiation, impart color to the leaves, prevent fat oxidation, and protect the various plant enzymes and vitamins. In plants, flavonoids are synthesized in certain parts of the plant. They are responsible for the color and aroma of flowers, which help them attract insects for pollination, thereby affecting seed/spore dispersion and germination. They also help the plants to protect from abiotic and biotic stresses. In plants, they act as signal molecules to filter UV rays, detoxification, and antimicrobial defense. They impart resistance to drought, frost hardiness, tolerance to freezing, and acclimatization to heat (Panche et al., 2016; Rodríguez De Luna et al., 2020; Wang et al., 2018a, b).

Flavonoids are low molecular weight compounds that possess a 2-phenylchromone nucleus. They are generally acetic acid or phenylalanine derivatives through the shikimic acid pathway. Based on the annularity of ring C, the connection position of ring B and oxidation degree classification of flavonoids are made. Flavonoids are generally found either as glycosylated or esterified forms. It consists of three rings, namely rings A (C_6) and B(C_6), linked by 3-carbon-ring C (Figure 3.1).

Flavonoids are generally obtained from natural sources especially through plants. More than 9,000 flavonoids are reported in the literature. Flavonoids (Keservani and Sharma, 2014) possess anticancer, anti-inflammatory, antioxidant, antimutagenic, antiviral, antithrombotic, antibacterial properties, etc. So, flavonoids are routinely used in cosmetic, food, and pharmaceutical industries. Humans have made several efforts to extract these flavonoids from plants by several techniques for these properties. A daily intake of up to 500 mg of flavonoids from various sources is recommended from dietary sources. Such increasing demands have made it possible to synthesize flavonoids, that is, synthetic flavonoids. But, when compared to synthetic flavonoids, the natural counterparts are relatively less toxic. So humans have recently developed a keen interest in efficiently extracting flavonoids from natural sources (Panche et al., 2016; Rodríguez De Luna et al., 2020; Wang et al., 2018a, b).

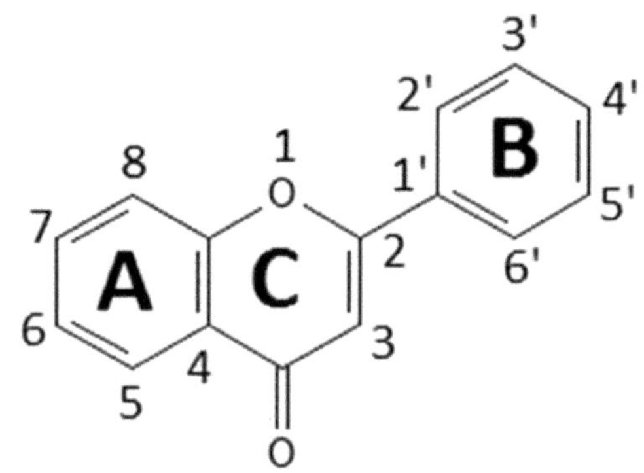

FIGURE 3.1 General structure of flavonoids.
Source: Adapted from Kawser et al. 2016.

To extract these flavonoids from plants, various techniques are employed. In ancient days, simple conventional methods such as heating plants in water were used to remove these plant components. Over time, variations in techniques were employed. Nowadays, with sophisticated instruments, improved efficiency of extraction of flavonoids was reported. Several unconventional methods include pressurized liquid extraction, ultrasound, mechanochemical, supercritical fluid, high hydrostatic pressure, intensification of vaporization by decompression to the vacuum, negative pressure cavitation microwave, and high-voltage electrical discharges, infrared, pulsed electric field, and enzyme-assisted extraction are being used currently for flavonoids extraction. Despite the availability of many techniques, there is no single technique that gives high yield and purity[2]. This chapter summarizes the conventional and unconventional methods employed to extract flavonoids from plants. This chapter also provides a comprehensive idea regarding the description of processes, parameters, equipment, the main parameters, operating conditions, and the main parameters used during the extraction of flavonoids.

3.2 FLAVONOIDS AND THEIR TYPES

Flavonoids are a group of phenolic compounds secreted by plants (Panche et al., 2016). They are predominantly present in fruit and vegetable pulp, seeds, barks, leaves, grains, stems, and roots. Flavonoids are also present in beverages like beer, wine, and green and black tea. Flavonoids incorporate rich colors in plants and shield against UV rays from the sun (Rodríguez De Luna et al., 2020). They are one of the abundant pigments next to chlorophyll and carotenoids. Intensive high-end research has been done on flavonoids to study their ample range of biological activities, namely, antioxidant, anticancer, anti-inflammatory, antimutagenic, antibacterial, antiviral, and vasodilator antithrombotic, antidiabetic, antiaging (Wang et al., 2018a, b).

Flavonoids are compounds with low molecular weight and are biosynthesized via the shikimic acid pathway from phenylalanine derivatives. They have a common skeleton (C6-C3-C6), which comprises a pair of phenyl rings linked through a pyran ring. The fundamental structure of flavonoids favors an array of substitution reaction patterns in the rings (Shibata et al., 2014). The functionality of flavonoids depends on a few principle characteristics: they are a) presence of an O-dihydroxy structure in the second ring, b) presence of a specific double bond in the pyran ring, hydroxyl groups in the first phenyl and pyran ring (de Lira Mota et al., 2009). Based on the chemical nature and position of phenyl and pyran rings, flavonoids are classified into 14 subgroups. The major types are mentioned below:

1. Flavones: Flavones are one among the most significant flavonoids. 2-phenyl-1-benzopyran-4-one is the general chemical structure of flavones. They constitute a ketone group in the fourth position of the pyran ring and a double bond in the phenyl ring in between the second and third positions. Flavones mainly occur in leaves, flowers, and fruits as glycosides. Primary flavone sources include mint, red peppers, parsley, honey, ginkgo biloba, celery, and grapes. Apigenin, luteolin, and tangeritin are flavones (Iwashina, 2013).
2. Flavonols: The skeletal structure of flavonol is 3-hydroxy flavone with ketone group. Flavonol is the elementary unit of proanthocyanins. Their primary sources include onion, apple, grapes, berries, kale, tomatoes, lettuce, tea, and wine. Regular consumption of flavanols provides antioxidants and reduces the risk of heart diseases. Quercetin belongs to this subgroup (Manach et al., 2004).
3. Flavanones: Flavanones are dyhydroflavones with a saturated pyran ring. They occur in both free and combined states as glycosides.

They usually are present in citrus fruits, including lemons, oranges, and grapes. The bitter flavor of the peel and juice of citrus is due to flavanones' presence. Naringenin, hesperidin, and eriodictyol belong to the flavanone subgroup. Flavanone has numerous health benefits associated with its free radical scavenging activity (Iwashina, 2013).

4. Isoflavonoids: Isoflavonoids are 3-phenyl chromen-4-one. They are formed by the migration of the benzene ring from the second position to the third position of the central ring. Their presence is limited to soybeans and legumes in the plant kingdom. Daidzein and genistein are common isoflavonoids (Matthies et al., 2008).
5. Aurones: In the chemical structure of aurones, a furan ring replaces the pyranic ring of the chalcones via conversion reaction.
6. Anthocyanidins: Anthocyanidins are a 2-phenyl benzopyrylium group of compounds. They are present in most wild fruits and incorporate a rich purple color to the fruits (Rodríguez De Luna et al., 2020).
7. Anthocyanins: Anthocyanins are structurally indistinguishable from anthocyanidins. Anthocyanins have a glycosylated region in the oxygen substituent in third position carbon. They are the chief pigment present in plants, flowers, and fruits responsible for color. They are generally spotted in the outer cell layer of colored fruits like red grapes, blueberries, blackberries, black currants, strawberries, raspberries, and cranberries. They are highly stable and possess enormous health benefits, enabling their wide usage in the food industry (Giusti and Wrolstad, 2003).
8. Chalcones: Flavonoids without a pyran ring are classified as chalcones and are also mentioned as open-chain flavonoids. Phloretin, phloridzin, chalconaringenin, and arbutin are a few chalcones. Mainly occur in wheat products, tomatoes, bearberries, strawberries, and pears (Hertog et al., 1993).

3.3 METHODS OF EXTRACTION

Flavonoids have broad applications for the welfare of humanity. Several attempts have been made to extract them. Generally, flavonoids are abundant in leaves and flowers, so they are the target region for conducting extraction. On the flip side, fresh leaves and barks are not preferred because wax and resin make this tissue hard to handle during the extraction process (Nurdalilah et al., 2018). In ancient times flavonoid extraction was carried out by boiling the target tissue in water. Currently, there are two methods of extraction: conventional and unconventional extraction methods. The traditional method utilizes low-cost equipment, a humongous amount of extraction solvent, high temperature, atmospheric pressure, and a higher extraction time. At the same time, unconventional extraction methods use sophisticated equipment which can operate at higher temperature and pressure conditions. The latter is ecological, modern and requires less extraction time (Rodríguez De Luna et al., 2020).

3.3.1 CONVENTIONAL EXTRACTION METHODS

The currently used conventional extraction methods for bioactive compounds are heat reflux, maceration, digestion, infusion, decoction, and percolation. The soxhlet extraction technique is a modernized approach to maceration and decoction, and this is extensively used today. The conventional methods require enormous amounts of solvents like water and ethanol. Acetone, acrylonitrile, methanol, petroleum ether, ethyl acetate, dichloromethane, and hexane are also used; they are highly hazardous. The temperature requirements of these processes are hardly above 100°C. Most of these processes occur in open vessels, so the ratio of solvent evaporated during the extraction is higher. A workable amount is recovered only in heat flux and the soxhlet extraction method. The extraction period can range from a few minutes to a few weeks, as in maceration. The inconveniences mentioned above are the significant disadvantages of conventional methods.

Along with these disadvantages, thermal degradation of biologically active compounds can also occur, and the usage of toxic chemicals and higher energy consumption is not a sustainable option. These problems restrict biologically active compounds in the pharmaceuticals, cosmetic, and food industries. Despite this, soxhlet extraction is used in the comparative study of recently developed, more sophisticated techniques (Zhang et al., 2011).

3.3.2 UNCONVENTIONAL EXTRACTION METHODS

The unconventional extraction methods are, on the whole, classified into mechanical force methods, electromagnetic forces, electric forces, and enzymatic extraction for specific reactions. Unconventional extraction methods have advantages over conventional methods: 1. Cutting of reduction time, 2. No thermal degradation of thermally labile compounds, 3. Less solvent requirement, 4. Reduction in energy consumption, 5. Enhanced yield with selectivity, 6. Reproducibility, 7. Environmental friendly and sustainable

(Luengo et al., 2013; Briones-Labarca et al., 2019). An overview of various unconventional methods is given in Figure 3.2.

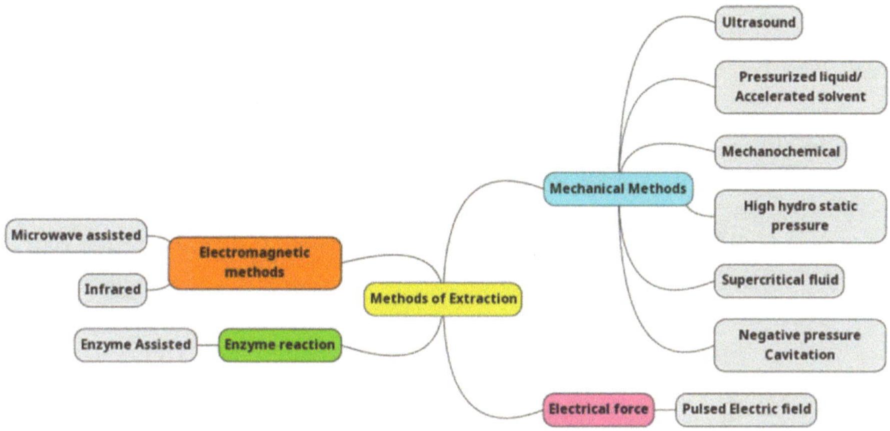

FIGURE 3.2 Flowchart representing the various unconventional methods of extraction of flavonoids.

3.4 MECHANICAL FORCE METHODS

3.4.1 ULTRASOUND EXTRACTION

Ultrasound extraction depends mostly on mechanical wave propagation which induces the movement of bubble cavitation because of temperature change and pressure change. At the point when the cavitation is formed, the bubbles break down through the pressure–rarefaction patterns of mechanical vibration. When bubbles explode, their heating rate and tension can reach 5000 K and 100 MPa. These circumstances cause the material present in the plant to break down, thereby allowing it to escape the extract (Sheng et al., 2014). The ultrasonic device generates waves using a probe or bath, and they are distinctive while delivering ultrasound waves. The most utilized gadget is an ultrasonic bath, a tempered steel tank that predominantly works at 40-kHz frequencies. However, when combined with a probe that operates primarily at 20 kHz, the ultrasound produces a much greater force than an ultrasonic bath. The ultrasonic probe is submerged straightly within the test container, usually composed of steel and glass. Force conveyed by the probe is typically greater than bath frameworks due to the vibrations provided via the test tip. This constructs the trial to be more repeatable than bath gear. In any case, probes are more modest, and for the most part, fewer measures of material of plant are treated (Esclapez et al., 2011).

3.4.1.1 OVERALL BENEFITS AND DRAWBACKS OF ULTRASOUND EXTRACTION

The benefits of ultrasound extraction are somewhat workable and straightforward, contrasted with the strategies whose strategies need to create a void space or build tension to be worked. In expansion, it likewise provides the benefit that tends to be utilized to supplement and promote the extraction of different procedures like microwaves or supercritical liquids. Ultrasound extraction is considered an energy-saving methodology because the extraction time is short and has greater precision and high yields. One of its weaknesses is the probing framework, where little weight measures can be dealt with. In any case, working in series can resolve this issue (Chemat et al., 2017). The bath framework has a fundamental limitation. It has limited repeatability due to the water present in the steel container or the glass constricting the energy that is not satisfactorily transferred to such a solvent used for the extraction.

3.4.2 PRESSURIZED LIQUID/ACCELERATED SOLVENT (PLE)

The first presentation of pressurized fluid extraction, also named accelerated dissolvent extraction, occurred in 1995. Over its temperature range, it is an interaction that connects great tension and temperatures with natural solvents (Richter et al., 1996). PLE seems to be a promising technology that works with supercritical fluids dissolvent at pressures ranging between 4 and 12 MPa and temperatures between 50°C and 300°C. There are a few marketed devices for extracting bioactive combinations using pressured fluids like the accelerated solvent extractor ASE200. This apparatus permits the plant extraction at a maximum tension (close to 13 MPa) and significant temperature (close to 200°C) (Hirondart et al., 2020). The ASE350 (Dionex, Thermo Fisher Scientific, USA) is a type that is additionally utilized in PLE extraction, working in pressure exceeding 10.3 MPa (Lama-Muñoz et al., 2020). The sample is enclosed in a container loaded with dissolvent and used to dynamically extricate the sample under an elevated range of temperature and tension settings for small durations (5–10 minutes) in this equipment. The compressed gas is utilized for washing the sample once it is removed from the cell extraction and placed in a flask (Tamkutė et al., 2020).

3.4.2.1 THE APPROACHES TO PRESSURIZED LIQUID

PLE is used to resolve bioactive combinations in plant lattices that are rapid, clean, and environmentally friendly. There are many market instruments for this extraction method. With these types of apparatus, the extraction methods may be customized and then performed, which is beneficial for the quality testing phase. The expensive cost of equipment may be an issue with this procedure (Hirondart et al., 2020; Pereira et al., 2020; Vigano et al., 2020).

3.4.3 MECHANOCHEMICAL EXTRACTION (MAE)

Korolev et al. (2003) used mechanochemical-assisted extraction (MAE) for the first time in 2003 to extract triterpene acids. From that point forward, this procedure has gotten a lot of consideration from the industries because of its small utilization of natural dissolvents (Xie et al., 2013). MAE involves three main steps: (1) raw substance pretreatment, (2) raw substance enactment by a mechanochemical process, and (3) extraction technique. The crude plant material is typically dehydrated and pounded to achieve the molecular mass dissemination of 0.5–2 mm during sample preparation. This substance is stored at room temperature in dry conditions. The following stage happens mainly in the ball mill plant, in which the material of the plant is blended with any accompanying reactants: $NaHCO_3$, NaOH, Na_2CO_3, SiO_2, or $Na_2B_4O_7.10H_2O$, commonly called alkaline specialists. In this subsequent stage, the goal is to change the flavonoids in the chemical form, present in the plant materials. Flavonoids are mildly acidic chemicals that cause a balance response when homogenized with alkaline inside the ball mill. In flavonoid salt, the result of the neutralization response is obtained. From a flavonoid, a salt prepared is chemically more water-soluble than the flavonoid itself. In the third step, the mechanochemical step of the process permits water to be utilized as an extraction solvent. Finally, the extraction step occurs. The flavonoid-developed salt is decomposed in H_2O and separated. The supernatant is then collected and fermented to transform the saline into the flavonoid. Finally, the extract solution is condensed in a vacuum evaporator to obtain the extract (Wu et al., 2017).

3.4.3.1 THE OUTLOOK OF MAE

The main advantages of MAE are the water used as a solvent, the reasonably short duration of extraction happens, with the ease with which the materials can be controlled during the process. Regardless of these benefits, the kinetics experiments are yet done to determine the essential information of weight exchange which helps to improve the method further. The tests are critical for increasing MAE in current contexts (Zhu et al., 2011). The reaction between the flavonoid and the soluble basicity agent in the ball mill should be thoroughly examined. Because this is a solvent-free interaction, it is fascinating to learn about the parameters which influence it. Scanning electron microscopy (SEM), a surface characterization technique, is an incredibly valuable representation method for studying the morphological alterations in chemicals that occur inside the ball mill and the mechanism involved in the reaction (Wang et al., 2009).

3.4.4 HIGH HYDROSTATIC PRESSURE EXTRACTION (HHPE)

This HHPE is considered a unique approach for obtaining flavonoids first used in 2005 (Shouqin et al., 2005). The method's standards are divided into hexad stages: (1) Shredded dehydrated materials of the plant are sieved through 40 or 60 sizes, (2) the relevant dissolvent is chosen, (3) the substances which are shredded are routinely combined with the dissolvent within a polyethylene sack, then they got sterilized and fastened, (4) after packing, it is placed in a pressure container containing valves to distribute tension; with that thermal regulator is placed inside the container, (5) this container then compressed using a liquid, regularly H_2O, and when the desired tension is obtained, the duration of solvent extraction is tailored, and lastly, (6) the crude particles are removed from the extract (Xi, 2017).

3.4.4.1 THE OVERVIEW OF HHPE

HHPE is a promising technology for releasing bioactive chemicals from plants in further years, even though it has not been used on a large scale. This unique technology requires more investigation, particularly identifying essential variables that influence extract yields and determining kinetic models (Briones-Labarca et al., 2015).

3.4.5 SUPERCRITICAL FLUID (SF) EXTRACTION

The SF extraction is a technique that normally utilizes CO_2 as dissolvable. The utilization of supercritical liquids for extracting chemicals from plants began in the 1980s, and it has since found widespread use in cosmetics, medicine, and food enterprises. The SF has proven to extract both polar and nonpolar combinations such as polyphenols, lipids, and carotenoids (Huang et al., 2012).

3.4.5.1 THE VIEWPOINT ON SF EXTRACTION

Compared to other standard and alternative procedures, the SF extraction procedure allows the plant extract as a sterilizer that recovers quickly because it does not concentrate the solvent during the process. The CO2 used in certain SFE methods is mainly employed as a product of manufacturing applications, such as brewing beer, which reduces outflows toward the environment (Raks et al., 2018). Another benefit of SFE is the ability to couple with liquid and gas chromatography at the end, which is commonly utilized to detect unpredictable, volatile mixes. The SFE approach is used on a modern industrial scale, demonstrating its broad applicability from milligrams in research organizations to tonnes in large companies. However, SFE flavonoid separation is not effectively used to know how to increase the yield by changing variables. The limitation of this technique is the early equipment cost is extremely high, and it takes a lot of ability to conduct SFE release strategies (Ibáñez et al., 2015).

3.4.6 NEGATIVE PRESSURE CAVITATION EXTRACTION (NPCE)

This NPCE entails introducing gas (usually nitrogen) into a vessel containing the plant material and solvent in a pressurized environment. The gas stream and negative tension can be controlled (Liu et al., 2009). Under vacuum, the gas contact with the solvent assists in making the cavitation interaction which produces vicious and turbulence flow developments of the solids and solvent. Therefore, cells get harmed because of the solid's collision and air pockets, and this technique works on the mass exchange of the flavonoids from within the cells to the solvents (Yao et al., 2015; Dong et al., 2011).

3.4.6.1 THE OUTLOOK ON NPCE

Along with NPCE, crossbreed or synergistic procedures were used to streamline prepared parameters. Consolidating using microwave and ultrasonic extractions improved the NPCE. Because of the increased burst of the cell wall, the synergistic effect of NPCE also helped to increase the extraction yield of total flavonoids (Jiang et al., 2018). These combinations of processes with NPCE are being investigated to improve extraction yield and update variables to expand with modern industrial applications (Wang et al., 2018a, b).

3.5 ELECTROMAGNETIC FORCE METHODS

3.5.1 MICROWAVE-ASSISTED EXTRACTION (MWAE)

Microwave ovens have recently obtained importance in the field of logical inquiry research and their widespread utilization as kitchenware. One of the most popular ways is microwave-assisted extraction (MWAE) which successfully separates chemicals associated with plant components (Alonso-Carrillo et al., 2017). Even though exceptional MW apparatus for a research facility and scientific applications is now economically accessible, few research centers utilize microwave ovens purchased as kitchen machines. On the other hand, this machine is frequently updated and adjusted to meet the needs of the situation. The insertion of a condenser outside the machine to retrieve the detergent loss and the installation of temperature regulation are some of the kitchen microwave's good features. Microwaves are a common power source, and most of the apparatus features good force regulation. MW equipment commercially available is specially used for research facilities and offers more extraordinary adaptability in time, temperature, and power than kitchen hardware. The extraction process to collect the compounds of interest follows a similar activity of the MW equipment independently (Zill-E-Huma et al., 2011).

3.5.1.1 PERSPECTIVES OF MWAE

In connection with using nontoxic detergents or even the chance to carry out productions without a detergent, MWAE outperforms traditional extraction methods. Additionally, MWAE permits the creation of numerous plans to change the selectivity for precise compounds. For instance, MWAE has been successfully used to upgrade the recovery of essential flavonoids such as quercetin, kaempferol, and their glucoside subordinates (Rodríguez-Pérez et al., 2016). Flavonoid production is a process that involves extracting flavonoids from plants. Total flavonoid content (TFC), total quercetin (TQ), or quercetin equivalent are the most common forms of MWAE (QE). Further to improve extraction, MWAE has been combined with various approaches. These synergistic cycles have resulted in higher flavonoid extraction yields. A portion of this process incorporates enzyme-assisted ultrasonic microwave (EAUMSE), vacuum microwave hydro dispersion gravity extraction (VMHG), and microwave and ultrasound-assisted extraction (MUAE) (Xu et al., 2018; Yedhu Krishnan et al., 2016; Jha et al., 2017).

3.5.2 INFRARED EXTRACTION

Infrared-assisted extraction equipment usually includes an IR light. The radiation source could be a single light with a few power modes or a cluster of lights with power ranges between 50 and 1000 W. The plant source was kept in a round-bottom flask vessel with varying contents based on the rigid thickness. Water, methanol, or other solvents are poured into this flask. The majority of IRAE instruments come with a condenser for detergent recovery. The flask is kept close to the lamp to expose the chemical to IR radiation. The IR radiation heats the detergent and plant source, improving flavonoid production. Usually, the extraction is accomplished (Mou et al., 2017).

3.5.2.1 THE OUTLOOK ON INFRARED EXTRACTION

The IRAE, MWAE, standard heat extraction, and ultrasonic extraction techniques were compared by Cai et al. (2011). When infrared warming was used, it resulted in shorter processing times and increased extraction. They concluded that regular warming time was longer because of heat loss in the flask and air. The most crucial advantage of infrared is that it warms the detergent uniformly, making it appropriate for extracting essential components from combinations like procyanidins and catechins. Several research work has indicated that IR innovation increased the extraction activity compared to MWAE and UE (Liang et al., 2017).

3.6 ELECTRICAL FORCE METHODS

3.6.1 PULSED ELECTRIC FIELD

A pulsed electric field (PEF) is a method of extraction in which vegetable matrices are exposed to the electrical potential to extract phytochemicals. It works on the basic principle, of generating electric pulses by a transformer and increasing the voltage's range from 140 or 220 to 1000 V; sometimes, it may rise over 25000 V. A capacitor is used to transform this higher voltage in a narrow pulse and is done in a chamber sealed with metallic electrodes. The size of this chamber is based on the substantial volume, and there is no singular weight, size, or shape of solids. Hence, PEF is a profitable method to treat whole fruits, even those differing in particle size, like grapes or oranges, or slices (El Darra et al., 2013). The perimeter of the electrode plate may range from 10 to 100 cm, and the two electrodes distance from 1 to 20 cm; however, the size of the chamber depends on the volume of the vegetable material that is to be treated.

3.6.1.1 PARAMETERS AND OPERATION CONDITIONS

3.6.1.1.1 Effects of Plant Material Energy and Electric Field Strength

The principal parameter of PEF is to control the energy per unit mass ratio. The energy is an action of electric current and electric potential, and here it is the adequate time of pulses that may range from a few microseconds to milliseconds. This method can be used at both the industrial and laboratory levels to extract flavonoids. Disruption of the cell wall is effectively enhanced by this method. The electric charges will interact to rupture the bovine cells within the cells with the electric field. In this case, cells are subjected to disruption, improving the internal pore formation. This pore size corresponds to the intensities of the electric field. Hence, lower intense electric fields produce tiny pores, and higher powers make larger pores and irreversible cell breakdown. This is known as electroporation, where the external electric field stimulates the permeability of the cell membrane. Cell damage increases the permeability and electroporation, which depends on the strength of the electric field, pulse number, and capacity of the discharge capacitor. Higher electric fields such as 40 kV/cm give more significant solid fragments like mango peels, improving extraction yield. Anyhow, extraction yields always correspond to the optimum rate of the strength of the electric field (Zimmermann et al., 1974; Gachovska et al., 2009).

3.6.1.1.2 Cell Disruption and Permeability

Measurement of cell damage can be defined as acoustic disintegration (Z_a), electrical conductivity disintegration (Z_c), and cutting disintegration (Z_f) indices. These indices are based on the electric field strength where the electric field rises and the integration index also increases. Peiro et al. (2019) did a preliminary study on the treatment of lemon residues. This indicated that PEF lower than 5 kV/cm shows no severe cell damage (Z_c<0.2), while increasing it to 7 kV/cm proportionally increases Z_c to 0.5. Maximum extraction was found on lemon residues treated with 7 kV/cm. Due to other causes, the solids with identical disintegration indices vary in yield. For example, the size and shape of apple fruit are linked to the essential differences in the kinetics of the fruit extract with identical acoustic integration indices (Z_a>0.8). In the case of apple slices, this method was more uniform than the whole apple, and the pulsed electric field affects the external surface alone. Furthermore, smaller particle sizes like sawdust give maximum extraction yield (Grimi et al., 2011).

PEF can also be carried out using solvents. In PEF, basic pH gives a higher yield than acidic pH. Usually, acidic pH shows a higher output of phenolic compounds in conventional methods. Permeability of plant cell membrane is achieved by centrifuging under dry conditions. Evaluation and comparison are made with non-PEF treated, and PEF treated identical parts for the permeability and enhancement of the extraction. Also, incrementing the PEF intensity causes higher permeability, and therefore intracellular compound extraction is improved (Redondo et al., 2018).

3.6.1.1.3 PEF Effects on Properties of Bioactive Compounds

The pulsed electric field can reshape the bioactive compound's property and quality. For example, the PEF technique increases the level of antioxidants in fruits and grains. When a vegetable material is exposed to PEF, it is observed to produce increased autoprotection molecules that are phenolic compounds and the chemical stabilization of flavonoids (Soliva-Fortuny et al., 2009). Also production of flavonoids is increased due to the auto-prevention process by chemical reagents like oxygen. PEF increased the flavonoid extraction by higher enzymatic oxidation associated with polyphenolic compounds (Redondo et al., 2018). This technique also influenced the activity of secondary metabolites because of phenolic compound degradation.

3.6.1.2 PERSPECTIVES OF PULSED ELECTRIC FIELD

A pulsed electric field is majorly used in the food industry to improve the cottonseed oil extraction, juice production, and life-span by inactivating the microbes in beverages such as fruit juice, milk, and beer. It also increases the sucrose yield from sugar beets. Even though it has enormous commercial application, it has been exposed to in-depth scientific research for a deeper understanding of its response to various vegetable matrices and optimizing the process parameters, mainly in the case of auto-protective mechanisms. Hence, it is necessary to do more exhaustive investigations on the PEF method in the flavonoid extraction process.

3.7 ENZYME REACTION

Enzymes have the biocatalytic activity to hydrolyze the cell walls of the plants, releasing the intracellular components. Substrates will bind to the active site of an enzyme where the catalysis process takes place. This enzyme–substrate reaction will lead to a conformational change in the structure and an intense interaction. Enzyme-assisted extraction is a widely used novel and efficient method of extraction from the target molecule depending on the biological catalytic action of various enzymes. The fundamental principle of this method is the breakage of the cell wall by pretreating the plant with multiple enzymes that undergo hydrolysis. Usually, a plant cell wall consists of polysaccharide molecules like hemicellulose, pectin, cellulose, lignin, and a few proteins that act as the substrate. These molecules give the cell wall stability and resist the intracellular molecules' extraction. Thus, EAE uses various carbohydrate hydrolyzing enzymes to break the cell walls. It may be a single or a combination of diverse enzymes. To achieve maximum extraction yield and permeability, enzymes like pectinases, celluloses, and hemicelluloses can hydrolyze the cell wall components and produce distinct compounds like polyphenols and polysaccharides oils, and pigments from various other medicinal compounds (Rodríguez De Luna et al., 2020).

This method is considered efficient for polyphenols extraction, flavonoids, and particularly anthocyanidins extractions from its corresponding glycosides (Gligor et al., 2019). In conventional methods, some phytochemicals present in the cell cytoplasm and a few more are maintained in the lignin-polysaccharide network using H-bonds. And they are consistently not able to be accessed by the solvents.

3.7.1 ENZYME TYPES

Several enzymes are employed in the EAE process, like amylase, xylanase, papain, and protease. They are categorized as hydrolyzation, oxidation-reduction, isomerization, carboxylation, and group transfer enzymes. Each enzyme has a specific characteristic that functions in a particular substrate based on its catalysis process (Gligor et al., 2019). Glycosides are the intracellular insoluble substances found in plant materials like phenolic compounds. They connected to the cell walls with the help of hydrogen bonds and hydrophobic interactions. Various bonds are formed with the cell wall through carboxylic groups like ether bonds are formed by phenolic acids with lignin via hydroxyl groups available in the phenolic aromatic ring, and ester bonds form with other acids. In plant material, some polyphenols are also present as glycosides linked covalently to the glucose modules. Here in flavonoids, glycosides are linked with glucose through hydroxyl groups or carbon–carbon bonds. Also, pectinase hydrolyzes the cellulose, and pectin solubilizes the cell wall components, improving the discharge of intracellular compounds (Nadar et al., 2018).

3.7.2 OPERATING CONDITIONS AND MAIN PARAMETERS

In addition to optimizing the extraction yield, certain parameters and operating conditions must be considered: size of the particle and moisture of the plant, enzyme type and concentration, extraction temperature, extraction time, and pH of the system (Nadar et al., 2018).

3.7.2.1 SIZE OF THE PARTICLE AND MOISTURE OF THE PLANT

Pretreatment of the plant is essential to reduce the moisture content and uniformization of the particle size. In small particle diameters, high content of antioxidant properties is present. Very small particle sizes will have higher bioactive compounds subjected to degradation during cell wall rupture (Santana and Macedo, 2019). Generally, dehydration eliminates the water present in the plant, thus improving the porosity of the cellular molecules. The enzyme–substrate connection is increased, enhancing the diffusion rate that recovers the maximum phytochemicals in the ultimate extract.

3.7.2.2 ENZYME TYPE AND CONCENTRATION

Enzyme type and their concentration play a significant role in efficient extraction. This parameter depends on the plant's nature and composition to achieve specificity. And it is determined based on the enzyme's active site and structure to fix the specific substrate. With higher substrate concentration, the reaction rate can be accelerated by adding the enzymes. The combination of enzymes assures complete cell wall fragmentation and results in higher yield. Ruviaro et al. made a combination with tanasse and beta-glucosidase for the flavonoid extraction from citrus juice by-products, producing a high output. Also, the restoration of hesperetin and naringenin was more significant when a mixture of enzymes was used. This indicates the synergy effect of tanasse and beta-glucosidase (Ruviaro et al., 2019).

3.7.2.3 EXTRACTION TIME

If the exposure time is expanded between the solvent and substrate-enzyme, it gradually increases the solubility of the cell wall components. On an industrial scale, extended time will reduce the quality of the extract and result in energy inefficiency.

3.7.2.4 EXTRACTION TEMPERATURE

The optimal temperature of enzymatic reaction takes place between 15 and 45°C. If the temperature rises above 60 C, it will lead to a reversible reaction of enzymes due to the action of heat. And at the same time, if the temperature is very low, the response will not be accelerated, reducing the extract efficiency. Thus, choosing the specific temperature for high yield extraction is essential.

3.7.2.5 pH OF THE SYSTEM

Maintaining the optimum pH of the extraction process is essential to achieving maximum yield. Most reactions occur under acidic conditions, thus destabilizing the hydrogen bonds linked to the bioactive compounds that help cell walls rupture.

3.7.3 SYNERGY EFFECTS BETWEEN EAE AND OTHER METHODS

Set combinations of various conventional techniques that produce maximum yield with the EAE method. Tchabo et al. used the combination of ultrasound and enzyme for flavonoid extraction from mulberry. It is observed that it yields about 378.02 mg/100 ml of total flavonoid content (Tchabo et al., 2014). Teles et al. (2021) used a combination of enzyme and high hydrostatic pressure (HHP) to extract the phenolic compounds of grape pomace. HHP improves the action of enzymes employed in the extraction process. Recovering hesperidin and naringenin of pomelo peels using ultrasound and enzyme-assisted methods showed effective extraction of bioactive compounds (Van Hung et al., 2020). Combining ultra-sonication, microwave with enzyme approach for the extraction of curcumin from turmeric powder also improves the antioxidant activity of extracted species. These factors significantly interfere with the extraction efficiency (Wang et al., 2021). Thus, the mixture of methods maximized the extraction yield and then separated techniques.

3.7.4 PERSPECTIVES OF THE EAE METHOD

The EAE method provides various advantages like specific substrate binding, greater extraction yield, and accelerates the phytochemical release from compact materials and resistance, such as bark and roots (Nadar et al., 2018). However, large-scale industries do not acquire it as the enzymes are relatively expensive. Small modulation in the parameters like pH, nutrients, oxygen concentration, and temperature will help alter the specificity of enzymes.

3.8 CONCLUSIONS

Plant-based natural products are invaluable, and nowadays, there is an increasing trend to extract and isolate such compounds to understand their full potential. Currently, scientists across the globe are working toward extraction processes that are eco-friendly and economically viable with increased efficiency employing short process time and energy and less consumption of solvents. These extraction techniques also depend on the material from which the bioactive compound is being extracted, the molecule of interest, and the selection criteria for the processes. Using a combination of such extraction techniques may help us overcome the limitations posed by the current approaches. Despite the ginormous amount of literature available on the extraction of flavonoids, the variability of results employing different techniques persists. Understanding the parameters and the variables utilized in these techniques will give an insight into developing new innovative extraction strategies to maximize the yield.

ACKNOWLEDGMENT

The authors would like to thank the Bannari Amman Institute of Technology for providing the necessary facilities to carry out this work.

KEYWORDS

- **flavonoids**
- **extraction**
- **efficiency**
- **conventional methods**
- **unconventional methods**

REFERENCES

Alonso-Carrillo, N.; de los Ángeles Aguilar-Santamaría, M.; Vernon-Carter, E. J.; Jiménez-Alvarado, R.; Cruz-Sosa, F.; Román-Guerrero, A. Extraction of Phenolic Compounds from *Satureja macrostema* Using Microwave-Ultrasound Assisted and Reflux Methods and Evaluation of Their Antioxidant Activity and Cytotoxicity. *Ind. Crops Prod.* **2017,** *103*, 213–221.

Briones-Labarca, V.; Giovagnoli-Vicuna, C.; Canas-Sarazua, R. Optimization of Extraction Yield, Flavonoids and Lycopene from Tomato Pulp by High Hydrostatic Pressure-Assisted Extraction. *Food Chem.* **2019,** *278*, 7517–7559.

Briones-Labarca, V.; Plaza-Morales, M.; Giovagnoli-Vicuña, C.; Jamett, F. High Hydrostatic Pressure and Ultrasound Extractions of Antioxidant Compounds, Sulforaphane and Fatty Acids from *Chilean papaya* (*Vasconcellea pubescens*) Seeds: Effects of Extraction Conditions and Methods. *LWT—Food Sci. Technol.* **2015,** *60*, 525–534.

Cai, Y.; Yu, Y.; Duan, G.; Li, Y. Study on Infrared-Assisted Extraction Coupled with High Performance Liquid Chromatography (HPLC) for Determination of Catechin, Epicatechin, and Procyanidin B2 in Grape Seeds. *Food Chem.* **2011,** *127*, 1872–1877.

Chemat, F.; Rombaut, N.; Sicaire, A. G.; Meullemiestre, A.; Fabiano-Tixier, A. S.; Abert-Vian, M. Ultrasound assisted extraction of food and natural products. Mechanisms, techniques, combinations, protocols and applications. A review. *Ultrason. Sonochem.* **2017,** *34*, 5405–5460.

de Lira Mota, K. S.; Dias, G. E. N.; Pinto, M. E. F.; Luiz-Ferreira, Â.; Monteiro Souza-Brito, A. R.; Hiruma-Lima, C. A.; Barbosa-Filho, J. M.; Batista, L. M. Flavonoids with Gastroprotective Activity. *Molecules* **2009,** *14* (3), 9791–012.

Dong, L. L.; Fu, Y. J.; Zu, Y. G.; Li, J.; Li, X. J.; Efferth, T. Negative Pressure Cavitation Accelerated Processing for Extraction of Main Bioactive Flavonoids from *Radix Scutellariae*. *Chem. Eng. Process. Process Intensif.* **2011,** *50*, 780–789.

El Darra, N.; Grimi, N.; Vorobiev, E.; Maroun, R. G.; Louka, N. Pulsed Electric Field Assisted Cold Maceration of Cabernet Franc and Cabernet Sauvignon Grapes. *Am. J. Enol. Vitic.* **2013,** *64* (4), 4764–4784.

Esclapez, M. D.; García-Pérez, J. V.; Mulet, A.; Cárcel, J. A. Ultrasound-Assisted Extraction of Natural Products. *Food Eng. Rev.* **2011,** *3*, 108–120.

Gachovska, T. K.; Adedeji, A. A.; Ngadi, M. O. Influence of Pulsed Electric Field Energy on the Damage Degree in Alfalfa Tissue. *J. Food Eng.* **2009,** *95*, 558–563.

Giusti, M. M.; Wrolstad, R. E. Acylated Anthocyanins from Edible Sources and Their Applications in Food Systems. *Biochem. Eng. J.* **2003,** *14* (3), 2172–2175.

Gligor, O.; Mocan, A.; Moldovan, C.; Locatelli, M.; Crişan, G.; Ferreira, I. C. Enzyme-Assisted Extractions of Polyphenols—A Comprehensive Review. *Trends. Food. Sci. Technol.* **2019,** *88*, 302–315.

Grimi, N.; Mamouni, F.; Lebovka, N.; Vorobiev, E.; Vaxelaire, J. Impact of Apple Processing Modes on Extracted Juice Quality : Pressing Assisted by Pulsed Electric Fields. *J. Food Eng.* **2011,** *103*, 52–61.

Hertog, M. G.; Hollman, P. C.; Van de Putte, B. Content of Potentially Anticarcinogenic Flavonoids of Tea Infusions, Wines, and Fruit Juices. *J. Agric. Food Chem.* **1993,** *41* (8), 1242–1246.

Hirondart, M.; Rombaut, N.; Fabiano-Tixier, A. S.; Bily, A.; Chemat, F. Comparison Between Pressurized Liquid Extraction and Conventional Soxhlet Extraction for Rosemary Antioxidants, Yield, Composition, and Environmental Footprint. *Foods.* **2020,** *9* (5), 584.

Huang, Z.; Shi, X. H.; Jiang, W. J. Theoretical Models for Supercritical Fluid Extraction. *J. Chromatogr. A.* **2012,** *1250*, 2–26.

Ibáñez, E.; Mendiola, J. A.; Castro-Puyana, M. Supercritical Fluid Extraction. *Encycl. Food Heal.* **2015,** *8*, 227–233.

Iwashina, T. Flavonoid Properties of Five Families Newly Incorporated into the Order *Caryophyllales* (Review). *Bull. Natl. Mus. Nat. Sci.* **2013,** *39*, 25–51.

Jha, P.; Das, A. J.; Deka, S. C. Optimization of Ultrasound and Microwave Assisted Extractions of Polyphenols from Black Rice (*Oryza sativa* cv. Poireton) Husk. *J. Food Sci. Technol.* **2017,** *54*, 3847–3858.

Jiang, Y.; Li, D.; Ma, X.; Jiang, F.; He, Q.; Qiu, S.; Li, Y.; Wang, G. Ionic Liquid–Ultrasound-Based Extraction of Biflavonoids from *Selaginella helvetica* and Investigation of Their Antioxidant Activity. *Molecules* **2018,** *23*, 1–18.

Kawser H. M., Abdal Dayem, A., Han, J., Yin, Y., Kim, K., Kumar Saha, S., Yang, G.M., Choi, H.Y. and Cho, S.G.,Molecular mechanisms of the anti-obesity and anti-diabetic properties of flavonoids. *International journal of molecular sciences*. **2016,** *17*, 4, 569.

Keservani, R. K.; Sharma, A. K. Flavonoids: Emerging Trends and Potential Health Benefits. *J. Chin. Pharma. Sci.* **2014,** *23* (12), 815.

Korolev, K. G.; Lomovskii, O. I.; Rozhanskaya, O. A.; Vasil'ev, V. G. Mechanochemical Preparation of Water-Soluble Forms of Triterpene Acids. *Chem. Nat. Compd.* **2003,** 3*9439*, 366–372.

Lama-Muñoz, A.; del Mar Contreras, M.; Espínola, F.; Moya, M.; Romero, I.; Castro, E. Content Of Phenolic Compounds and Mannitol in Olive Leaves Extracts from Six Spanish Cultivars: Extraction with the Soxhlet Method and Pressurized Liquids. *Food Chem.* **2020,** *320*, 126626.

Liang, Q.; Chen, H.; Zhou, X.; Deng, Q.; Hu, E.; Zhao, C.; Gong, X. Optimized Microwave-Assistant Extraction Combined Ultrasonic Pretreatment of Flavonoids from *Periploca forrestii* Schltr.; Evaluation of Its Anti-Allergic Activity. *Electrophoresis* **2017,** *38*, 1113–1121.

Liu, W.; Fu, Y.; Zu, Y.; Kong, Y.; Zhang, L.; Zu, B.; Efferth, T. Negative-Pressure Cavitation Extraction for the Determination of Flavonoids in Pigeon Pea Leaves by Liquid Chromatography-Tandem Mass Spectrometry. *J. Chromatogr. A* **2009,** *1216*, 3841–3850.

Luengo, E.; Álvarez, I.; Raso, J.; 2013. Improving the Pressing Extraction of Polyphenols of Orange Peel by Pulsed Electric Fields. *Innov. Food. Sci. Emerg. Technol.* **2013,** *17*, 798–794.

Manach, C.; Scalbert, A.; Morand, C.; Rémésy, C.; Jiménez, L. Polyphenols: Food Sources and Bioavailability. *Am. J. Clin. Nutr.* **2004,** *79* (5), 7277–7247.

Matthies, A.; Clavel, T.; Gütschow, M.; Engst, W.; Haller, D.; Blaut, M.; Braune, A. Conversion of Daidzein and Genistein by an Anaerobic Bacterium Newly Isolated from the Mouse Intestine. *Appl. Environ. Microbiol.* **2008,** *74* (15), 4847–4852.

Mou, Q.; He, J.; Yin, R.; Yang, B.; Fu, M.; Fang, J.; Li, H. Response Surface Optimized Infrared-Assisted Extraction and UHPLC Determination of Flavonoid Types from *Flos sophorae*. *Molecules* **2017,** *22*, 1–12.

Nadar, S. S.; Rao, P.; Rathod, V. K. Enzyme Assisted Extraction of Biomolecules as an Approach to Novel Extraction Technology: A Review. *Food. Res. Int.* **2018,** *108*, 309–330.

Nurdalilah, O.; Teoh, Y. P.; Ooi, Z. X.; Sam, S. T. Comparative Study on the Extraction of Bioactive Secondary Metabolites from Pomelo and Pineapple Peels Extract Comparative Study on the Extraction of Bioactive Secondary Metabolites from Pomelo and Pineapple Peels Extract. *IOP Conf. Ser., Mater. Sci. Eng.* **2018,** *429* (1), 012040.

Panche, A. N.; Diwan, A. D.; Chandra, S. R. Flavonoids: An Overview. *J. Nutr. Sci.* **2016,** *5* (e47), 11–15.

Peiró, S.; Luengo, E.; Segovia, F.; Raso, J.; Almajano, M. P. Improving Polyphenol Extraction from Lemon Residues by Pulsed Electric Fields. *Waste. Biomass. Valori.* **2019,** *10* (4), 8898.

Pereira, D. T. V.; Marson, G. V.; Barbero, G. F.; Tarone, A. G.; Cazarin, C. B. B.; Hubinger, M. D.; Martinez, J. Concentration of Bioactive Compounds from Grape Marc Using Pressurized Liquid Extraction Followed by Integrated Membrane Processes. *Sep. Purif. Technol.* **2020,** *250*, 117206.

Raks, V.; Al-Suod, H.; Buszewski, B. Isolation, Separation, and Preconcentration of Biologically Active Compounds from Plant Matrices by Extraction Techniques. *Chromatographia.* **2018,** *81*, 189–202.
Redondo, D.; Venturini, M. E.; Luengo, E.; Raso, J.; Arias, E. Pulsed Electric Fields as a Green Technology for the Extraction of Bioactive Compounds from Thinned Peach by-Products. *Innov. Food Sci. Emerg. Technol.* **2018,** *45*, 335–343.
Richter, B. E.; Jones, B. A.; Ezzell, J. L.; Porter, N. L.; Avdalovic, N.; Pohl, C. Accelerated Solvent Extraction: A Technique for Sample Preparation. *Anal. Chem.* **1996,** *68*, 1033–1039.
Rodríguez De Luna, S. L.; Ramírez-Garza, R. E.; Serna Saldívar, S. O. Environmentally Friendly Methods for Flavonoid Extraction from Plant Material: Impact of Their Operating Conditions on Yield and Antioxidant Properties. *Sci. World J.* **2020,** 2020 (6792069), 13–18.
Rodríguez-Pérez, C.; Gilbert-López, B.; Mendiola, J. A.; Quirantes-Piné, R.; Segura-Carretero, A.; Ibáñez, E. Optimization of Microwave-Assisted Extraction and Pressurized Liquid Extraction of Phenolic Compounds from *Moringa oleifera* Leaves by Multiresponse Surface Methodology. *Electrophoresis* **2016,** *37*, 1938–1946.
Ruviaro, A. R.; Paula, P. De, Barbosa, M.; Macedo, G. A. Enzyme-Assisted Biotransformation Increases Hesperetin Content in Citrus Juice by-Products. *Food Res. Int.* **2019,** *124*, 2132.
Santana, Á. L.; Macedo, G. A. Effects of Hydroalcoholic and Enzyme-Assisted Extraction Processes on the Recovery of Catechins and Methylxanthines from Crude and Waste Seeds of Guarana (*Paulliniacupana*). *Food Chem.* **2019,** *281*, 222–230.
Sheng, Z.; Wang, Y.; Wan, P.; Li, Y. Ultrasound-Assisted Extraction of Total Flavonoids from Leaves of *Syringa Oblata* Lindl. *Lat. Am. Appl. Res.* **2014,** *44*, 131–135.
Shibata, C.; Ohno, M.; Otsuka, M.; Kishikawa, T.; Goto, K.; Muroyama, R.; Kato, N.; Yoshikawa, T.; Takata, A.; Koike, K. The Flavonoid Apigenin Inhibits Hepatitis C Virus Replication by Decreasing Mature microRNA122 Levels. *Virology* **2014,** 462–*463*, 42–48.
Shouqin, Z.; Jun, X.; Changzheng, W. High Hydrostatic Pressure Extraction of Flavonoids from Propolis. *J. Chem. Technol. Biotechnol.* **2005,** *80*, 50–54.
Soliva-Fortuny, R.; Balasa, A.; Knorr, D.; Martín-Belloso, O. Effects of Pulsed Electric Fields on Bioactive Compounds in Foods: A Review. *Trends. Food. Sci. Technol.* **2009,** *20* (111–2), 5445–5456.
Tamkutė, L.; Liepuoniūtė, R.; Pukalskienė, M.; Venskutonis, P. R. Recovery of Valuable Lipophilic and Polyphenolic Fractions from Cranberry Pomace by Consecutive Supercritical CO_2 and Pressurized Liquid Extraction. *J. Supercrit. Fluids.* **2020,** *159*, 104755.
Tchabo, W.; Ma, Y.; Engmann, F. N.; Zhang, H. Ultrasound-Assisted Enzymatic Extraction (UAEE) of Phytochemical Compounds from Mulberry (*Morus nigra*) Must and Optimization Study Using Response Surface Methodology. *Ind. Crop. Prod.* **2014,** *63*, 2142.
Teles, A. S. C.; Chávez, D. W. H.; Coelho, M. A. Z.; Rosenthal, A.; Gottschalk, L. M. F.; Tonon, R. V. Combination of Enzyme-Assisted Extraction and High Hydrostatic pressure for Phenolic Compounds Recovery from Grape Pomace. *J. Food Eng.* **2021,** *288*, 110128.
Van Hung, P.; Yen Nhi, N. H.; Ting, L. Y.; Lan Phi, N. T. Chemical Composition and Biological Activities of Extracts from Pomelo Peel By-Products Under Enzyme and Ultrasound-Assisted Extractions. *J. Chem.* **2020,** *2020*, 1043251.
Vigano, J.; de Paula Assis, B. F.; Nathia-Neves, G.; Dos Santos, P.; Meireles, M. A. A.; Veggi, P. C.; Martinez, J. Extraction of Bioactive Compounds from Defatted Passion Fruit Bagasse (*Passiflora edulis* sp.) Applying Pressurized Liquids Assisted by Ultrasound. *Ultrason. Sonochem.* **2020,** *64*, 104999.

Wang C, Yang H, Li J. Combination of Microwave, Ultrasonic, Enzyme Assisted Method for Curcumin Species Extraction from Turmeric (*Curcuma Longa* L.) and Evaluation of their Antioxidant Activity. *eFood.* **2021,** 2 (2):738–0.

Wang, P.; Tang, J. Solvent-Free Mechanochemical Extraction of Chondroitin Sulfate from Shark Cartilage. *Chem. Eng. Process. Process Intensif.* **2009,** *48*, 1187–1191.

Wang, T.; Guo, N.; Wang, S. X.; Kou, P.; Zhao, C. J.; Fu, Y. J. Ultrasound-Negative Pressure Cavitation Extraction of Phenolic Compounds from Blueberry Leaves and Evaluation of Its DPPH Radical Scavenging Activity. *Food Bioprod. Process.* **2018,** *108*, 69–80.

Wang, T. Y.; Li, Q.; Bi, K. S. Bioactive Flavonoids in Medicinal Plants: Structure, Activity and Biological Fate. *Asian J. Pharm. Sci.* **2018,** *13* (1), 12–23.

Wu, K.; Ju, T.; Deng, Y.; Xi, J. Mechanochemical Assisted Extraction: A Novel, Efficient, Eco-Friendly Technology. *Trends. Food. Sci. Technol.* **2017,** *66*, 166–175.

Xi, J. Ultrahigh Pressure Extraction of Bioactive Compounds from Plants—A Review. *Crit. Rev. Food Sci. Nutr.* **2017,** *57*, 1097–1106.

Xie, J.; Lin, Y. S.; Shi, X. J.; Zhu, X. Y.; Su, W. K.; Wang, P. Mechanochemical-Assisted Extraction of Flavonoids from Bamboo (*Phyllostachys edulis*) Leaves. *Ind. Crops Prod.* **2013,** *43*, 276–282.

Xu, L.; He, W.; Lu, M.; Yuan, B.; Zeng, M.; Tao, G.; Qin, F.; Chen, J.; Guan, Y.; He, Z. Enzyme-Assisted Ultrasonic-Microwave Synergistic Extraction and UPLC-QTOF-MS Analysis of Flavonoids from Chinese Water Chestnut Peels. *Ind. Crops Prod.* **2018,** *117*, 179–186.

Yao, X. H.; Zhang, D. Y.; Luo, M.; Jin, S.; Zu, Y. G.; Efferth, T.; Fu, Y. J. Negative Pressure Cavitation-Microwave Assisted Preparation of Extract of *Pyrola incarnata* Fisch. Rich in Hyperin, 2′-O-galloylhyperin and Chimaphilin and Evaluation of Its Antioxidant Activity. *Food Chem.* **2015,** *169*, 270–276.

Yedhu Krishnan, R.; Neelesh Chandran, M.; Vadivel, V.; Rajan, K. S. Insights on the Influence of Microwave Irradiation on the Extraction of Flavonoids from *Terminalia chebula. Sep. Purif. Technol.* **2016,** *170*, 224–233.

Zhang, D. Y.; Zu, Y. G.; Fu, Y. J.; Luo, M.; Gu, C. B.; Wang, W.; Yao, X. H. Negative Pressure Cavitation Extraction and Antioxidant Activity of Biochanin A and Genistein from the Leaves of *Dalbergia odorifera* T. Chen. *Sep. Purif. Technol.* **2011,** *83*, 91–99.

Zhu, X. Y.; Lin, H. M.; Chen, X.; Xie, J.; Wang, P. Mechanochemical-Assisted Extraction and Antioxidant Activities of Kaempferol Glycosides from *Camellia oleifera* Abel. Meal. *J. Agric. Food Chem.* **2011,** *59*, 3986–3993.

Zill-E-Huma, Abert-Vian, M.; Elmaataoui, M.; Chemat, F. A Novel Idea in Food Extraction Field: Study of Vacuum Microwave Hydrodiffusion Technique for by-Products Extraction. *J. Food Eng.* **2011,** *105*, 351–360.

Zimmermann, U.; Pilwat, G.; Riemann, F. Dielectric Breakdown of Cell Membranes. *Biophys. J.* **1974,** *14* (11), 8818–8899.

CHAPTER 4

EXTRACTION METHODS OF POLYPHENOL DERIVATIVE (FLAVONOID)

TARKESHWAR[1], MANISHA ARORA PANDIT[1], KAPINDER[2], and TAHIRA ANSARI[2]

[1]Department of Zoology, Kalindi College, University of Delhi, Delhi, India

[2]Department of Zoology, DDU Gorakhpur University, Gorakhpur, Uttar Pradesh, India

ABSTRACT

Flavonoids are secondary plant metabolites, having polyphenol group, found in many natural products and are responsible for the flavor, color, and pharmacological activities. They function as antioxidants which help the plant to protect against adverse environment and also exhibit immuno-modulatory, noninflammatory, antimicrobial, and anticancer properties. The flavonoids extraction methods greatly influence their biological characteristics. The process of extraction and separation of requisite flavonoids is the most critical phase in the processing of plant raw materials. The conventional method of extraction process includes maceration and soxhlet extraction which are much slower, less ecologically friendly, and have lower levels of automation than recently used more advanced techniques. The alternatives techniques of extraction are currently being used includes microwave, ultrasound, and pressured fields. These alternative non-conventional methods are simplified procedure that is inexpensive in cost and great in efficiency. When compared to other extraction processes, it is more environmentally friendly and has a greater potential for use in industrial production. Furthermore, because

The Flavonoids: Extraction and Applications. Deepika Saini, Rajesh K. Kesharwani, & Raj K. Keservani (Eds.)

this approach is carried out at ambient temperature, it has the potential to decrease or avoid the degradation of thermo-sensitive substances.

4.1 INTRODUCTION

Humans obtain innumerable benefits through the nutritional and medicinal properties of cereals, legumes, fruits, vegetables, and their products. The nutritional properties of plant products are to some extent dependent on the presence of phytochemicals in their tissues. Phytochemicals are plant-based compounds that are known to have advantageous biological and pharmacological properties. One of the most well-known categories of phytochemicals present in the plants and their products are the flavonoids (Perlatti et al., 2014). Nobel Laureate Szent-György in 1930, identified these chemicals after isolating citrine from lemon peels, which modulates capillary permeability. They were first identified as vitamin P (due to its capacity to improve capillary permeability) and vitamin C_2 (due to properties comparable to vitamin C) (Guo et al., 2019). Flavonoids (Keservani and Sharma, 2014; Keservani et al., 2016a,b) are secondary plant metabolites that may be found in nearly every aspect of a plant's developing system and control their flavor, color, and pharmacological properties (Nabavi et al., 2018; Scarano et al., 2018).

In addition to chlorophyll and carotenoids, flavonoids are known to be the most prevalent pigment in plants (Stalikas, 2007). The 1,3-diphenylpropane is the most elemental architecture of flavonoids and is composed of two aromatic rings connected by a three-carbon structure. Unless chalcones and dihydrochalcones are present, this molecule produces an oxygenated heterocycle. Based on their amount in food and their construction, flavonoids are classified into several categories such as flavones, flavonols, flavanones, isoflavones, flavandiols, proanthocyanidins, anthocyanins, and catechins (Bravo and Mateos, 2008). Other known flavonoids include anthocyanidins, chalcones, and flavan-3-ols along with metabolic intermediates like dihydrochalcones, biflavonoids, and aurones produced during flavonoid synthesis (Bravo and Mateos, 2008; Stalikas, 2007). Flavonoids can also be synthesized using aromatic amino acids like malonate, phenylalanine, and tyrosine (Stalikas, 2007). They function as antioxidants that aid in plant protection against damaging environmental factors and exhibit minimal toxic effects as compared to synthetic antioxidants (Nabavi et al., 2018). Their antioxidant capacity is due to the presence of a range of phenolic hydroxyl groups, iron chelation abilities, and various transition metals. They have also been reported to have immuno-modulatory, noninflammatory (Yahfoufi et al., 2018) antimicrobial, and anticancer properties (Chirumbolo et al., 2018; Abotaleb et al., 2018; Rodriguez-Garcia and Sanchez-Quesada, 2019). They can protect the organisms from oxidizing agents like UV light, pollution and synthetic food additives, among other things. Since the human body is not capable of synthesizing these compounds in sufficient quantities to defend itself; therefore, these compounds must be received from external sources such as food or supplementation.

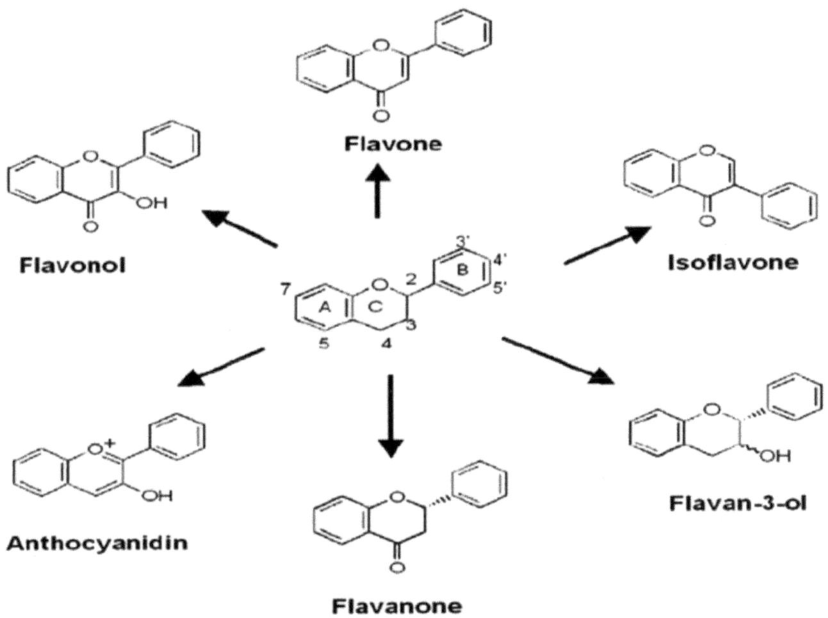

Apart from their above-mentioned benefits, they have curative effects in a wide range of diseases such as cancer and heart disease (Miao et al., 2017; Moon et al., 2018; Diwan et al., 2017). They are also reported to minimize diseases arising due to oxidative-stress and geriatric human ailments (Santi et al., 2018). Flavonoids also exhibit other attributes such as activation of enzymes, inducing communication across gap junctions, as well as detoxification enzymes such as cytochrome P-450 dependent monooxygenase and regulation of cell growth, etc. (Stahl et al., 2002). Due to these benefits associated with flavonoids, it is imperative to integrate them in drugs, herbal medicines, food and health products, bio-preservatives, and cosmetics.

Characterization of bioactive flavonoids from extracts of vegetables or fruits and plant parts such as flowers and seeds, etc., is mainly carried out via chromatography or separation techniques due to quality control purposes. Bioactive flavonoids can also be obtained from other sources such as green and black tea, wine, and beer (Martínez-Flórez et al., 2002). Due to availability of flavonoids in the leaf and flower at higher concentration, these plant parts are most commonly employed for extraction. In contrast, fresh leaves and bark are not very good sources for flavonoids extraction mainly due to the presence of resins and waxes in them (Geissman, 1962). The manner of flavonoids extraction greatly influences their biological characteristics. Biotechnological approach reduces or eliminates the use of hazardous chemicals, saves processing time as well as retains bioactivity of the compounds. The process of extraction and separation of requisite flavonoids is the most critical phase in the processing of plant raw materials. Other important criteria are, the kind of plant matter used, sample preparation, type of solvent, extraction strategy, and physicochemical environment. Together these factors influence the extract quality and effectiveness of a particular procedure.

In research labs and in companies where the scale of manufacturing is less, classical methods of flavonoid extraction like Maceration and Soxhlet extraction are favored while other methods such as Enzyme and Microorganism-assisted Extraction, Microwave-Assisted extraction (MAE), Supercritical Fluid Extraction (SFE), Matrix Solid-Phase Dispersion (MSPD), Ultrasound-Assisted Extraction (UAE), Pressurized Liquid Extraction, Negative Pressure Cavitation Extraction (NPCE), etc., are the more modern methods that give better yield at a lower cost (Azwanida, 2015).

4.2 EXTRACTION METHODS OF FLAVONOIDS

Flavonoids are a type of polyphenol found in many natural products and are responsible for the flavor, color, and medicinal properties of foods. Their extraction is defined as a separation process wherein the requisite flavonoid is disassociated from a matrix like in the case of liquid–liquid extraction or solid-phase extraction. It is based on the partition theory wherein the solute is initially in equilibrium within the two phases followed by its movement into the extraction solvent from the initial solvent. Washing is an extraction procedure which entails the removal of contaminants from the solvent to obtain the required component (Alara et al., 2018). Some of the commonly used extraction methods are as follow:

4.2.1 CLASSICAL OR CONVENTIONAL OR TRADITIONAL METHODS

a) **Maceration:** Is a quick and easy method that is used for plant material that is thermolabile. It involves the extraction of the compound by covering the source material (stem bark, leaves, or root bark) with the menstruum. Menstruum is the solvent that is utilized in the extraction from medicinal plants. Once the source drug material is completely immersed in the menstruum, filtration, or decantation is carried out to remove the micelle from the marc followed by evaporation to separate the micelle from the menstruum (Ingle et al., 2017; Azwanida, 2015; Pandey and Tripathi, 2014; Doughari, 2019; Majekodunmi, 2015; Ujang et al., 2013).

b) **Soxhlet extraction:** It is an uninterrupted hot extraction using a Soxhlet glass extractor on which the method is named. The setup consists of an extraction chamber, a siphon tube, a round bottom flask, and a condenser. A thimble or a porous bag is filled with finely ground plant material that is tightly closed. The extraction solvent is held in the bottom flask, with the closed thimble in the extraction chamber. Upon heating the solvent, it evaporates and enters the extraction chamber where it undergoes condensation while extracting the plant botanical from the thimble. The process of evaporation, condensation, and extraction continues till the entire botanical is obtained from the raw material. The extraction is deemed to be complete when no residue is found in the extraction chamber. Soxhlet method of extraction is more suitable for thermo stable, partially soluble plant material and materials containing insoluble adulterants (Hossain et al., 2014; Harborne, 1998).

4.2.2 ADVANCED OR NON-CONVENTIONAL OR MODERN OR GREEN EXTRACTION METHODS

Alternatives techniques of extraction are currently being used, including microwaves, ultrasound, and pressured fields. Traditional extraction processes are slower, less ecologically friendly, and have lower levels of automation than these newer techniques. The numerous flavonoid extraction methods accessible from natural sources, as well as the key parameters involved, will be discussed in this chapter like temperature, solvent, sample volume, extraction time, etc. Recent trends are addressed in depth, as well as their industrial significance, to provide insight into their potential. As a result, the purpose of this study is to look at how compound extrapolation has progressed. Some of the methods used are given as follows:

a) Microwave-Assisted Extraction (MAE)
b) Ultrasound-Assisted Extraction (UAE)
c) Supercritical Fluid Extraction (SFE)
d) Enzyme & Microorganism-Assisted Extraction
e) Pressurized Liquid Extraction
f) Matrix Solid-Phase Dispersion (MSPD)
g) Negative Pressure Cavitation Extraction (NPCE)

a) **Microwave-assisted extraction (MAE):** It is one of the more complex extraction approaches mostly used for preparing medicinal herbs. It makes use of dipole rotation and ionic transfer methods to displace charged ions in both the solvent and the drug. This method can be used to extract flavonoids satisfactorily. It uses electromagnetic radiation of wavelength ranging from 1 cm to 1 m with a frequency range of 300 MHz to 300 GHz. Microwaves produce 600 to 700 W of energy at a frequency of 2450 Hz. The bombardment of an item with microwave radiation causes it to absorb electromagnetic energy and convert it into heat.

   The heat produced in turn moves the solvent easily in the drug matrix. The use of a polar solvent causes ion migration and dipole rotation, further improving solvent penetration and helping extraction. A nonpolar solvent, on the other hand, produces inadequate microwave radiation with just a little quantity of heat; hence, this strategy does not encourage utilizing nonpolar solvents (Lakhssassi et al., 2017; Bhan, 2017).

b) **Ultrasound-assisted extraction (UAE):** This technique is used to disintegrate cell walls of plants and amplify the surface area for solvent movement. Sound waves of high frequency, more than 20 kHz are used in the process that leads to the formation of secondary metabolites. Sample preparation is carried out by drying the plant material which is then ground into a fine powder and then sieved. The appropriate extraction solvent is added to the prepared powder and put in the extractor. The high sound energy accelerates the extraction process by reducing the quantity of heat required.

c) **Supercritical fluid extraction (SFE):** A material that exists at a temperature and pressure above its thermodynamic critical point is known as a supercritical fluid. Such fluids induce an increased diffusivity and lowered viscosity of the solvents employed in optimizing matter transfer (Dasso and Li, 2019). As a result, the SFE methodology has shown substantially higher flavonoid extraction yields than previous procedures (Román Páez et al., 2016). The most beneficial solvent in supercritical extraction procedures is carbon dioxide gas (CO_2) that is used for its many favorable characteristics such as flammability, volatility, nontoxicity, low cost, and ease of removal (Belbaki et al., 2017; Kavoura et al., 2019). Low temperatures preserve product integrity, high solvent volatility reduces waste, and extraction does not exhibit phase shifts, and easy separation of volatile and nonvolatile substances are some of the benefits of this extraction method. However, there are certain drawbacks associated with this method which include maintenance of equilibrium between solute and solvent, high running and maintenance costs, difficulty in adding solids to the extract due to high pressure that leads to heavy investment (Song et al., 2019; Panja, 2018).

One of the most crucial aspects of the procedure is temperature. Reduced temperatures are used in this process to keep the product as stable as feasible. Value of expellers from pressing of soybean oil was examined with extraction at 40 MPa and 35°C or 40°C using CO_2 gas as the solvent. The expellers were permeated with ethanol and high flavonoid content values of 65.0 and 31.3 QE/100 gdm were obtained (Alvarez et al., 2019). The flavonoids in *Odontonema strictum* leaves were extracted using supercritical CO_2 and ethanol in another study. Temperature (55–65°C) was used to optimize the total flavonoid content. Dried extracts had a total flavonoid level of 99.33 to 247.78 mg/g, whereas dried leaves powder had a total flavonoid content of 10.68 to 18.92 mg/g (Ouédraogo et al., 2018). A study using supercritical CO_2 extraction with or without ethanol from *Citrus unshiu* peels at a pressure of 30 MPa and temperature of 59.85°C was carried out. The extracts obtained, which included nobiletin, enhanced the presence of ethanol in supercritical carbon dioxide but also lengthen extraction duration. Pressure also plays a vital part in these procedures and is dependent on the solvent utilized. Another study, for example, discovered that utilizing enzyme-assisted supercritical fluid; it was possible to harvest plants effectively. Because of its design, the Box-Behnken was chosen to maximize extraction parameters like pressure, which was kept between 100 and 300 bar. For total polyphenol content, the most advantageous parameters were temperature of 68°C, pressure at 205 bar, and co-solvent content of 15.5%. This method led to the highest polyphenol component release and efficient enzyme-assisted supercritical fluid extraction (Krakowska et al., 2018). Therefore, supercritical fluid techniques are more secure, more effective, and eco-friendly for the extraction and recovery of flavonoids.

d) **Enzyme-assisted extraction (EAE):** This technique involves pre-treatment with enzymes wherein specialized hydrolyzing enzymes are added at the extraction stage (Vongsak et al., 2018) leading to the destruction of the cell wall, structural polysaccharides, and lipid bodies by hydrolysis under moderate circumstances, resulting in higher yields. This method is thought to be an innovative and successful way to extract a wide range of secondary plant metabolites having antioxidant characteristics, such as flavonoids (Selvamuthukumaran and Shi, 2017).

Various aspects have to be taken into consideration for an effective extraction. These include the temperature of the reaction, time taken for extraction, pH, amount of enzyme required, and size of the substrate particles (Maier et al., 2008). Grape pomace was treated with a blend of pectinolytic and cellulolytic enzymes in a 2:1 ratio for 2-hrs at a temperature of 40°C and 4.0 pH to obtain bioactive compounds such as anthocyanins and non-anthocyanin flavonoids. In comparison to sulfite-assisted extraction, the yield was greater with the procedure resulting in the separation of phenol-rich fractions. By using the hydrolyzing enzymes of the plants treated, Wallace and Burong developed a selective EAE-procedure for isoflavone extraction from chickpeas, soybeans, *Lupinus albus* seeds, and leaves from *Trifolium subterraneum* L. and *Mentha piperita.* The ground material is first suspended in water, then warmed (up to 62°C for leaves of *Trifolium subterraneum* L) and then hydrolysis is allowed to complete. The time period can range from 10 min to a few hours to overnight.

Following this procedure, pH was raised to 9 (pH range between 6 and 12) in order to achieve a isoflavone fraction that is soluble. Precipitation of the isoflavones was carried out by lowering the pH in the range of 3.5–5.6 followed by filtration. The precipitate can be further dissolved in ethanol then acetone in order to precipitate all dissolved proteins, sugars, and saponins to some extent. This methodology helps extract a variety of bioactive chemicals after filtration from plants which are usually greater in small molecular compounds such as water-soluble flavonoids. Post extraction, the flavonoids can be isolated in stages by varying the pH and addition of a variety of solvents (Wallace and Burong, 2006). However, in terms of flavonoid extraction from plant materials, EAE is a non-selective approach.

e) **Pressurized liquid extraction (PLE):** This method was first introduced by the Dionex Corporation in 1995 (Benthin et al., 1999) and different research groups have referred to pressurized liquid extraction (PLE) as "accelerated solvent extraction." Because of the high pressure utilized by PLE, the solvents continue to be in a liquid state even above their boiling point. The results observed were substantial solvent penetration into the matrix, as well as greater solubility and rate of diffusion of hydrophobic substances within the solvent (Zhang et al., 2018). PLE involves shorter extraction time and solvent usage than the other procedures discussed in this review, and it is more repeatable and easily automated, allowing for extraction in less time and with less solvent.

According to Gizir et al., high temperatures and pressure used in this method do not destroy the thermolabile substances (Gizir et al., 2008). The solvents used in PLE are crucial and should be easy to remove, non-toxic, and non-flammable while the extraction by itself should be as selective as possible (Ameer et al., 2017). The use of acidified water as the solvent has been beneficial in obtaining anthocyanin-rich extract from black carrots and lyophilized red grape skins. It was seen that the rate of anthocyanin degradation is time dependent and in order to avoid molecular degradation, temperatures up to 100°C, the pressure of 50 bar, and a short-time period of a few minutes only, gave better results. To obtain the highest return rates of anthocyanins and phenolics in the extracts, acidified methanol was selected from among other solvents with the conditions of 333.15 K and 10.1 MPa. These optimizations also reduced the solvent utilization and extraction time to 3 min (Ju and Howard, 2003). Santos et al. found that at an optimization of 5 MPa, 553 K & 9 mins, in static mode, the recovery rate of anthocyanins and phenolic compounds improved in comparison to low-pressure percolation approaches (Santos et al., 2012). Another study by Luthria, 2008 also demonstrated similar outcomes. PLE can be used to optimally extract flavones from parsley flakes by utilizing two binary solvent arrangements: ethanol:water (50:50 v/v) and acetone:water (50:50 v/v). The flush volumes used in the solvent systems reduced the amount of extractant significantly along with the waste amounts. PLE discovered greater flavonol and falvan-3-ol recoveries from *Rheumpalmatum* using 80% aqueous methanol (Tan et al., 2019). Therefore, this methodology is usually suggested for use in quality control. The influence of physiochemical parameters such as rate of flow of the solvent and loading of the feed on the efficiency of PLE of iso-flavones from soybean flakes, as well as the ethanol/water ratio was studied by Chang et al. (2007) who found that the extraction process was 95% recovered using 80% aqueous ethanol. Another work used PLE to isolate isoflavones from *Trifolium* (L) in a micro-preparative manner (Zgórka, 2009). The effects of various parameters like the type of solvent used, temperature, and cycle count needed for extraction were explored and at 125°C, good extraction capacity was achieved using a 75:25 (v/v) methanol–water mixture. In this investigation, PLE outperformed traditional solvent extraction and UAE due to its low cost, higher return of isoflavone aglycones, and superior efficiency of extraction. Other research studies also claim obtaining success using PLE to obtain flavonoids from various plant sources. Another form of PLE is Pressurized Hot Water Extraction (PHWE).

Supercritical water extraction and pressurized low-polarity water extraction are other names used for PHWE. In this method, the operational properties of the pressurized water used are the temperature range above the boiling point and below the critical point of water, which are 273 K at 0.1 MPa and 647 K at 22.1 MPa, respectively (Ameer et al., 2017). Water's intermolecular hydrogen bonds break down in subcritical circumstances, lowering its dielectric constant. This constant is lowered to 27 at a temperature of 250°C (Vergara-Salinas et al., 2015). The dielectric constant of methanol and ethanol is comparable. Using water instead of ethanol or methanol lowers expenses making the process more environment friendly. The exclusivity of this method is based on the temperature-dependent solubility of flavonoids. The aqueous solubility of a variety of hydrophobic substances increases at temperatures over 373.15 K, while water remains incompressible as the working pressure rises. The breakdown of phenolic compounds that are thermolabile and the need for additional cleansing are two undesirable effects of high temperature. In conclusion, careful temperature optimization is critical for PHWE. At 150°C, PHWE produced improved flavonol extraction from Seabuckthorn (Kumar et al., 2011) at 100–110°C, and anthocyanins from dried red grape skin (Ju and Howard, 2003) while more polar flavonoids, such as anthocyanins, flavonols, flavonones, and flavones can be preferentially obtained at temperature range from 100°C to 120°C (Rodriguez–Meizoso et al., 2006). The flavonoids are destroyed at temperatures above 120°C and after prolonged exposure. Industrial applications of PHWE were tested by Ko et al. (2016) on a small scale before deciding on its commercialization. The impact of PHWE extraction on flavonoids from satsuma mandarin peel (dried) was investigated and the following parameters were determined to be optimal, temperature of 130°C, extraction time of 15 mins and a ratio of 1/34 for the solute/solvent. Under these conditions, flavonoids yielded 117.8 and 113.4 mg/g, respectively. In the pilot plant, flavonoids recovered from PHWE were 96.3%, demonstrating the method's ability for future industrial use. Quantitative extraction of isoflavones does not require PHWE (Blicharski and Oniszczuk, 2017). Even at high temperatures and moderate pressures, water does not isolate isoflavones as well as methanol, ethanol, and their mixtures with water (Li-Hsun et al., 2004). Furthermore, during PHWE, the breakdown of malonyl and glucoside forms was also observed at higher temperatures (Jung et al., 2008; Moras et al., 2017).

f) **Matrix solid-phase dispersion (MSPD):** MSPD involves a prominent solid–liquid extraction technique alternative. The target chemicals are extracted, separated, and produced using this method (Blicharski and Oniszczuk, 2017). Most of the issues associated with traditional procedures are eliminated during the process, which includes simultaneous homogenization, extraction, and purification. Samples of dried leaves along with hydrophobic C18-resin were put in tiny columns, and extraction of isoflavones was carried out using dichloromethane and methanol mixture (Visnevschi-Necrasov et al., 2009). This was done as per an MSPD method designed for isoflavones in *Trifolium pratense* L. MSPD, unlike traditional extraction processes, requires no special laborious equipment, uses reduced quantities of solvent, is faster, cheaper, and more eco-friendly. The eluent used determines its selectivity (Hong et al., 2015). An MSPD approach was developed for the simultaneous extraction and isolation of kaempferol and other polyphenols such as hydroxysafflor yellow-A from *Carthamus tinctorius L* using silica gel as a dispersing sorbent (sorbent/sample ratio 3:1) and methanol: water as a solvent. Graphene-encapsulated silica when used as an adsorbent in MSPD of polymethoxylated flavonoids from dried *Murraya paniculata (L.),* its recovery rate was more than 93.60% (Sun et al., 2015). Graphene-encapsulated silica outperformed the other five sorbents (graphene, silica gel, C18-resins, diatomaceous earth, and neutral alumina) in terms of target analyte extraction efficiency.

g) **Negative pressure cavitation extraction (NPCE):** NPCE (negative pressure cavitation) extraction is a new, efficient, and environmentally friendly method. This method uses negative pressure to create cavitation, which is then continually delivered into the liquid–solid system to improve turbulence, collision, and mass transfer between the extracting solvent and solid matrix. This method was found to be effective in terms of optimizing substrate and enzyme mixing, enhancing the rate of enzymatic degradation of plant cell wall components and the release of intracellular contents into the extraction solvent. NPCE is a novel form of cavitation that may be used to isolate thermo-unstable plant derivatives (Blicharski and Oniszczuk, 2017). The technique revolves around the continuous input of nitrogen into the extraction method. Lower pressure causes N2-bubbles to form in the liquid–solid system, forming a very unstable gas–liquid–solid phase. These activities encourage turbulence, collision, and mass transfer between the extractant and the plant matrix, resulting in successful target analyte extraction. The procedure is kept at a low temperature, making it effective, easy, low-cost, and environmentally benign. NPCE has the potential to be used in industry (Roohinejad et al., 2016). Another advantage of this approach is that thermo-responsive chemicals like isoflavonoids can be quantitatively removed at ambient temperature (Zhang et al., 2011; Zhao et al., 2011; Dong et al., 2011). Radix Scutellariae flavones (**the dried root of the medicinal plant Scutellariae baicalensis**) were separated using NPCE with the following ideal technological parameters: 40 mL/g solid, 75% ethanol in water, 60 min extraction period, and 0.07% MPa vacuum. In comparison to other traditional extraction procedures, the researchers have found that NPCE had high extraction efficiency. When NPCE was combined with ultrasonication, its efficiency increased.

Wang et al. (2020) coupled negative pressure compression with a UAE strategy for extracting flavonols from Flos Sophorae immaturus (NPC-UAE). The scientists found that the yield of six chemicals extracted by NPC-UAE was 1.27–1.62 and 1.17–1.40 times higher than that of NPCE and UAE, respectively; 72% ethanol, 16 min of treatment duration, 25:1 mL/g liquid-to-solid ratio, ultrasonic intensity 0.347 W/cm^2, negative pressure 0.07 MPa, and temperature 60°C were the best extraction conditions.

4.3 ADVANTAGES OF NON-CONVENTIONAL METHODS

This approach is a simplified procedure that is inexpensive in cost and great in efficiency. When compared to other extraction processes, it is more environmentally friendly and has a greater potential for use in industrial production. Furthermore, because this approach is carried out at ambient temperature, it has the potential to decrease or avoid the degradation of thermo-sensitive substances. Extraction process such as microwave-assisted extraction provides several advantages including reduced time of extraction, minimum requirement of solvents, and increased efficiency of extraction. The primary method for extracting flavonoids consists of maceration of fresh or dry raw material in different solvents such as methanol, acidified water, ethyl alcohol, ethyl acetate, and hexane. Similarly, separation of flavonoids with ultrasound-assisted extraction method is limited to small sample size and also caused reduction in extraction time, minimum utilization of solvent with maximum yield of the desired product. The technique revolves around the continuous input of nitrogen into the system. Negative pressure in the extraction system results in bubbling of N_2 gas to form a very unstable gas–liquid–solid phase. These activities encourage turbulence, collision and mass flow between plant matrix and the extractant, resulting in successful target analyte extraction. The process is reduced to the lowest temperature, making it effective, easy, low-cost, and environmentally benign. NPCE has the potential to be used in industry. Thermo-sensitive chemicals like iso-flavonoids can be removed quantitatively at room temperature.

4.4 ADVANTAGES OF CONVENTIONAL METHODS

1. Large amount of drug can be extracted with smaller amount of solvent.
2. It is also applicable to plant materials that are heat stable.

3. No filtration is required and high amount of heat could be applied.
4. Rapid analysis can also take place.
5. Required minimal investment in the equipment that analyze the sample with the help of traditional method.
6. With less solvent, a larger amount of medication may be extracted. Plant materials that are heat stable can also be used.

4.5 APPLICATION OF CONVENTIONAL METHODS

Maceration has long been a popular and inexpensive home technique for preparing tonics. In addition, this technology has been used for the extraction of essential oils and active compounds from plant materials. Maceration process helps in breakdown of cell structure of plant materials that release chemical components from cell such as different enzymes, polyphenols, lipids, amino acids, etc. In case of tea plant, catechins released from cell when comes into contact with the enzyme, polyphenol oxidase, initiates "tea oxidation," erroneously considered as "fermentation."

In the process of maceration, intact tissues are turned into a suspension of cells, resulting in the pulp product which is used as the base material for pulp juices, nectar, baby food, and dairy products such as puddings and ingredients for yoghurts (Pilnik and Vorgen, 1993; Khatri et al., 2015). Maceration releases components from pomace (seeds, skins and pulp) after crushing, facilitated by the release and activation of hydrolytic enzymes. A Soxhlet extractor used to extract lipids and other compounds from solid samples. A condenser, a Soxhlet extractor, and a round-bottomed flask make up a Soxhlet extraction device.

4.6 APPLICATIONS FOR SOXHLET EXTRACTION

The Soxhlet is very useful for food testing, biofuel, soil, sludge and environmental waste analysis. The Soxhlet extraction process ensures that the sample matrix is in direct contact with the extraction solvent. This approach may be used to extract a wide range of semivolatile organic molecules, which can subsequently be evaluated using the proper chromatographic procedure. When other compounds or contaminants are insoluble in that solvent, the Soxhlet device allows isolation of desired oil even if it has limited solubility in that solvent. What happens is that the oil is continuously extracted by contact with the solvent. Of course, Soxhlet extraction necessitates a volume of solvent that reaches the syphon level, as well as time to separate nearly all of the oil.

This may appear to be a negative; however, it is more efficient than extraction through a separatory funnel. Soxhlet extraction is helpful since it is a thorough extraction process. It guarantees that all analytes from the plant matrix are removed. On the other hand, the high temperatures required in its functioning (60–80 degrees) may result in the loss of several heat labile phytochemicals.

4.7 CONCLUSIONS

Plant derived flavonoids isolation and their applications are attracting a lot of attention. The most important aspect in the development of novel and effective extraction procedures that are both economically and ecologically sound, allowing for a larger recovery of active substances from natural materials in a shorter period of time. The benefits of such procedures have been enormous in terms of producing high-quality and quantity extracts, as well as reduced solvent expenditure, energy consumption, and extraction time. The animals are not able to synthesized flavonoids within their body, however, flavonoids are important food constituents of mammals and their intake generally ranged between 55 and 850 mg (Routray and Orsat, 2012; Kumar and Pandey, 2013). Due to the property of flavonoid to remove free radicals from the animal body and also check their production, these are considered as potential natural antioxidants. Moreover, flavonoids have the ability to chelate metal ions to prevent lipid peroxidation. Synthetic antioxidants are reported to exhibit certain toxicity to mammalian cells. As a result, there has been a surge of interest in extraction of natural antioxidants from plants. Natural antioxidant-based medications have been demonstrated to be relatively benign, safe, and free of major adverse effects (Fernando and Soysa, 2014).

Unconventional and conventional methods of flavonoids extraction were analyzed using several techniques including Soxhlet extraction, MAE, NPCE, UAE, MSPD, and several others. These techniques have gained attention of researchers for using them for improving the quality of life of human beings and also conservation of environment as well as generating the innovative processes for plant-based flavonoids extraction. Using traditional extraction methods, flavonoids and polyphenols are usually extracted together. Plants that are high in flavonoids produce extracts that are high in flavonoids. When the plant material is ground and alcohols are used as biosorbents, the extraction efficiency is improved. Temperature and duration of treatment affect the extraction of thermo unsteady flavonoids. The yields of flavonoids and polyphenols are substantial when solvent extraction procedures like MAE,

UAE, and MSPD are used. With the right extractant, these procedures could be effective for selective flavonoid extraction. By implementing a methanol–water or ethanol–water medium, PLE was more successful than MAE and UAE. At temperatures of 100–120°C, PHWE may preferentially extract more polarity of flavonoids including flavonols, flavones, flavonones, and anthocyanins. The best extraction procedure depends on the plant matrix and compost type, and specific screening process should be applied. Sustainable extraction technologies have recently been proved to be great alternatives to conventional extraction techniques in latest researches. Several studies are still going on to enhance these new green extraction methods, with the goal of lowering extraction cost, time it takes, the quality of the extract, environment protection, and wellness (Chaves et al., 2020). The Soxhletation and Meceration methods are useful for extraction of substrates that weak of heat. Both of these methods take more time for extracting and purifying plant materials. So, overall finding is that the unconventional method of extraction is more efficient than conventional methods of flavonoid extraction.

ACKNOWLEDGMENTS

The authors thank the Kalindi College, University of Delhi and Deen Dayal Upadhyay University Gorakhpur, U.P. for their continuation support and providing conductive environment to complete this assignment.

KEYWORDS

- **flavonoids**
- **extraction**
- **plant botanical**
- **secondary plant metabolite**

REFERENCES

Altemimi, A.; Lakhssassi, N.; Baharlouei, A.; Watson, D. G.; Lightfoot, D. A. Phytochemicals: Extraction, Isolation, and Identification of Bioactive Compounds from Plant Extracts. *Plants (Basel)* **2017,** 6.

Alvarez, M.; Cabred, S.; Ramírez, C.; Fanovich, M. Valorization of an Agroindustrial Soybean Residue by Supercritical Fluid Extraction of Phytochemical Compounds. *J. Supercrit. Fluids* **2019,** *143*, 90–96.

Ameer, K.; Shahbaz, H. M.; Kwon, J. H. Green Extraction Methods for Polyphenols from Plant Matrices and Their Byproducts: A Review. *Compr. Rev. Food Sci. Food Saf.* **2017,** *16*, 295–315.

Azwanida, N. N. A Review on the Extraction Methods Use in Medicinal Plants, Principle, Strength, and Limitation. *Med. Aromat. Plants* **2015,** *4*, 196.

Belbaki, A.; Louaer, W.; Menial, A. H. Supercritical CO2 Extraction of Oil from Crushed Algerian Olives. *J. Supercrit. Fluids* **2017,** *130*, 165–171.

Benthin, B.; Danz, H.; Hamburger, M. Pressurized Liquid Extraction of Medicinal Plants. *J. Chromatogr. A* **1999,** *837*, 211–221.

Bhan M. Ionic Liquids as Green Solvents in Herbal Extraction. *Int. J. Adv. Res. Dev.* **2017,** *2*, 10–2.

Blicharski, T.; Oniszczuk, A. Extraction Methods for the Isolation of Isoflavonoids from Plant Material. *Open Chem.* **2017,** *15*, 34–45.

Chang, L. C.; Chang, C. M. J. Continuous Hot Pressurized Fluids Extraction of Isoflavones and Soyasaponins from Defatted Soybean Flakes. *J. Chin. Inst. Chem. Eng.* **2007,** *38*, 313–319.

Chaves, J. O.; de Souza, M. C.; da Silva, L. C.; Lachos, P. D.; Torres, M. P. C.; da Fonseca, M. A. P.; Forster, C. T.; Vázquez, E. M.; Ana Velasco González, P. A. V.; Barbero, G. F.; Rostagno, M. A. Extraction of Flavonoids from Natural Sources Using Modern Technologies. *Front. Chem.* **2020,** 8.

Dasso, E. S.; Li, Y. O. Mechanisms and Effects of Ultrasound-assisted Supercritical CO_2 Extraction. *Trends Food Sci. Tech.* **2019**, *86*, 492–501.

Diwan, V.; Brown, L.; Gobe, G. C. The Flavonoid Rutin Improves Kidney and Heart Structure and Function in an Adenine-induced Rat Model of Chronic Kidney Disease. *J. Funct. Foods* **2017**, *33*, 85–93.

Dong, L. L.; Fu, J. Y.; Zu, Y. G.; Li, J.; Li, X. J.; Efferth, T. Negative Pressure Cavitation Accelerated Processing for Extraction of Main Bioactive Flavonoids from Radix Scutellariae. *Chem. Eng. Process* **2011**, *50*, 780–789.

Doughari, J. H. Phytochemicals: Extraction Methods, Basic Structures, and Mode of Action as Potential Chemotherapeutic Agents, Phytochemicals—A Global Perspective of Their Role in Nutrition and Health. In *A Global Perspective of Their Role in Nutrition and Health*; Venketeshwer, R., Ed.; InTech, 2012.

Santi, F.; Obem, C.; Caetano, V.; et al. Analysis of Polyphenols in Brewer's Spent Grain and its Comparison with Corn Silage and Cereal Brans Commonly used for Animal Nutrition. *Food Chem.* **2018,** *239*, 385–401.

Gizir, A. M.; Turker, N.; Artuvan, E. Pressurized Acidified Water Extraction of Black Carrot (Daucus Carota ssp. Sativus var. Atrorubens Alef.) Anthocyanins. *Eur. Food Res. Technol.* **2008**, *226*, 363–370.

Guo, L.; Gao, L.; Ma, X.; Guo, F.; Ruan, H.; Bao, Y.; Wang, Y. Functional Analysis of Flavonoid 30-hydroxylase and Flavonoid 30,50-hydroxylases from Tea Plant (Camellia Sinensis), Involved in the B-ring Hydroxylation of Flavonoids. *Gene* **2019**, *717*, 144046.

Harborne, J. B. *Phytochemical Methods: A Guide to Modern Techniques of Plant Analysis*, 3rd ed.; New York, NY; London, UK: Thomson Science, 1998; p. 219.

Hong, B.; Wang, Z.; Xu, T.; Li, C.; Li, W. Matrix Solid-phase Dispersion Extraction Followed by High Performance Liquid Chromatography-diode Array Detection and

Ultra-performance Liquid Chromatography–quadrupole–time of Flight–mass Spectrometer Method for the Determination of the Main Compounds from Carthamus Tinctorius L. (Hong-hua). *J. Pharm. Biomed. Anal.* **2015,** *107*, 464–472.

Hossain, M. A.; Al-Hdhrami, S. S.; Weli, A. M.; Al-Riyami, Q.; Al-Sabahi, J. N. Isolation, Fractionation and Identification of Chemical Constituents from the Leaves Crude Extracts of *Mentha Piperita* L Grown in Sultanate of Oman. *Asian Pac. J. Trop. Biomed.* **2014**, *4*, 368–72.

Ingle, K. P.; Deshmukh, A. G.; Padole, D. A.; Dudhare, M. S.; Moharil, M. P.; Khelurkar, V. C. Phytochemicals: Extraction Methods, Identification, and Detection of Bioactive Compounds from Plant Extracts. *J. Pharmacogn. Phytochem.* **2017,** *6*, 32–6.

Ju, Z. Y.; Howard, L. R. Effects of Solvent and Temperature on Pressurized Liquid Extraction of Anthocyanins and Total Phenolics from Dried Red Grape Skin. *J. Agric. Food Chem.* **2003,** *51*, 5207–5213.

Jung, S.; Murphy, P. A.; Sala, I. Isoflavone Profiles of Soymilk as Affected by High-pressure Treatments of Soymilk and Soybeans. *Food Chem.* **2008**, *111*, 592–598.

Kavoura, D.; Kyriakopoulou, K.; Papaefstahiou, G.; Spanidi, E.; Gardikis, K.; Louli, V.; Aligiannis, N.; Krokida, M.; Magoulas, K. Supercritical CO2 Extraction of *Salvia Fruticosa. J. Supercrit. Fluids.* **2019**, *146*, 159–164.

Keservani, R. K.; Sharma, A. K. Flavonoids: Emerging Trends and Potential Health Benefits. *J Chin Pharma Sci.* **2014,** *23* (12), 815.

Keservani, R. K.; Sharma, A. K.; Kesharwani, R. K. Nutraceutical and Functional Foods for Cardiovascular Health. In *Food Process Engineering*. Apple Academic Press, CRC Press, 2016a; pp 291–312. ISBN: 9781771884020.

Keservani, R. K.; Sharma, A. K.; Kesharwani, R. K. Medicinal Effect of Nutraceutical Fruits for the Cognition and Brain Health. *Scientifica* **2016b,** 10. Article ID 3109254. doi: 10.1155/2016/3109254.

Ko, M. J.; Kwon, H. L.; Chung, M. S. Pilot-scale Subcritical Water Extraction of Flavonoids from Satsuma Mandarin (Citrus Unshiu Markovich) Peel. *Innov. Food Sci. Emerg. Technol.* **2016,** *38*, 175–181.

Krakowska, A.; Rafinska, K.; Walczak, J.; Buszewski, B. Enzyme-assisted Optimized Supercritical Fluid Extraction to Improve Medicago Sativa Polyphenolics Isolation. *Ind. Crop. Prod.* **2018**, *124*, 931–940.

Kumar, M. S. Y.; Dutta, R.; Prasad, D.; Misra, K. Subcritical Water Extraction of Antioxidant Compounds from Seabuckthorn (*Hippophae Rhamnoides*) Leaves for the Comparative Evaluation of Antioxidant Activity. *Food Chem.* **2011,** *127*, 1309–1316.

Li-Hsun, C.; Ya-Chuan, C.; Chieh-Ming, C. Extracting and Purifying Isoflavones from Defatted Soybean Flakes using Superheated Water at Elevated Pressures. *Food Chem.* **2004,** *84*, 279–285.

Luthria, D. L. Influence of Experimental Conditions on the Extraction of Phenolic Compounds from Parsley (*Petroselinum Crispum*) Flakes using a Pressurized Liquid Extractor. *Food Chem.* **2008,** *107*, 745–752.

Maier, T.; Goppert, A.; Kammerer, D. R.; Schieber, A.; Carle, R. Optimization of a Process for Enzyme-assisted Pigment Extraction from Grape (*Vitis Vinifera* L.) Pomace. *Eur. Food Res. Technol.* **2008,** *227*, 267–275.

Majekodunmi, S. O. Review of Extraction of Medicinal Plants for Pharmaceutical Research. *Merit Res. J. Medicine Med. Sci.* **2015,** *3*, 521–7.

Miao, M.; Cao, L.; Xu, K.; Xin, W.; Zheng, Y. Intervention Action of Total Flavonoids from Root of Ilex Pubescensin Cerebral Ischemic Tolerance with Blood Stasis. *Saudi J. Biol. Sci.* **2017,** *24*, 729–736.

Moon, K. M.; Lee, B.; Cho, W. K.; Lee, B. S.; Kim, C. K.; Ma, J. Y. *Swertiajaponin* as an Anti-browning and Antioxidant Flavonoid. *Food Chem.* **2018**, *252*, 207–214.

Moras, B.; Rey, S.; Vilarem, G.; Pontalier, P. V. Pressurized Water Extraction of Isoflavones by Experimental Design from Soybean Flour and Soybean Protein Isolate. *Food Chem.* **2017**, 214: 9–15.

Ouédraogo, J. C.; Dicko, C.; Kini, F.; Bonzi-Coulibaly, Y.; Dey, E. Enhanced Extraction of Flavonoids from *Odontonema Strictum* Leaves with Antioxidant Activity Using Supercritical Carbon Dioxide Fluid Combinedwith Ethanol. *J. Supercrit. Fluids.* **2018,** *131*, 66–71.

Pandey, A.; Tripathi, S. Concept of Standardization, Extraction, and Pre-phytochemical Screening Strategies for Herbal Drug. *J. Pharmacogn. Phytochem.* **2014,** *2*, 115–9.

Panja, P. Green Extraction Methods of Food Polyphenols from Vegetable Materials. *Curr. Opin. Food Sci.* **2018,** *23*, 173–182.

Perlatti, B.; Forim, M. R.; Zuin, V. G. Green Chemistry, Sustainable Agriculture and Processing Systems: A Brazilian Overview. *Chem Biol Technol Agric.* **2014,** *1*, 1–9.

Rodríguez-Meizoso, I.; Marin, F. R.; Herrero, M.; Señorans, F. J.; Reglero, G.; Cifuentes, A.; Ibáñez, E. Subcritical Water Extraction of Nutraceuticals with Antioxidant Activity from Oregano. Chemical and Functional Characterization. *J. Pharm. Biomed. Anal.* **2006,** *41*, 1560–1565.

Román Páez, M.; Rivera Narváez, C.; Cardona Bermúdez, L.; Muñoz, L.; Gómez, D.; Passaro Carvalho, C.; Quiceno Rico, J. *Guía de Extracción por Fluidos Supercríticos: Fundamentos y Aplicaciones*; Servicio Nacional de Aprendizaje—SENA: Rionegro, Colombia, 2016.

Roohinejad, S.; Kouba, M.; Barba, F. J.; Greiner, R.; Orlien, V.; Lebovka, N. I. Negative Pressure Cavitation Extraction: A Novel Method for Extraction of Food Bioactive Compounds from Plant Materials. *Trends Food Sci. Technol.* **2016,** *52*, 98–108.

Kumar, S.; Pandey, A. K. Chemistry and Biological Activities of Flavonoids: An Overview. *Sci. World J.* **2013,** *2013*, 16. Article ID 162750.

Fernando C. D.; Soysa, P. Total Phenolic, Flavonoid Contents, *In-vitro* Antioxidant Activities and Hepatoprotective Effect of Aqueous Leaf Extract of *Atalantia ceylanica*. *BMC Complement. Altern. Med.* **2014,** *14*, 1–8.

Martínez-Flórez, S.; González-Gallego, J.; Culebras, J. M.; Tuñón, M. J. Flavonoids: Properties and Anti-oxidizing Action. *Nutrición Hospitalaria.* **2002,** *17*, 271–278.

Santos, D. T.; Veggi, P. C.; Meireles, M. A. Optimization and Economic Evaluation of Pressurized Liquid Extraction of Phenolic Compounds from Jabuticaba Skins. *J. Food Eng.* **2012,** *108*, 444–452.

Selvamuthukumaran, M.; Shi, J. Recent Advances in Extraction of Antioxidants from Plant by-products Processing Industries. *Food Qual. Saf.* **2017,** *1*, 61–81.

Song, L.; Liu, P.; Yan, Y.; Huang, Y.; Bai, B.; Hou, X.; Zhang, L. Supercritical CO_2 Fluid Extraction of Flavonoid Compounds from *Xinjiang Jujube* (*Ziziphus Jujube* Mill.) Leaves and Associated Biological Activities and Flavonoid Compositions. *Ind. Crop. Prod.* **2019,** *139*, 111508.

Stahl, W.; Ale-Agha, N.; Polidori, M. C. Non-antioxidant Properties of Carotenoids. *J. Biol. Chem.* **2002**, *383*, 553–558.

Sun, T.; Li, X.; Yang, J.; Li, L.; Jin, Y.; Shi, X. Graphene-encapsulated Silica as Matrix Solid-phase Dispersionextraction Sorbents for the Analysis of Poly-methoxylated Flavonoids in the Leaves of *Murraya Panaculata* (L.) Jack. *J. Sep. Sci.* **2015,** *38*, 2132–2139.

Geissman, T. A. *Fe Chemistry of Flavonoids Compounds*; Oxford, UK: Pergamon Press, 1962.

Tan, L.; Ji, T.; Jiang, G.; Hu, F. Simultaneous Identification and Quantification of Five Flavonoids in the Seeds of *Rheum Palmatum* L. by using Accelerated Solvent Extraction and HPLC-PDA-ESI/MSn. *Arab. J. Chem.* **2019,** *12*, 1345–1352.

Ujang, Z. B.; Subramaniam, T.; Diah, M. M.; Wahid, H. B.; Abdullah, B. B.; Rashid, A. A.; Appleton, D. Bioguided Fractionation and Purification of Natural Bioactive Obtained from *Alpinia Conchigera* Water Extract with Melanin Inhibition Activity. *J. Biomater. Nanobiotechnol.* **2013,** *4*, 265–72.

Vergara-Salinas, J. R.; Cuevas-Valenzuela, J.; Perez-Correa, J. R. Pressurized Hot Water Extraction of Polyphenols from Plant Material. In *Biotechnology of Bioactive Compounds: Sources and Applications*; Kumar, V. G.; Tuohy, M. G.; Lohani, M.; O'Donovan, A., Eds.; John Wiley & Sons Inc.: Hoboken, NJ, USA, 2015; pp 63–101.

Visnevschi-Necrasov, T.; Cunha, C. S.; Nunes, E.; Oliveira, M. B. P. P. Optimization of Matrix Solid-phase Dispersion Extraction Method for the Analysis of Isoflavones in *Trifolium pretense. J. Chromatogr. A.* **2009**, *1216*, 3720–3724.

Vongsak, B.; Sithisarn, P.; Mangmool, S.; Thongpraditchote, S.; Wongkrajang, Y.; Gritsanapan, W. Maximizing Total Phenolics, Total Flavonoids Contents and Antioxidant Activity of *Moringa Oleifera* Leaf Extract by the Appropriate Extraction Method. *Ind. Crops Prod.* **2013,** *44*, 566–571.

Routray, W.; Orsat, V. Microwave-assisted Extraction of Flavonoids: A Review. *Food and Bioprocess Technology*, **2012,** *5* (2), 409–424.

Wallace, R. G. W.; Burong, G. W. B. Extraction of Flavonoids. DE Patent DE60114151T2, July 20, 2006.

Wang, G.; Cui, Q.; Yin, L. J.; Li, Y.; Gao, M. Z.; Meng, Y.; Li, J.; Zhang, S. N.; Wang, W. Negative Pressure Cavitation based Ultrasound-assisted Extraction of Main Flavonoids from Flos Sophorae Immaturus and Evaluation of its Extraction Kinetics. *Sep. Purif. Technol.* **2020,** *244*, 115805.

Zgórka, G. Pressurized Liquid Extraction Versus other Extraction Techniques in Micro-preparative Isolation of Pharmacologically Active Isoflavones from *Trifolium* L. Species. *Talanta* **2009**, *79*, 46–53.

Zhang, D. Y.; Zu, Y. G.; Fu, Y. J.; Luo, M.; Gu, C. B.; Wang, W.; Yao, X. H. Negative Pressure Cavitation Extraction and Antioxidant Activity of Biochanin A and Genistein from the Leaves of *Dalbergia Odorifera T. Chen. Sep. Purif. Technol.* **2011,** *83*, 91–99.

Zhang, Q. W.; Lin, L. G.; Ye, W. C. Techniques for Extraction and Isolation of Natural Products: A Comprehensive Review. *Chin. Med.* **2018,** *13*, 20.

Zhao, B. S.; Fu, Y. J.; Wang, W.; Zu, Y. G.; Gu, C. B.; Luo, M.; Efferth, T. Enhanced Extraction of Isoflavonoids from *Radix Astragali* by Incubation Pretreatment Combined with Negative Pressure Cavitation and its Antioxidant Activity. *Innov. Food Sci. Emerg. Tech.* **2011,** *12*, 577–585.

CHAPTER 5

BIOCHEMISTRY AND APPLICATIONS OF FLAVONOIDS

AKANSHA MEHRA[1], ANUPRIYA CHATTERJEE[2], and NITESH KUMAR[1]

[1]*Amity Institute of Advanced Research and Studies (Materials and Devices) Amity University, Noida, India*

[2]*Health Sciences, University of the People, USA*

ABSTRACT

Flavonoids are a collection of natural compounds with diverse phenolic structures. They are a group of plant secondary metabolites present in various fruits (complete or in part) and vegetables. Several epidemiological studies have revealed their multitude of health benefits, leading to the surge in research on the structural, biological, and functional diversity of flavonoids. The bioavailability, metabolic activity, and biological properties of flavonoids are determined by their structures, total count of hydroxyl groups, and replacement of functional groups. Flavonoids are well-recognized as an essential element in a wide range of nutraceutical, therapeutic, pharmacological, and cosmetic uses. This is attributed to their health-promoting, anticarcinogenic, anti-inflammatory, and immune-modulatory properties. Microbial biotechnology has opened up avenues for the cost-effective large-scale production of many classes of flavonoids for therapeutic uses. The recent developments in the field of flavonoids focuse upon their extraction, detection, characterization, and scrutinizing their applicability for health benefits.

The Flavonoids: Extraction and Applications. Deepika Saini, Rajesh K. Kesharwani, & Raj K. Keservani (Eds.)

5.1 INTRODUCTION

Flavonoids are a naturally occurring highly diverse class of polyphenolic compounds ubiquitously present in plants. Flavonoids are found as pigments that color most flowers, fruits, and seeds. They are abundantly present in the spermatophytes (seed-producing vascular terrestrial plants: gymnosperms and angiosperms) but have also been reported in primitive taxa, namely, bryophytes (nonvascular terrestrial plants, including liverworts, mosses, and hornworts), pteridophytes (non-seed producing vascular terrestrial plants, that is, lycophytes, horsetails, and all ferns) and algae (Santos et al., 2017). To date, 10,000 flavonoids have been identified which have been highlighted as the third largest group of naturally occurring products following alkaloids (12,000) and terpenoids (30,000) (Dixon and Pasinetti, 2010).

They are an essential cellular component of all the higher plants. They are encapsulated in the mesophyll cell's nucleus and impregnated in the centers of reactive oxygen species (ROS) generation (Havsteen, 2002). Plants have evolved to synthesize flavonoids to protect themselves against fungal pathogens, herbivores, and ultraviolet (UV) radiation. They are recognized by pollinating agents, such as insects, birds, and animals, which help the dispersion of seeds (Falcone Ferreyra et al., 2012). Research has revealed that flavonoids (Keservani and Sharma, 2014) exhibit biological and pharmacological activities, which include antioxidant, antiviral, antibacterial, anti-inflammatory, antiallergic, antithrombotic, cytotoxic, anticancer, cardioprotective (Keservani et al., 2016a), hepatoprotective, neuroprotective (Keservani et al., 2016b), antimalarial, antileishmanial, antitrypanosomal, and antemedial properties (Mierziak et al., 2014). They can enhance the health-promoting properties with high antioxidant capacity in both in vivo and in vitro systems which have made them a significant component of our diet. They have the ability to instigate the human protective enzyme systems along with protective effects against several infectious (bacterial and viral diseases) and degenerative diseases, such as cardiovascular diseases, cancers, and other age-related diseases (Santos et al., 2017). They also stand as a secondary antioxidant defense system in tissues of plants exposed to various abiotic and biotic stresses. They also regulate the plant growth factors for example auxin. Biosynthetic genes have been identified in numerous bacteria and fungi for the enhanced production of flavonoids (Havsteen, 2002). Flavonoids are classified into six major subgroups namely chalcones, flavones, flavonols, flavandiols, anthocyanins, and proanthocyanidins or condensed tannins (Figure 5.1), and a seventh group, auronesis studied in some species. Different flavonoids have different biological functions, such as protection against ultraviolet (UV) radiation and phytopathogens, signaling during nodulation, male fertility, auxin transport, as well as the flower colors, the visual signal that attracts pollinators. Flavonoids are also responsible for displaying faded color in several plants, protecting leaf cells from photooxidative damage, and enhancing the efficiency of nutrient retrieval during senescence. Flavonols are probably the most vital flavonoids participating in stress responses; they are the most ancient and abundantly found flavonoids with a wide range of potent physiological properties.

The antioxidant property of flavonoids is often much higher than those of vitamin E and vitamin C. Hence, they can prevent oxidative stress by suppressing and scavenging ROS while upregulating the antioxidant enzyme systems, as per the respective type of flavonoid. While they can directly scavenge the free radicals by donating the hydrogen atom, they are also able to inhibit some enzymes, including NADH oxidase (NOX), glutathione-S-transferase, mitochondrial succinoxidase, and microsomal monooxygenase, which are important sources of ROS production (Dixon and Pasinetti, 2010). They chelate free metal ions, enhancing the formation of highly reactive hydroxyl radicals. Additionally, they can restrict the lipid peroxidation and oxidative damage of membrane lipids. Epidemiological studies have also revealed that flavonoid-rich diets add less risk to the incidence of cardiovascular diseases (CVD); they display hepatoprotective; neuroprotective including therapeutic options for neurodegenerative diseases and improvement of cognitive function; broad-spectrum antimicrobial including against the human immunodeficiency virus (HIV-1) and various other effects (Panche et al., 2016).

5.2 FLAVONOIDS AS NUTRACEUTICALS

"Nutraceutical" is a term given by Stephen DeFelice in 1979 (DeFelice, 1992). It is known as "a food or food component that provides medical or health benefits, including the prevention and treatment of disease." Nutraceuticals constitute isolated nutrients, dietary supplements, and diets to genetically engineered "designer" food, herbal products, and processed food, such as cereals, soups, and beverages. Any nontoxic food extract supplement that has been scientifically proven with health benefits for the treatment and prevention of disease can be termed nutraceuticals (Keservani et al., 2010a,b; Keservani et al., 2017; Keservani et al., 2020; Dillard and German, 2000). The increasing interest in nutraceuticals highlights the fact regarding epidemiological studies signifying that a specific diet or component of the diet is associated with a lower risk for a certain disease. The majorly active nutraceutical components in a plant are flavonoids. It is typical for phenolic compounds, which can act as powerful antioxidants and metal chelators. They have long been recognized to harbor anti-inflammatory, antiallergic, antithrombotic, antiviral, hepatoprotective and anticarcinogenic activities as discussed in the following subsections:

5.2.1 ANTIOXIDANT ACTIVITY

The best-studied property of almost all the flavonoid groups is their ability to act as an antioxidant. The flavones and catechins seem to be the most vital flavonoid for protecting the ROS. The cells and tissues of the body are continuously threatened by the damage made by free radicals and ROS which are synthesized during normal oxygen metabolism or are induced by exogenous damage (De Groot, 1994; Grace, 1994). Free radicals and ROS have been found to be responsible for huge number of human diseases. Kaempferol, morin, myricetin, and rutin, quercetin, acts absolutely well as antioxidants with beneficial effects such as anti- antiallergic, antiviral, anticancer, and inflammatory, activity (Ares and Outt, 1998; Wegener and Fintelmann, 1999). They have also been found to be protective against hepato diseases, cataracts, and heart diseases. Quercetin and silybin act as free radical scavengers and is found to restrict the liver reperfusion ischemic tissue damage (Halliwell, 1994). The scavenging activity of flavonoids has been studied and reported to be in the given order: Myrcetin > quercetin > rhamnetin > morin > diosmetin > naringenin > apigenin > catechin > 5,7-dihydroxy-3′,4′,5′-trimethoxyflavone > robinin > kaempferol > flavone (Ratty and Das, 1988).

5.2.2 ANTIMICROBIAL ACTIVITY

Flavonoids and the esters of phenolic acids have been investigated and studied for their antibacterial, antifungal as well as antiviral activities.

5.2.2.1 ANTIBACTERIAL ACTIVITY

The antibacterial activity is demonstrated by quite a good a number of flavonoids. Quercetin is studied to completely arrest the growth of *Staphylococcus aureus*. Majority of the flavonones without sugar moiety exhibited antimicrobial activities, unlike the flavonols and flavonolignans which showed antagonistic activity on microorganisms (Havsteen, 1983).

5.2.2.2 ANTIFUNGAL ACTIVITY

A large number of flavonoids isolated from the tangerine orange, when analyzed for fungistatic behavior against *Deuterophoma tracheiphila* were found to be active; nobiletin and langeritin exhibited strong and weak activities both, respectively, while hesperidin could promote fungal growth slightly. Chlor-flavonin was the first chlorine-containing flavonoid-type antifungal substance produced by the strains of *Aspergillus candidus* (Ververidis et al., 2007).

5.2.2.3 ANTIVIRAL ACTIVITY

The naturally occurring flavonoids display antiviral activity that has been recognized since the 1940s. Recent studies report the attempts made to make synthetic modifications in natural compounds to improve their antiviral activity. The flavonoids such as apigenin, catechin, dihydroquercetin (taxifolin), morin, rutin, hesperidine, and quercetin exhibit antiviral properties against some of the 11 types of viruses (Xu et al., 2014). The antiviral activity studied is associated with the non-glycosidic compounds, and hydroxylation at the third position is a prerequisite for antiviral behavior. Flavonols have been demonstrated to be much more active than the flavones against Herpes simplex virus type 1 and the order of importance was galangin > kaempferol > quercetin 26. Recently, a naturally occurring plant flavonoid polymer with a molecular weight of 2,100 daltons was found to exhibit antiviral properties against two strains of type 1 Herpes simplex virus and type 2 Herpes simplex viruses (Hong et al., 2007). The investigation of the antiviral activity of flavonoids has mainly focused on HIV which is spread globally since the 1980s (Singh et al., 2005). Several recent studies on the anti-AIDS activity of flavonoids have been reported. Out of the 28 flavonoids tested, flavans were found to be highly effective than flavones and flavonones (Hämäläinen et al., 2007).

5.2.3 HEPATOPROTECTIVE ACTIVITY

The liver is subject to acute and potentially lethal injury by numerous substances including phalloidin (the toxic constituent of the mushroom,

Amanita phalloides), CCl4, galactosamine, ethanol, and other compounds. Studies have revealed that flavonoids possess hepatoprotective activity. A study carried out to investigate the flavonoid derivatives silymarin, apigenin, quercetin, and naringenin, as putative therapeutic agents against microcrystin LR-induced hepatotoxicity, concluded that silymarin was the most effective one. Flavonoids, rutin, and venoruton showed regenerative and hepatoprotective effects in experimental cirrhosis (Bose et al., 2018).

5.2.4 ANTI-INFLAMMATORY ACTIVITY

The anti-inflammatory nature of flavonoids is found in several animal models. Flavone/flavonol glycosides, flavonoid aglycons, has been found to possess LO and COX inhibitory activities (Wu and Yen, 2005).

5.2.5 ANTIDIABETIC EFFECT

Flavonoids, especially quercetin, have been studied to harbor antidiabetic activity. It was reported that quercetin promotes the regeneration of cells of pancreatic islets and probably enhances the release of insulin in streptozotocin-induced diabetic rats (Vessal et al., 2003). Also in another study, Hif and Howell unveiled that quercetin stimulates the release of insulin and enhanced Ca^{2+} uptake from isolated islets cell which the flavonoids a substance of choice in noninsulin-dependent diabetes (Soares et al., 2017).

5.3 BIOSYNTHESIS OF FLAVONOIDS

5.3.1 THE CHEMICAL STRUCTURE OF FLAVONOIDS

The basic chemical structure of flavonoids comprises 15 carbon atoms, arranged in the manner C6-C3-C6, which corresponds to two aromatic rings A and B linked by a unit of three carbon atoms, which may or may not give rise to the third ring. The rings are labeled as A, B, and C (Mierziak et al., 2014). The presence of hydroxylation pattern, conjugation between the aromatic rings, glycosidic moieties, methoxy groups, and other substituents influences the chemical nature of the flavonoids. The conjugated double bonds and groups (hydroxyl or other substituents) in the flavonoids donate electrons through resonance in order to stabilize the free radicals, which come into being from the electronic spectra of flavonoids (Gupta et al., 2016). Studies on flavonoids by UV spectroscope have revealed that most of the flavonoids consist of two major absorption maxima: Band II (240–285 nm) which corresponds to the benzoyl system of the A ring, while band I (300–400 nm) exhibits the cinnamoyl system of the B-ring (Figure 5.1) (Fossen and Andersen, 2006; Markham and Mabry, 1975).

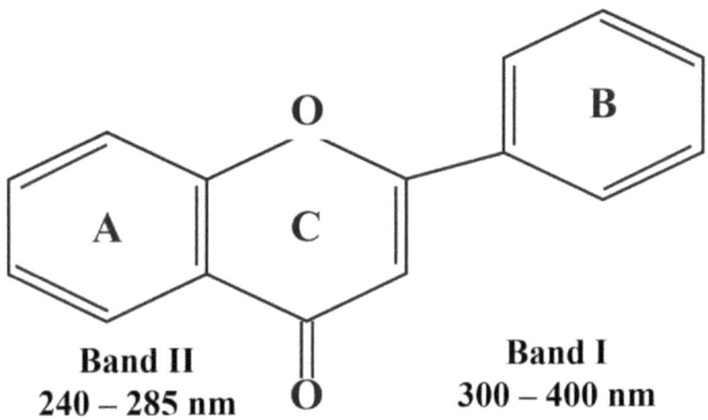

FIGURE 5.1 Band II absorption (originated from A-ring benzoyl system) and band I (from the B-ring cinnamoyl system).

Source: Adapted from Croft (1998).

UV (or UV-vis) spectroscopy has been used for a long time for the analyses of flavonoids (Falcone Ferreyra et al., 2012). Functional groups associated with the flavonoid skeleton may influence the absorption. Flavonoids are natural antioxidants derived from plants that can sequester free radicals. They are widely present in foods and beverages (De Groot, 1994). The three fundamental structural features of flavonoids required for antioxidant activity can be determined by: (1) a 3′,4′-dihydroxy (catechol) structure in the B ring which favors the electron delocalization (**A**), (2) an unsaturated 2-3 bond in association with a 4-keto group which provides electron delocalization from the B-ring (**B**) and (3) hydroxyl groups at positions 3 and 5 forms intramolecular hydrogen bonding along with the keto group (**C**) (Figure 5.2). These effects increase the radical scavenging by delocalizing the electrons or by donating the hydrogen (Ares and Outt, 1998).

Flavonoids have different activation mechanisms, such as the scavenging of free radicals, peroxide inactivation and other ROS, chelation of metals, and quenching of secondary lipid oxidation products. The radical scavenging properties associated with the structure of flavonoids defend against oxidative stress and in doing so reduce heart disease, prevent cancer, and slow down the aging processes in cells responsible for degenerative diseases (Croft, 1998).

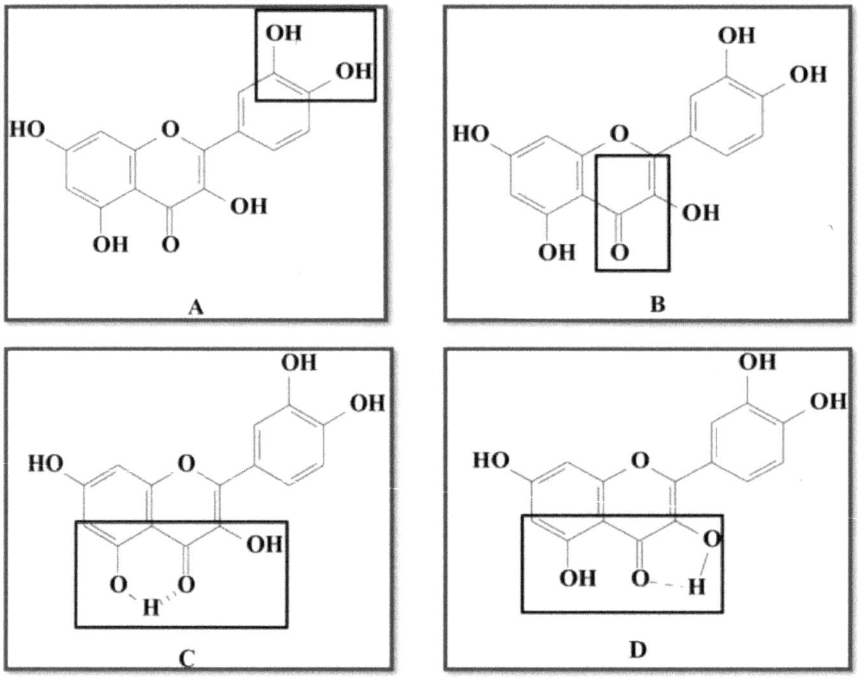

FIGURE 5.2 Structural groups for radical scavenging.

Source: Adapted from Croft (1998).

5.3.2 BIOSYNTHETIC PATHWAY

In plants, there are two distinct biosynthetic pathways responsible for the synthesis of flavonoid-based compounds. They are namely, the shikimic acid pathway producing the phenyl propanoids (C6-C3) skeleton and the acetate pathway serving as a building block for polymeric 2-carbon units (Nabavi et al., 2020). Plants synthesize flavonoids to fulfill several actions, such as allelopathy, antioxidants, antimicrobials, feeding repellents, oviposition stimulants, photoreceptors, phytoalexins, and visual attractors. It has been studied that flavonoid biosynthesis is stimulated by oxidative stress, which is triggered by ROS generated during the metabolic function of plants. Flavonoids inhibit the production of ROS and also protect against damage induced by UV radiation (Santos et al., 2017). The main backbone structure as well as a substitution at the various sites is responsible for the antioxidant effect and UV absorbability of flavonoids. Although B-ring dihydroxylated flavonoids are studied to show higher antioxidant activity, the monohydroxylated ones exhibit a higher potential to absorb UV wavelengths. Glycosylation of flavonoids further makes it water-soluble, preventing autooxidation of hydroxyl groups, and thus facilitates their transportation to the cell membrane. In maize, flavonoids have been found to be involved in aluminum toxicity, where their production is stimulated in aluminum-exposed roots. Flavonoids regulate the movement and catabolism of auxins, creatin, and an auxin gradient that causes different morpho-anatomical characters of plants. Moreover, flavonoids regulate indole-3-acetic acid oxidase effect and modulate protein activity implicated in cell growth. Plant flavonoids do also serve as active ingredients in a variety of pharmaceutical, nutraceutical, and cosmetic preparations for human uses. This aspect is also associated with their broad-spectrum health-promoting actions, including anti-allergenic and antimicrobial (Havsteen, 2002).

The first step in the flavonoid biosynthesis is marked by the condensation of one *p*-coumaroyl-CoA molecule (shikimate derived, B-ring) with three malonyl-CoA (polyketid origin, A ring) molecules producing chalcone (2′, 4′, 6′, 4-tetrahydroxychalcone). This reaction takes place in the presence of the enzyme chalcone synthase (CHS) (Santos et al., 2017). Chalcone flavanone isomerase (CHI) subsequently isomerizes chalcone to flavanone. Apart from the central intermediates formed, the pathway bifurcates into several side branches, each yielding a different class of flavonoids (Figure 5.3) (Alzand and Mohamed, 2012).

Although the biosynthesis pathway is found to be conserved in plants, a group of enzymes, such as isomerases, reductases, hydroxylases, has been studied to modify the basic flavonoid skeleton, producing different flavonoid classes. Chalcones and flavanones are also produced as intermediary compounds in the biosynthesis pathway. Anthocyanins, flavones, flavonols and proanthocyanidins, and other classes of flavonoids are the end products of the biosynthesis pathway. Another important class includes isoflavonoids, which are synthesized by the migration reaction of 2-aryl side chain to 3-position catalyzed by the enzyme isoflavone synthase (Santos et al., 2017). The retrochalcones are known to be unusual flavonoids that are characterized by reversed A and B-rings. Their biosynthesis is not yet clearly understood but are most likely to be produced from the common C_{15} intermediate of the general biosynthesis pathway of the flavonoids, more precisely from the reduction of dibenzoyl methane or by 2-hydroxylation of a flavanone. They are restricted to a few plant species, some species of the families Leguminosae, Annonaceae, and Basellaceae have been found to harbor these compounds. Several species are reported to harbor the chalcone dimers linked by cyclobutane (Figure 5.4). For the first time, dimerization in retrochalcones was observed in the roots of *Dahlstedtia grandiflora*. The mechanisms of [2 + 2] cycloaddition led to the formation of these compounds.

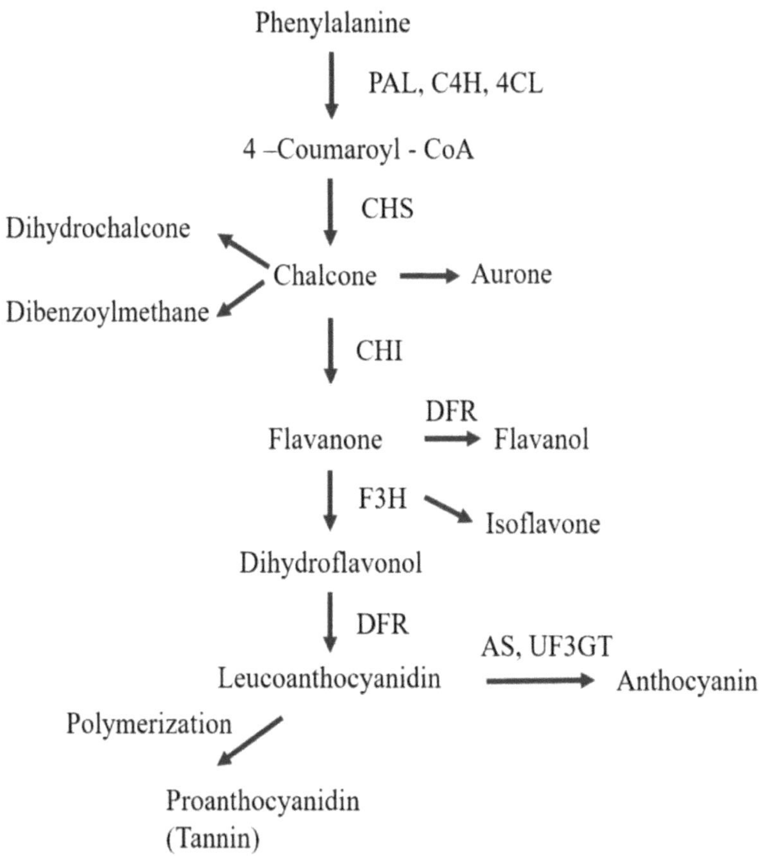

FIGURE 5.3 Flavonoid biosynthetic pathway schematic representation. The key enzymes involved in catalysis: PAL, phenylalanine ammonialyase; C4H, cinnamate 4-hydroxylase; 4CL, 4-coumaroyl-coenzyme A ligase; CHS, chalcone synthase; CHI, chalcone flavanone synthase; F3H, flavanone 3β-hydroxylase; DFR, dihydroflavonol 4-reductase; FLS, flavonol synthase; IFS, isoflavonoid synthase; AS, anthocyanin synthase and AUF3GT, UDP glucose: Flavonoid 3-O-glucosyltransferase.

Source: Adapted from Alzand and Mohamed (2012).

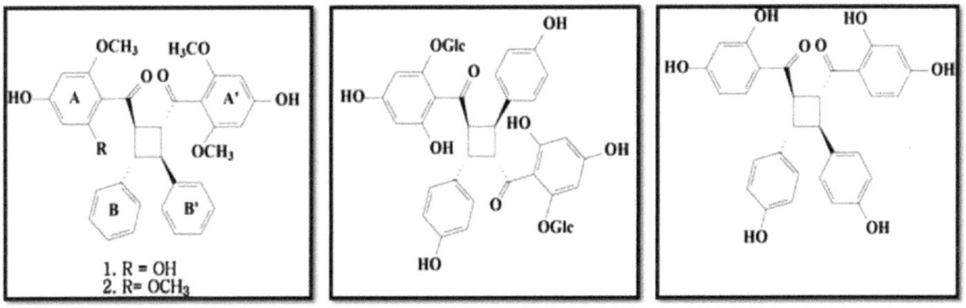

FIGURE 5.4 Chemical structures of the chalcone dimmers present in *Combretum albopunctatum, Helichrysum zivojinii*, and *Agapanthus africanus.*

Source: Adapted from Santos et al. (2017).

Flavonoids found in the conjugated forms with sugar are known as glycosides and the ones occurring without sugar are known as aglycones. They are generally witnessed to be hydroxylated in the positions, namely, 3, 5, 7, 3′, 4′, and 5′. Some of the hydroxyl groups are often methylated, acetylated, or sulfated. The occurrence of prenylation is usually found in the carbon atoms of the aromatic rings. When glycosides are synthesized, the glycosidic linkage is normally located in the third and seventh position and the resulting carbohydrates are D-glucose, glucose rhamnose, galactose, L-rhamnose or arabinose (Alzand and Mohamed, 2012; Santos et al., 2017). These changes often influence their reactivity, solubility, and stability. Flavonoids are majorly present in the form of glycosides under natural conditions.

5.4 APPLICATIONS OF FLAVONOIDS

5.4.1 BIOLOGICAL APPLICATIONS OF FLAVONOIDS

A number of phenylpropanoid superstructure compounds have key roles in plant mechanical stability, UV protection, fertility, and growing regulatory physiology and signaling of a plant cell. Phenylpropanoids are also essential chemical regulators of plant interaction with insects and microorganisms, acting as either attractants or insect repellents, as phytoalexins that protect plants from illnesses and herbivores, and as attractants for pollinators through petal color. Whenever secreted by symbiotic nitrogen-fixing rhizobacteria, they also cause root nodulation (Mandal et al., 2010).

The biological roles of flavonoids are coupled to their significant intrinsic cytotoxicity and ability to communicate with enzymes via protein binding. Certain flavonoids promote stress resistance, for instance, functioning as scavenging free radicals including ROS and chelating metals which create ROS via the Fenton reaction. Flavonoids have an important role to play in protecting maize from aluminum damage. The root system of maize plants exposed to aluminum secreted huge levels of phenolics including catechin and quercetin, suggesting that their ability to chelate metals could be an in vivo method to mitigate aluminum toxicity (Emamverdian et al., 2015).

5.4.2 FLAVONOIDS IN LEGUME–RHIZOBIAL INTERACTIONS DURING ROOT NODULATION

Flavonoids are engaged in the implementation of nodulation. Flavonoid-deficient roots of transformants generated via chalcone synthase RNA interference seem unable to produce nodules (Wang et al., 2018). Extracellular naringenin and liquiritigenin, progenitors transformed to the *Medicago truncatula* and soybean products, supplemented the nutrient uptake and flavonoid deficiency in the roots, implying that flavonoid deficit was the cause of deficiency of nodulation. Furthermore, the flavonoid-deficient roots displayed enhanced auxin transfer and local suppression of auxin movement at the nodulation region was absent.

5.4.3 FLAVONOIDS IN PLANT DEFENSE

5.4.3.1 THE ACTIVITY OF FLAVONOIDS AGAINST UV RADIATION

UV-absorbing properties of flavonoids have long been thought to constitute evidence for their involvement in UV protection. UV radiation causes the production of flavonol molecules in a variety of species, including *Ligustrum vulgare, Arabidopsis, petunia,* and *Vitis vinifera*, according to recent studies (Falcone Ferreyra et al., 2012). Because the existence of the hydroxyl group in the 3-position of flavonoid is the major structural property necessary for metal chelation, such as iron, copper, zinc, and aluminum, this limits the generation of free radicals as well as reduces the ROS once produced.

5.4.3.2 ACTIVITY OF FLAVONOIDS AGAINST INFECTION

Plants are protected by flavonoids from diseases and herbivores. Bioactive compounds are probably the most crucial facilitators of plant–insect interactions, as per the phytochemical coevolution hypothesis. Plants and insect herbivores have diversified resulting in plant defense (i.e., secondary metabolites) and herbivore counterattack (i.e., detoxification ability). Alterations in abiotic ecological factors like CO_2 in the atmosphere and ozone (O_3) patterns, Ultraviolet light, rainfall patterns, or temperature may all have a serious influence on the concentrations of secondary compounds in plants, which in return may alter herbivory or pathogenic infection thresholds (Shin et al., 2005).

5.4.4 APPLICATIONS IN INDUSTRY

In industries, the regulation of the flavonoid system for the deliberate development of chemicals has been widely exploited. More than decades have gone since the first plant color was altered via genetic manipulation: The red color petunia stores pelargonidin and expresses the maize dihydroflavonol reductase gene. A large number of genetic engineering endeavors have been presented to develop innovative flower colors across several species of plants, including petunia, rose, carnation, and torenia by tweaking the flavonoid biosynthesis pathway, either through transcriptional down-regulation, inactivation of enzymes responsible to carry out the anthocyanin pathway, or by hexavalent chromosome insertion (Kumar et al., 2018).

5.4.5 APPLICATIONS IN FOOD

In the past few years, numerous flavonoids have been used as antioxidant compounds in fats and oils to avoid the oxidation process, to safeguard meals and drinks from exposure to light, to extend the shelf life of food, as a substitute for animal nutrition to maintain their well-being, to safeguard animal related products, as an antibacterial property in food products, and as a health functional food excipient in nutritional supplements; these uses can be attributed primarily to antioxidant and antimicrobial activities (Ruiz-Cruz et al., 2017). The use of artificial ingredients in the food service industry has been falling in past years, owing mostly to man's desire to lower the probability of contracting disorders related to synthetic ingredients (Holst and Williamson, 2008).

Countless vegetables and fruits contain considerable amounts of flavonoids; bioactive molecules and flavonoids have been identified as two of the greatest micronutrients that can be used in industries to minimize the use of synthetic substances in foods and enhance human health being promising candidates to reduce several illnesses. Because of their antibacterial and antioxidant capabilities, these bioactive chemicals can be employed to extend the shelf life and preserve several food products (Karak, 2019).

Further to that, in the past few years, there has been a surge in interest in supporting more organic products, owing to constraints about the use of chemical additives in food and a boost in the use of herbal (Sharma et al., 2020; Rane et al., 2020a,b,c) extracts, such as antioxidants or dietary components, which do have lower toxic effects when compared with synthetic ingredients (Saeed et al., 2012). Flavonoids are appropriate substances that can be used as food preservatives due to their previously proven beneficial properties, which include protection of fat as well as oil oxidation, performance enhancer for livestock feed, enzymes and vitamins preservation, repression of bacterial activity in food products, and wellness responsive component in foods and dietary supplements. According to a study, flavonoids are more commonly found in red meats and poultry to suppress the oxidation process, utilization of natural extracts in meals as functional components, and the prevention of spoilage microbe growth in meat. According to certain research, these chemicals have a preventive role in meat quality and other items (Ruiz-Cruz et al., 2017).

5.4.6 MISCELLANEOUS APPLICATIONS

Plant bioactive chemicals (mainly flavonoids) have long been employed in dermatology and cosmetic formulations due to their antioxidant, antibacterial, anti-inflammatory, and curative activities. Flavonoids are getting more popular as cosmetic compounds due to their ability to shield skin from UV light and heal the epidermis from oxidative stress, hence improving skin look (Jang, 2021). Because of the presence of phenolic groups, the activity of flavonoids on the skin is connected with their antiradical characteristics, however, these activities on the skin have received little attention. The use of herbal and biodegradable products to protect the skin from UV radiation has improved; personal care products are being acknowledged for their gains in the treatment and prevention of diseases and in the advancement of several dermatologic circumstances, which portray a feasible solution for industrial usage of natural extracts (Huang et al., 2020).

Skin protection by flavonoids is indicated by their anti-inflammatory effect. This activity is provided by the impact of bioactive substances on enzymatic reactions and other conditions that encourage the progression of inflammatory phases. The suppression of enzymes associated with cellular stimulation that fosters skin degeneration and the production of regulation of chemicals for their proliferation is the other mode of action. Flavonoids also have an affinity for protein complexes and estrogen receptors, and also anti-irradiating properties; as a result of all these properties, they are utilized in the cosmetic sector to alleviate congestion issues, with the intention of lowering the severity of inflammatory symptoms (Figure 5.5).

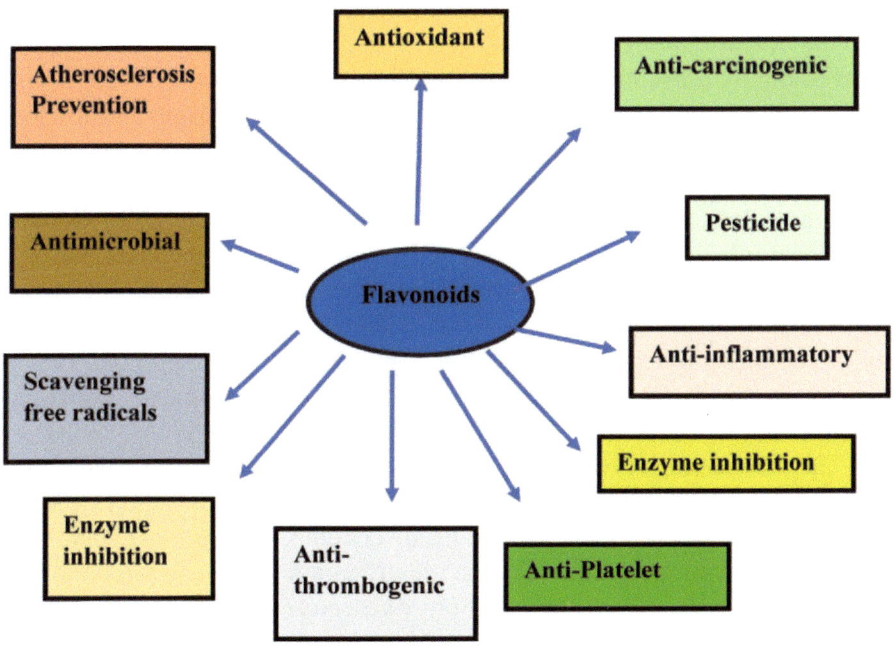

FIGURE 5.5 Representation of roles of flavonoids in several bioactivities, human health, and agriculture-related activities.

5.5 FUTURE CHALLENGES

The majority of the main enzymes and genes in the flavonoid system have been discovered. Several aspects of flavonoid biochemistry, however, remain unknown. Because of the variety of the diverse molecular configurations and the shortage of information on bioavailability, the research of flavonoids is challenging. Some of the transcriptional factors that govern this stemmed pathway, for example, have yet to be discovered in terms of expression profiles and functions. Besides, there is minimal evidence that protein–protein associations generate metabolic pathways that boost the efficacy of this route, and there is limited information about the transit of flavonoids into the extracellular environment.

The proportion of nutritional flavonoids ingestion is still being debated, with the total consumption of dietary phenolic and flavonoid compounds typically considered to be around 1 g/d, with phenolic acids and complex polymers accounting for a large share of this estimated amount (Kay, 2010). This estimate has remained virtually constant since Kuhnau et al. released it (Agrawal, 2011). As acknowledged in a recent review by Spencer et al. (2008), quantitatively establishing the benefits offered by flavonoids has proved highly difficult due to their great variability in foods, the constrained data available on their presence in food items, and inadequate knowledge regarding their metabolic consequences.

Furthermore, there are limited ways for measuring oxidative damage in vivo, and determining objective endpoints remains a concern. It is necessary to develop analytic procedures in order to acquire more information on intake and elimination. Data on the long-term effects of chronic flavonoid use are particularly sparse. Several studies have highlighted the importance of homology modeling in identifying prospective flavonoid compounds for use in the treatment of numerous human health conditions (Badhan and Penny, 2006).

The immense diversification of plants that emerged during development has resulted in a commensurate range of flavonoid compounds that are known to date and many that are still to be discovered. Further investigation of various plant species will lead to the identification of novel compounds, and maybe, new metabolic pathways. Future research will also help to enhance the applicability of flavonoids in the floricultural, foodstuff, medicinal, and chemical sectors.

5.6 CONCLUSIONS

Flavonoids are secondary plant metabolites found in plants or their parts. They have a wide diversity of structures and ecological relevance, for example, the colorful pigments in many flower petals operate as chemotaxonomic signal chemicals and have a wide range of biological applications. Because of their numerous health advantages, flavonoids are one of the most researched phytochemicals. Flavonoids are available in a multitude of chemical structures. Their biological behavior and interactions with other flavonoids and non-flavonoid antioxidants found in the same whole fresh produce as fruits are influenced by structural differences. The bioactivities of flavonoids are determined not only by their natural form either aglycones or glycosides but also by their metabolites. The knowledge gained above on the structure–bioactivity relationships and the interaction effects of flavonoids was mainly predicated on flavonoids that have been extracted or isolated. These findings may only provide some indications of a more fundamental character but are unlikely to be transferable to real food matrices required in interactions with bioactive substances, including flavonoid–non-flavonoid or flavonoid–flavonoid interactions, in a relatively close state to have a much better understanding and knowledge of the health-related benefits with reference to consuming a diet rich in flavonoids. The new advancements in nutritional supplements or enriched foodstuffs must pick out the aggregate of flavonoids with the maximum applicable properties.

KEYWORDS

- **flavonoids**
- **plants secondary metabolites**
- **structure**
- **and composition and biological properties**

REFERENCES

Agrawal, A. Pharmacological Activities of Flavonoids: A Review. *Int. J. Pharm. Sci. Nano.* **2011,** *4* (2), 1394–1398.

Alzand, K. I.; Mohamed, M. A. Flavonoids: Chemistry, Biochemistry and Antioxidant Activity. *J. Pharm. Res.* **2012,** *5* (40134012), 37.

Ares, J. J.; Outt, P. E. Gastroprotective Agents for the Prevention of NSAID-induced Gastropathy. *Curr. Pharm. Des.* **1998,** *4* (1), 17–36.

Badhan, R.; Penny, J. In Silico Modelling of the Interaction of Flavonoids with Human P-glycoprotein Nucleotide-binding Domain. *Eur. J. Med. Chem.* **2006,** *41* (3), 285–295.

Bose, S.; Sarkar, D.; Bose, A.; Mandal, S. C. Natural Flavonoids and its Pharmaceutical Importance. *Pharm. Rev.* **2018,** 61–75.
Croft, K. D. The Chemistry and Biological Effects of Flavonoids and Phenolic Acids A. *Ann. N. Y. Acad. Sci.* **1998,** *854* (1), 435–442.
De Groot, H. Reactive Oxygen Species in Tissue Injury. *Hepato-Gastroenterology* **1994,** *41* (4), 328–332.
DeFelice, S. Nutraceuticals: Opportunities in an Emerging Market. *Scrip. Mag.* **1992,** *9*, 14–15.
Dillard, C. J.; German, J. B. Phytochemicals: Nutraceuticals and Human Health. *J. Sci. Food Agric.* **2000,** *80* (12), 1744–1756.
Dixon, R. A.; Pasinetti, G. M. Flavonoids and Isoflavonoids: From Plant Biology to Agriculture and Neuroscience. *Plant Physiol.* **2010,** *154* (2), 453–457.
Emamverdian, A.; Ding, Y.; Mokhberdoran, F.; Xie, Y. Heavy Metal Stress and Some Mechanisms of Plant Defense Response. *Sci. World J.* **2015,** *2015.*
Falcone Ferreyra, M. L.; Rius, S.; Casati, P. Flavonoids: Biosynthesis, Biological Functions, and Biotechnological Applications. *Front. Plant Sci.* **2012,** *3*, 222.
Fossen, T.; Andersen, Ø., Eds.; Spectroscopic Techniques Applied to Flavonoids. In *Flavonoids: Chemistry, Biochemistry and Applications*, Taylor and Francis CRC Press, 2006; pp 37–142.
Grace, P. Ischaemia-reperfusion Injury. *J. Brit. Surg.* **1994,** *81* (5), 637–647.
Gupta, J.; Gupta, A.; Gupta, A. Flavonoids: Its Working Mechanism and Various Protective Roles. *Int. J. Chem. Stud.* **2016,** *4* (4), 190–198.
Halliwell, B. Free Radicals, Antioxidants, and Human Disease: Curiosity, Cause, or Consequence? *Lancet* **1994,** *344* (8924), 721–724.
Hämäläinen, M.; Nieminen, R.; Vuorela, P.; Heinonen, M.; Moilanen, E. Anti-inflammatory Effects of Flavonoids: Genistein, Kaempferol, Quercetin, and Daidzein Inhibit STAT-1 and NF-κB Activations, Whereas Flavone, Isorhamnetin, Naringenin, and Pelargonidin Inhibit only NF-κB Activation Along with Their Inhibitory Effect on iNOS Expression and NO Production in Activated Macrophages. *Mediators Inflamm.* **2007,** *2007.*
Havsteen, B. Flavonoids, a Class of Natural Products of High Pharmacological Potency. *Biochem. Pharmacol.* **1983,** *32* (7), 1141–1148.
Havsteen, B. H. The Biochemistry and Medical Significance of the Flavonoids. *Pharm. Therap.* **2002,** *96* (2–3), 67–202.
Holst, B.; Williamson, G. Nutrients and Phytochemicals: From Bioavailability to Bioefficacy Beyond Antioxidants. *Curr. Opin. Biotechnol.* **2008,** *19* (2), 73–82.
Hong, E.-Y.; Choi, S.-I.; Kim, G.-H. Determination of Flavonoids from Allium Victorialis var. Platyphyllum and their Effect on Gap Junctional Intercellular Communication. *Food Sci. Biotechnol.* **2007,** *16* (5), 747–752.
Huang, K.; Jinzhong, Z.; Zhu, T.; Morsi, Y.; Aldalbahi, A.; El-Newehy, M.; et al. PLCL/Silk Fibroin based Antibacterial Nano Wound Dressing Encapsulating Oregano Essential Oil: Fabrication, Characterization and Biological Evaluation. *Colloids Surf. B* **2020,** *196*, 111352.
Jang, A.-R. Composition of Phenolic Acids and Flavonoids and Skin Care Cosmetic Antioxidant Activity of Akebia quinata Fruit Extracts. *J. Digit. Converg.* **2021,** *19* (4), 365–371.
Karak, P. Biological Activities of Flavonoids: An Overview. *Int. J. Pharm. Sci.* **2019,** *10* (4), 1567–1574.
Kay, C. D. The Future of Flavonoid Research. *Br. J. Nutr.* **2010,** *104* (S3), S91–S95.
Keservani, R. K., Sharma, A. K., Kesharwani, R. K. (Eds.). *Nutraceuticals and Dietary Supplements: Applications in Health Improvement and Disease Management*; CRC Press, 2020. ISBN: 9781771888738.

Keservani, R. K.; Kesharwani, R. K.; Vyas, N.; Jain, S.; Raghuvanshi, R.; Sharma, A. K. Nutraceutical and Functional Food as Future Food: A Review. *Der. Pharmacia. Lettre* **2010a,** *2* (1), 106–116.

Keservani, R. K., Kesharwani, R. K.; Sharma, A. K.; Vyas, N.; Chadoker, A. Nutritional Supplements: An Overview. *Int. J. Curr. Pharm. Rev. Res.* **2010b,** *1* (1), 59–75.

Keservani, R. K.; Sharma, A. K. Flavonoids: Emerging Trends and Potential Health Benefits. *J. Chin. Pharm. Sci.* **2014,** *23* (12), 815.

Keservani, R. K.; Sharma, A. K.; Kesharwani, R. K. Nutraceutical and Functional Foods for Cardiovascular Health. In *Food Process Engineering*, Apple Academic Press, CRC Press, 2016a; pp 257–278, 291–312. ISBN: 9781771884020.

Keservani, R. K.; Sharma, A. K.; Kesharwani, R. K. Medicinal Effect of Nutraceutical Fruits for the Cognition and Brain Health. *Scientifica* 2016b, 10. Article ID 3109254. doi: 10.1155/2016/3109254.

Keservani, R. K.; Sharma, A. K.; Kesharwani, R. K. An Overview and Therapeutic Applications of Nutraceutical and Functional Foods. *Rec. Adv. Drug Deli. Tech.* **2017,** 160–201. ISBN:9781522507543.

Kumar, A.; Mosa, K. A.; Ji, L.; Kage, U.; Dhokane, D.; Karre, S.; et al. Metabolomics-assisted Biotechnological Interventions for Developing Plant-based Functional Foods and Nutraceuticals. *Crit. Rev. Food Sci. Nutr.* **2018,** *58* (11), 1791–1807.

Mandal, S. M.; Chakraborty, D.; Dey, S. Phenolic Acids Act as Signaling Molecules in Plant-microbe Symbioses. *Plant Signal. Behav.* **2010,** *5* (4), 359–368.

Markham, K. R.; Mabry, T. J. *Ultraviolet-visible and Proton Magnetic Resonance Spectroscopy of Flavonoids (The Flavonoids)*; Springer, 1975; pp 45–77.

Mierziak, J.; Kostyn, K.; Kulma, A. Flavonoids as Important Molecules of Plant Interactions with the Environment. *Molecules* **2014,** *19* (10), 16240–16265.

Nabavi, S. M.; Šamec, D.; Tomczyk, M.; Milella, L.; Russo, D.; Habtemariam, S.; et al. Flavonoid Biosynthetic Pathways in Plants: Versatile Targets for Metabolic Engineering. *Biotechnol. Adv.* **2020,** *38*, 107316.

Panche, A.; Diwan, A.; Chandra, S. Flavonoids: An Overview. *J. Nutr. Sci.* **2016,** *5*.

Rane, B. R.; Patil, A. S.; Keservani, R. K.; Jain, A. S. Novel Approaches in Herbal Formulation. In *Enhancing the Therapeutic Efficacy of Herbal Formulations through Novel Drug Delivery Systems* Kesharwani, R. K., Keservani, R. K., Sharma, A. K., Eds.; IGI Global International Publisher: Pennsylvania, USA, Chapter 2, 2020a; pp 43–68. ISBN13: 9781799844532.

Rane, B. R.; Bharath, M. S.; Patil, R. R.; Keservani, R. K.; Jain, A. S. Novel Approaches in Nutraceuticals. In *Enhancing the Therapeutic Efficacy of Herbal Formulations through Novel Drug Delivery Systems*, Kesharwani, R. K.; Keservani, R. K.; Sharma, A. K., Eds.; IGI Global International Publisher: Pennsylvania, USA, Chapter 11, 2020b; pp 241–266. ISBN13: 9781799844532.

Rane, B. R.; Tadavi, S. A.; Keservani, R. K. Naturopathy. In *Herbal Product Development*, Anil, K. S., Raj, K. K., Surya, P. G., Eds.; Apple Academic Press, CRC Press, Taylor & Francis Group, Chapter 12, 2020c; pp 321–347. ISBN: 9781771888776.

Ratty, A.; Das, N. Effects of Flavonoids on Nonenzymatic Lipid Peroxidation: Structure-activity Relationship. *Biochem. Med. Metab. Biol.* **1988,** *39* (1), 69–79.

Ruiz-Cruz, S.; Chaparro-Hernández, S.; Hernández-Ruiz, K. L.; Cira-Chávez, L. A.; Estrada-Alvarado, M. I.; Gassos Ortega, L.; et al. Flavonoids: Important Biocompounds in Food. In *Flavonoids—From Biosynthesis to Human Health*; Justino, G. C. Ed.; IntechOpen: London, UK, 2017; 353–369.

Saeed, N.; Khan, M. R.; Shabbir, M. Antioxidant Activity, Total Phenolic and Total Flavonoid Contents of Whole Plant Extracts Torilis Leptophylla L. *BMC Complement. Altern. Med.* **2012,** *12* (1), 1–12.

Santos, E. L.; Maia, B.; Ferriani, A. P.; Teixeira, S. D. Flavonoids: Classification, Biosynthesis and Chemical Ecology. In *Flavonoids—From Biosynthesis to Human Health*; Justino, G. C. Ed.; IntechOpen: London, UK, 2017, 1–14.

Sharma, A. K.; Keservani, R. K.; Gautam S. P. *Herbal Product Development*; Apple Academic Press, CRC Press, Taylor & Francis Group, 2020; ISBN: 9781771888776.

Shin, J.-E.; Kim, J.-M.; Bae, E.-A.; Hyun, Y.-J.; Kim, D.-H. In Vitro Inhibitory Effect of Flavonoids on Growth, Infection and Vacuolation of Helicobacter Pylori. *Planta Med.* **2005,** *71* (03), 197–201.

Singh, I. P.; Bharate, S. B.; Bhutani, K. Anti-HIV Natural Products. *Curr. Sci.* **2005,** 269–290.

Soares, J. M. D.; Leal, A. E. B. P.; Silva, J. C.; Almeida, J. R.; de Oliveira, H. P. Influence of Flavonoids on Mechanism of Modulation of Insulin Secretion. *Pharmacogn. Mag.* **2017,** *13* (52), 639.

Spencer, J. P.; Abd El Mohsen, M. M.; Minihane, A.-M.; Mathers, J. C. Biomarkers of the Intake of Dietary Polyphenols: Strengths, Limitations and Application in Nutrition Research. *Br. J. Nutr.* **2008,** *99* (1), 12–22.

Ververidis, F.; Trantas, E.; Douglas, C.; Vollmer, G.; Kretzschmar, G.; Panopoulos, N. Biotechnology of Flavonoids and other Phenylpropanoid-derived Natural Products. Part I: Chemical Diversity, Impacts on Plant Biology and Human Health. *Biotechnol. J.: Health. Nutri. Tech.* **2007,** *2* (10), 1214–1234.

Vessal, M.; Hemmati, M.; Vasei, M. Antidiabetic Effects of Quercetin in Streptozocin-induced Diabetic Rats. *Comp. Biochem. Physiol. C: Toxicol. Pharmacol.* **2003,** *135* (3), 357–364.

Wang, F.; Ren, G.; Li, F.; Qi, S.; Xu, Y.; Wang, B.; et al. A Chalcone Synthase Gene AeCHS from Abelmoschus Esculentus Regulates Flavonoid Accumulation and Abiotic Stress Tolerance in Transgenic Arabidopsis. *Acta Physiol. Plant* **2018,** *40* (5), 1–13.

Wegener, T.; Fintelmann, V. Flavonoids and Bioactivity. *Wein. Med. Wochem. Schr.* **1999,** *149*, 241–247.

Wu, C.-H.; Yen, G.-C. Inhibitory Effect of Naturally Occurring Flavonoids on the Formation of Advanced Glycation Endproducts. *J. Agric. Food Chem.* **2005,** *53* (8), 3167–3173.

Xu, J.-J.; Wu, X.; Li, M.-M.; Li, G.-Q.; Yang, Y.-T.; Luo, H.-J.; et al. Antiviral Activity of Polymethoxylated Flavones from "Guangchenpi", the Edible and Medicinal Pericarps of Citrus Reticulata 'Chachi'. *J. Agric. Food Chem.* **2014,** *62* (10), 2182–2189.

PART III

Applications of Flavonoids in Disease Management

CHAPTER 6

POTENTIAL PHARMACOLOGICAL HEALTH BENEFITS OF FLAVONOIDS

RICHA SAXENA[1], ABHIRUP MITRA[2], SONIYA JOSHI[2], RICHA SHARMA[3], PANKAJ KUMAR RAI[1], SANA ANSARI[1], NAIMA MIRZA[1], NISHTHA SRIVASTAVA[4], and VINEET K. MAURYA[5]

[1]*Department of Biotechnology, Invertis University, Bareilly, Uttar Pradesh, India*

[2]*Department of Agriculture, Invertis University, Bareilly, Uttar Pradesh, India*

[3]*Department of Microbiology, Mahatma Gandhi University of Medical Science and Technology, Jaipur, India*

[4]*Department of Applied Science, Invertis University, Bareilly, Uttar Pradesh, India*

[5]*Department of Botany and Microbiology, H.N.B. Garhwal University, Srinagar Garhwal, India*

ABSTRACT

Flavonoids, natural bioactive secondary metabolites, are found in medicinal plants, vegetables, grains, fruits, and beverages. These compounds are widely distributed in a group of polyphenols and have potential health benefits because of their therapeutic effects. Many researchers have revealed that they have numerous significant activities like sedative, antioxidant, antimicrobial, anti-inflammatory, antidiabetic, antidepressant, anticonvulsant, antiproliferative, antihypertensive, cardioprotective, hepatoprotective, antiulcerogenic, and anticancer. Recent research and developmental activities on flavonoids relate to human health benefits. Mostly, all pharmacological activities are found due to their antioxidant-rich properties and these are

The Flavonoids: Extraction and Applications. Deepika Saini, Rajesh K. Kesharwani, & Raj K. Keservani (Eds.)

associated with human health dietary ingredients, that is, tea, vegetables, fruits, plants, and wine, which are the major nutritional sources of flavonoids. Different research is being started on flavonoids due to their resourceful health benefits. In medicinal plants, flavonoids act as growth regulators and help to reduce oxidative stress. For pharmaceutical purposes, commercial mass production of various flavonoids has been potentially done by using microbial technology. This chapter aims to express the prospective health benefits of flavonoids from various sources and also provides insights on flavonoids as a natural bioactive compound for the improvement of effective pharmacological remedies in the future.

6.1 INTRODUCTION

Flavonoids are the foremost vital class of bioactive molecules that are extensively present in plants along with fruits and vegetables. They have a variety of positive biochemical properties attributed to their antioxidative effect associated with various diseases, like mutagenic, inflammatory, microbial, carcinogenic, and antidiabetic activity (Jasuja et al., 2013; Panche et al., 2016; Yahfoufi et al., 2018). Flavonoids (Keservani and Sharma, 2014; Keservani et al., 2016a, 2016b) are used as an indispensable component for various pharmaceutical, therapeutic, cosmetic, and nutraceutical purposes. They have significant anti-inflammatory, antioxidant, antimutagenic, and anticarcinogenic activities that integrate for the balance of important cellular enzyme actions. Flavonoids are also distinguished as effective inhibitors for various enzymes, that is, cyclooxygenase (COX), xanthine oxidase (XO), lipoxygenase (LOX), nitric oxide (NO), phosphoinositide 3-kinase, collagenase, and hyaluronidase (Walker et al., 2000; Lee et al., 2009; Rodriguez-Garcia et al., 2019).

Flavonoids are natural products extracted from various parts of plants, and more than 4000 flavonoids are present in nature. Flavonoids standardize plant growth and development and also organize other biological cells by several techniques (Havsteen, 2002). They consist of a large group of bioactive compounds widely scattered in the plant system. Flavonoids also play a significant role in the attractive colors of flowers, leaves, and fruits (Sangeetha et al., 2016). Though their appearances are not restricted to flowers, they are also found in the entire plant (Dewick, 2001). Indeed, flavonoids are the vital ingredient for human diet. Daily ingestion of flavonoids is based on the utilization of fruits and vegetables. The flavonoids classes are divided into different major classes, that is, flavones, isoflavones, flavonols, flavan-3-ols, flavanones, flavanonols, chalcones, aurones, and anthocyanidins. Several types of flavonoids are present in foods and beverages (tea, cocoa, and wine). Onions and tea are foremost nutritional sources of flavones and flavanols (Pietta, 2000). The reported clinical and preclinical data recommend cytotoxic and chemopreventive effects of flavonoids against different cancer cell lines.

Polyphenolic bioactive constituents (i.e., flavonoids) are low molecular weight and show an imperative function in plants, animals, and microorganisms. They provide protection for plants against several abiotic and biotic stresses and reveal a diverse variety of biological functions. They are also responsible for crucial functions in relation amid their environment and plants (Pourcel et al., 2007). The study of Samanta et al. (2011) reported that flavonoids participate in the development, growth, propagation, and protection of plants, beside numerous adverse conditions in the biosynthetic pathway. Consequently, flavonoids are used in fruit dispersal to facilitate the germination in spores and seeds and the development of seedlings (Panche et al., 2015).

Flavonoids act as a plant defense system and protect plants from different biotic (herbivores and pathogens) and abiotic stresses (UV radiation and heat). Flavonoids function as distinctive UV filters, signal molecules, detoxifying agents, phytoalexins, antimicrobial protective agents, and allopathic compounds (Mierziak et al., 2014). Flavonoid compounds are employed against drought resistance, frost hardiness, and can contribute in freezing tolerance and plant heat acclimatization (Alhaithloul et al., 2021). The quercetin is ubiquitous in various medicinal plants and works as an antifeedant against some insects and pests (Morimoto et al., 2000). Furthermore, some other flavonoids, like isoflavones and rotenoids, also act as antifeedants against various phytophagous insect pests (Palma-tenango et al., 2017). Flavonoids derived from medicinal plants are utilized by all humans and organisms because of their lots of health benefits in fighting diseases. A wide variety of flavonoid compounds is used in agriculture, medicine, pharmacy, and cosmetology. They are employed to improve acne, skin hydration, blackheads, wrinkles, and baldness, restore the skin's antibacterial and antifungal barriers, and reduce the speed of the aging progression (Bogdan Allemann and Baumann, 2008).

Most scientists have widely examined the techniques and conditions, particularly for the isolation of flavonoids in foods and foodstuffs, in addition to preparative, analytical, and industrial uses based on their potential applications (Chaves et al., 2020). Current research has discovered the herbal (Sharma et al., 2020) (green) extraction techniques recommending excellent substitutes for traditional techniques. In this field, numerous studies are still being performed to progress the extract quality, the time consumed, the ecological shelter, and health (Chaves et al., 2020). Jorgensen (1995) explained that the early development of flower genetics was mainly due to genetic variants in flavonoid-induced flowering, and indicated that genetic mutation in plants was allied by flavonoid biosynthesis (Jorgensen, 1995). Presently, there are about 8000 flavonoids, which have a significant role in human and animal health (Ahmed et al., 2016). Dixon and Pasinetti (2010) analyzed flavonoids and their types in brief and illustrated their agricultural performance and neuroscience in humans (Dixon and Pasinetti, 2010). Recently Panche et al. (2015) discussed the use of flavonoids as a secondary metabolite for their protective roles against various diseases. In the present study, efforts have been made to explain the current summary of flavonoids' therapeutic and pharmaceutical potential from the current published works, in addition to providing several remarkable guidelines for future investigation.

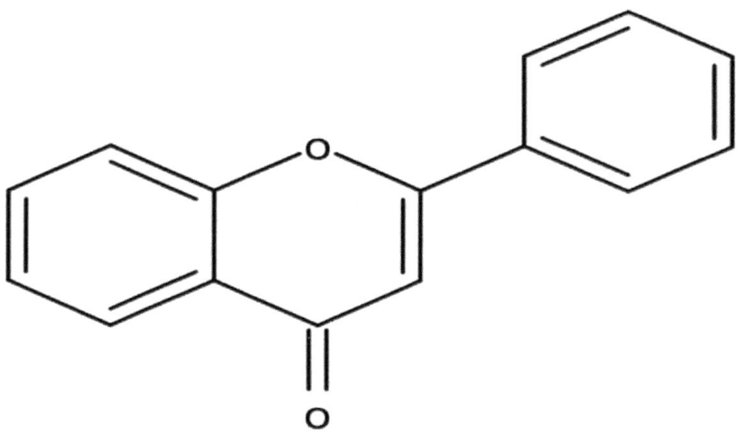

FIGURE 6.1 Basic structure of flavonoids.

6.2 CLASSIFICATION OF FLAVONOIDS

Flavonoids are phenolic compounds that are naturally found in plants and structurally derived from flavones, the parent substances occurring in nature (Ferreyra et al., 2012). The classification of flavonoids depends on the position of phenyl C-ring carbon and pyran B-ring carbon, apart from the degree of oxidation and unsaturation of the C-ring. For example, when the pyran B-ring is attached in position 3 of the phenyl C-ring carbon, it is known as isoflavonoids (3-benzopyrans). Similarly, when the pyran B-ring is attached to position 4 of the phenyl C-ring, it is known as 4-benzopyrans (neoflavonoids). Therefore, they are subdivided into various groups depending on the association of the pyran "B" ring at carbon position 2 of the phenyl "C" ring. These subclasses are also known as flavones, flavonols, flavanones, flavanonols, flavanols or catechins, anthocyanins, and chalcones (Iwashina, 2013; Panche et al., 2016; Ekalu and Habila, 2020; Subbaraj et al., 2021). Major classes of flavonoids have been shown in Figure 6.2. The classes of flavonoids along with their subclasses and sources have been shown in Table 6.1.

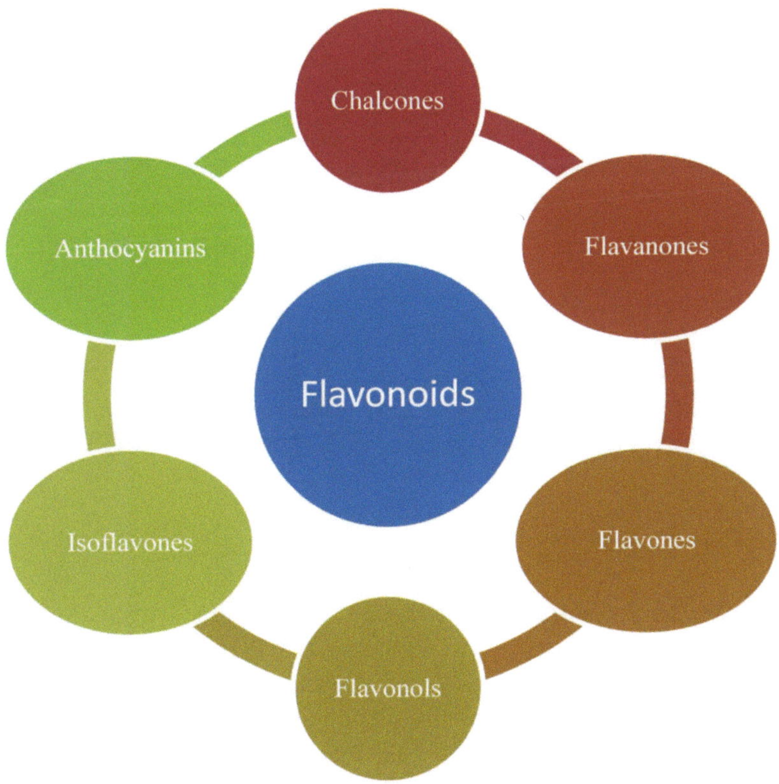

FIGURE 6.2 Major classes of flavonoids (Iwashina, 2013; Panche et al., 2016).

Source: Adapted from Iwashina, 2013; Panche et al., 2016

6.2.1 FLAVONES

Generally, flavones have a double bond between positions 2 and 3 and a ketone in position 4 of the ring. Although most of the flavones of fruits and vegetable fruits have a hydroxyl group in position 5 of the A-ring, the hydroxylation in other positions, for the majority in positions 3′ and 4′ of the B-ring and position 7 of the A-ring, can fluctuate according to the taxonomic classification of the definite fruit and vegetable (Gould and Lister, 2006; Kumar et al., 2009; Santos et al., 2017). Celery, garlic, and chamomile tea are rich sources of flavones (Clark et al., 2015). Various researches proved that luteolin has great beneficial properties among all (Lugasi and Hovari, 2002). It was also found that luteolin has the capability to minimize the blood pressure in the body of hypertensive rats (Milbury et al., 2006; Panche et al., 2016).

6.2.2 FLAVONOLS

Flavonols are also an important class of flavonoids found in broccoli, onions, fruits, and tea. It is represented by glycosides like kaempferol and quercetin (Kumar and Pandey, 2013).

6.2.3 FLAVANONES

Naringenin and hesperetin are the main representatives of this class. They are mostly found in fruit peels and citrus fruits (Mahmoud et al., 2019). They help to prevent the activity of free radicals (Khan et al., 2018).

6.2.4 ISOFLAVONES

3-aryl-4H-chromen-4-ones (isoflavones) are produced from flavanones by the relocation that supports 2,3-aryl migration, pursued through dehydrogenation. Even though it is generally assumed in the literature that the term isoflavonoids is obtained from the isolation of other phenolic compounds, that is, isoflavans, isoflavanones, and isoflavones, they are commonly found in nature for their health benefits (Vitale et al., 2013; Pinto et al., 2020). Isoflavones have similarity in structure with mammalian estrogens; hence they are expected to work like estrogen receptor agonists. Diadzein and genistein are the main representatives among the class (Jiang et al., 2005; Rodríguez-García et al., 2019).

6.2.5 *ANTHOCYANIDINS*

Anthocyanidins are much admired pigments responsible for color in vegetables and fruits. They are accountable for blue, red, and purple colors of fruits (Clark et al., 2015). They have great beneficial effects on the cardiovascular system that are employed by dropping the risk of severe myocardial infarction and endothelium-dependent vasodilation (Mckay et al., 2009; Ciumarnean et al., 2020).

6.2.6 *CHALCONES*

Chalcones are also known as 1,3-diaryl-2-propen-1-ones. In the aryl rings, chalcones are polyhydroxylated and have huge biological properties, including antimicrobial, antioxidant, anti-inflammatory, antifungal, and antitumor properties (Gupta and Jain, 2015).

6.3 SOURCES OF FLAVONOIDS

Flavonoids are mostly found in different agricultural commodities like fruits, vegetables, and grains, along with different herbs. Apart from the major sources, they are also found in different plant-based foods including cocoa, tea, wines, etc. In daily diet, flavonoids are present in several raw vegetables and fruits, along with other processed foodstuff products, apart from the different food ingredients including food colors and other additives, and health foods derived from natural resources like different herbs, berries, green tea, and ginkgo leaves. One of the major sources of flavonoids is grapes including wines prepared from it. Around 80% of the grapes are utilized for wine raw materials throughout the world. Grapes are usually categorized into two types, namely European and American. The European species are more popular for raw winemaking. In the major types of the European species, Merlot, Cabernet Sauvignon, and Pinot Noir are widely used for red wine, and Chardonnay, Semillon, and Sauvignon Blanc are used for making of white wines. Apart from the abovementioned sources, there are a huge number of sources for flavonoids (Tungmunnithum et al., 2018). Some of the major sources as per the classes and subclasses of flavonoids have been shown in Table 6.1.

TABLE 6.1 Different Classes of Flavonoids with Subclasses and Sources (Panche et al., 2016; Dias et al., 2021).

Sl. No.	Class of flavonoids	Subclasses	Sources
1.	Anthocyanins	a) Cyanidin b) Malvidin c) Delphinidin d) Peonidin	Red and purple grapes, red wine, cranberries, blueberries, strawberries, and blackberries
2.	Chalcones	a) Phloretin b) Arbutin c) Phlioridzin d) Chalconaringenin	Tomatoes, shallots, bean sprouts, and potatoes
3.	Flavanones	a) Hesparitin b) Naringin c) Naringenin d) Eriodictyol e) Hesperidin	Lemons, limes, oranges, and grapefruit
4.	Flavones	a) Apigenin b) Tangeretin c) Baicalein d) Rpoifolin	Parsley, red peppers, celery, chamomile, and peppermint
5.	Flavonols	a) Quercetin b) Myricetin c) Rutin d) Morin e) Kaempferol	Tea, red wine, peaches, berries, grapes, kale, tomatoes, onions, lettuce, broccoli, and scallions
6.	Isoflavones	a) Genistin b) Genistein c) Daidzein d) Glycitein e) Daidzin	Soy, soy products, and legumes such as fava beans

Source: Adapted from Panche et al., 2016; Dias et al., 2021

6.4 METABOLIC PROCESS OF FLAVONOIDS

Mainly two compartments are considered for the metabolism process of flavonoids in human body. One compartment usually consists of different tissues, like liver, small intestine, and kidneys, and the second one is colon (Figure 6.3). Flavonoids cannot be absorbed through the small intestine; they are soaked up and discharged with bile, and then outstretched to the colon. The benefits of bile secretion in rats to help in almost about 40% of total absorbed (+)-catechin was discharged through the bile toward the small intestine (Hollman and Katan, 1999; Spencer et al., 2000).

TABLE 6.2 Classes of Flavonoids with Backbone Structure and Health Benefits (Dias et al., 2021).

Class of flavonoids	Backbone structure	Health benefits
Anthocyanins	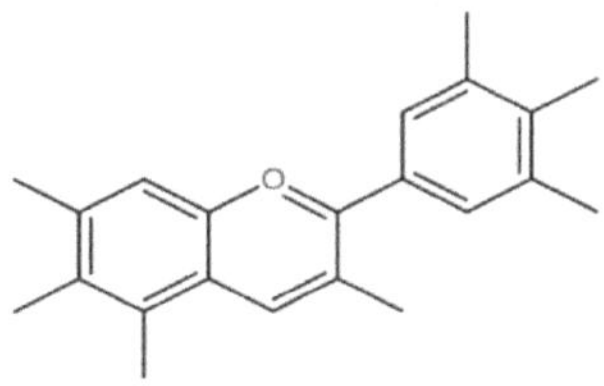	Antidiabetic, anticancer, anti-inflammatory, antimicrobial, and antiobesity activity
Chalcones	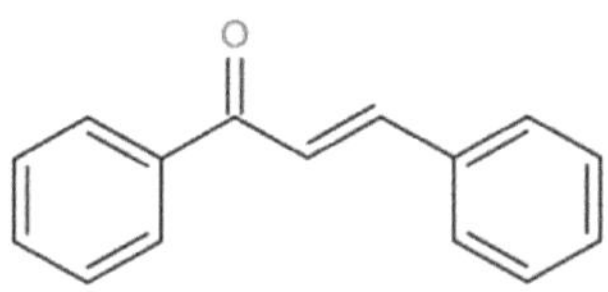	Anti-inflammatory, antimicrobial, antioxidative, cytotoxic, anticancer, and immunosuppressive activity
Flavanones	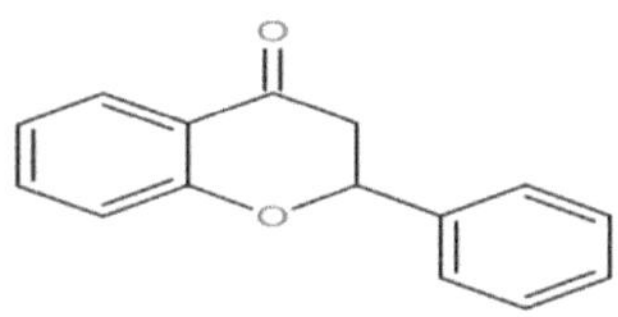	Anti-inflammatory properties, help to manage your weight and cholesterol
Flavones	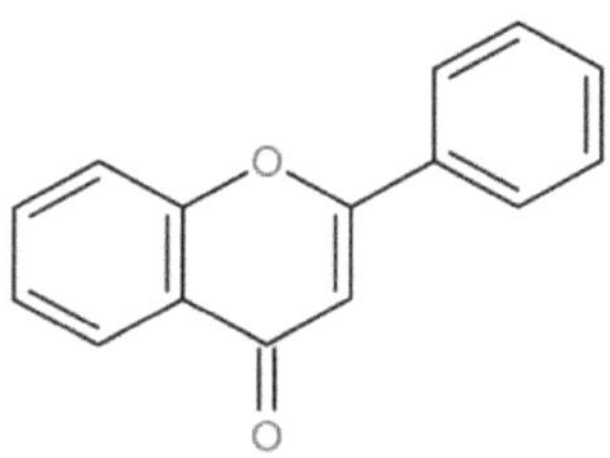	Protection against inflammation
Flavonols	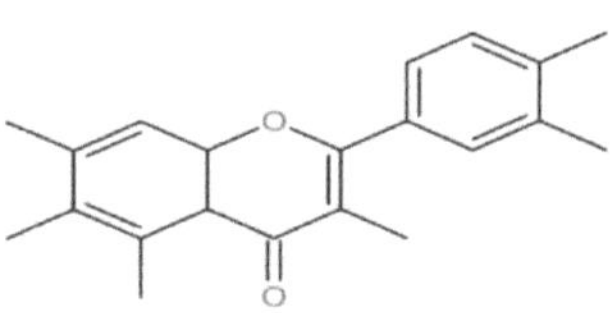	Antioxidant properties, help to protect against cardiovascular disease
Isoflavones	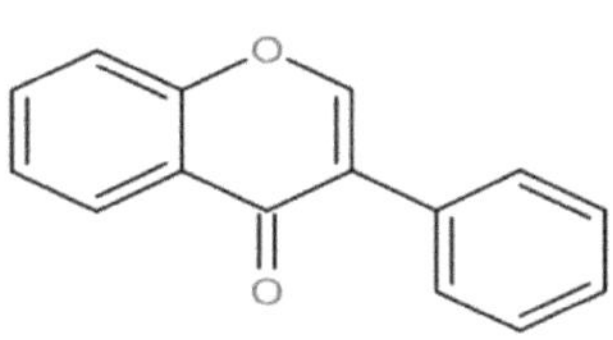	Help to maintain the balance of hormones

Source: Adapted from Dias et al., 2021

Different compartments for the metabolism of phenols are shown in Figure 6.3. Flavonoid is easily soaked up in the small intestine, for example, quercetin-3-glucoside.

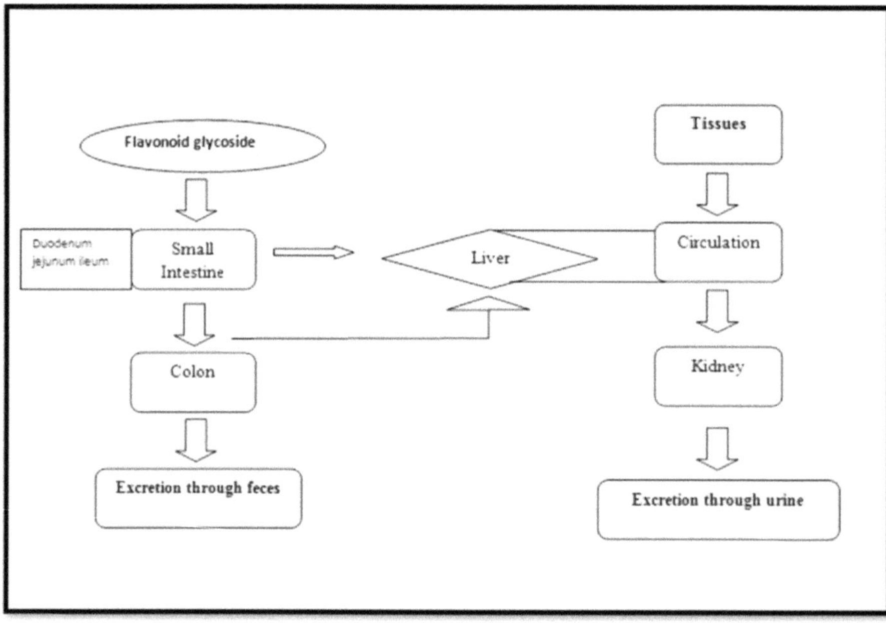

FIGURE 6.3 Metabolic process of flavonoids (Kumar and Pandey, 2013).
Source: Adapted from Kumar and Pandey, 2013

Maximum flavonoids, excluding the subgroups of catechins, are found in plants bound with sugars as b-glycosides. Aglycans can be effortlessly absorbed through the small intestine, while flavonoid glycosides should be transformed into the configuration of aglycan (Bae et al., 2020).

The flavonoid glucosides (hydrophilic) inclusive of quercetin are conveyed throughout the small intestine with help of intestinal Na+-based glucose cocarrier (SGLT1) (Hollman et al., 1999). Another mechanism indicates lactase phloridzin hydrolase (LPH) hydrolyzed the flavonoid glucosides, a β-glucosidase on the outer side of the comb outer layer of the small intestine. Glycosides, which are not substrates for the enzymes, are conveyed in the direction of the colon, where flavonoid glycosides are hydrolyzed by micro-organisms, which additionally degrade the aglycones (Walle et al., 2005). The absorption capability of the colon is somewhat less as compared to the small intestine; due to this reason, the most effective and insignificant absorption of those glycosides must be anticipated (Rowland et al., 2018).

After this entire process of absorption of flavonoids, they are combined with the liver through glucuronidation and sulfation, alternatively metabolized otherwise, methylation to phenolic components (Bravo, 1998). Because of this process of combination, urine or plasma does not contain any free aglycones, excluding the catechins (Hollman, 2004). Relying on different food sources, bioavailability of some flavonoids differs noticeably; as an instance, the quercetin absorption from tea or apple is about four times lower as compared with the onions (Hollman et al., 1997; Dabeek and Marra, 2019). The flavonoids, which are secreted through bile in the intestine and which cannot be absorbed by the small intestine, are disintegrated in the colon with the help of microflora present in the intestine by breaking the ring structure of flavonoids (Thilakarathna and Rupasinghe, 2013). Oligomeric flavonoids are also hydrolyzed into monomers and dimers in acidic environment inside the abdomen. Large molecules arrive at the colon, wherever they may be despoiled by microorganisms (Spencer et al., 2000; Kawabata et al., 2019).

6.5 PHARMACOLOGICAL PROPERTIES OF FLAVONOIDS

Flavonoids are the central bioactive agents for drug development having extensive pharmacological properties. The exclusive and enormous pharmacological health uses of flavonoids have prepared them for research interest. Flavonoids are effective against several diseases, such as antioxidant, anti-inflammatory, digestive complications, blood glucose, hypertension, nervous system disorders, cholesterol irregularities, cardioprotective, hepatoprotective, reproductive disorders, microbial infections, antiviral, antiallergic, anticancer, and tumorigenesis. Herein, the role of flavonoid compounds in fighting acute diseases and the processes associated with health-promoting procedures have been demonstrated in detail. The use of flavonoids that are isolated from medicinal plants, fruits, and vegetables has become a unique trend in clinical research and possesses a wide range of pharmacological and biochemical effects.

6.5.1 FLAVONOIDS AS ANTIOXIDANT AGENTS

Antioxidants are bioactive compounds that are used to prevent oxidation in living cells and act as protective agents against free radicals (Rao et al., 2011). Polyphenolic compounds are also known as free-radical scavengers and almost all flavonoids have the capacity to act as antioxidants. Antioxidant ability is mainly endorsed by the highest reactivities of the hydroxyl substituents involved in the results. The two most potent flavonoids, that is, flavones and catechins, are used to defend the body against ROS (reactive oxygen species). Antioxidant defense mechanism prevents the damage caused via free radicals and oxidative stress. The word "reactive oxygen species," that is, ROS represents oxygen-centered radicals, for example, hydroxyl (–OH) and superoxide ($–O_2$), in addition to nonradical species resulting from oxygen, for example, hydrogen peroxide, hypochlorous acid, and singlet oxygen. Several researchers have investigated the antioxidant potential of different flavonoids (Krumova and Cosa, 2016; Li et al., 2018). In the accomplishment of numerous xenobiotic compounds, reactive oxygen species (ROS) plays a significant role. Many scientists have investigated the increasing production of ROS utilized in the most forms of tissue injuries, and consequently, it has been concerned in a huge amount of ailment conditions ranging from cancer and immune inflammatory injury to myocardial infarction (Halliwell, 1991, 1992). The study of Tanwar and Modgil (2012) and Sharma et al. (2015) reported that highly potent antioxidant compounds, that is, rutin, kaempferol, quercetin, myricetin, and morin revealed remedial effects, for example, antiallergic, antiviral, anti-inflammatory, antimicrobial as well as anticancer activity. Some other flavonoids, for example, epicatechin, epigallocatechin, and gallocatechin have been shown to have antioxidant action against free radicals (Bae et al., 2020). The central purpose of antioxidant compounds is to defend the human being from free radicals oxidation (Saxena et al., 2017).

6.5.2 FLAVONOIDS AS ANTIMICROBIAL AGENT

Resistance against antimicrobial drugs has become a progressively essential and urgent worldwide trouble (Tim Cushnie and Lamb, 2005). Recently, flavonoids or polyphenolic compounds are a rapidly growing term for antimicrobial study (Yang et al., 2014). Several subclasses of flavonoids possess potent antiviral, antifungal, and antibacterial activity. Flavonoids are polyphenolic secondary metabolites naturally occurring in plants and food products (Hollman et al., 1999; Pandey and Rizvi, 2009; Nabavi et al., 2020). Moreover, they are found in beverages, food items, and numerous plant sources, such as huge variety of vegetables, fruits, cereals, tea, cocoa, and wine (Górniak et al., 2019; Uddin et al., 2020). Flavonoids are used as therapeutic agents and have their physiologically active components in traditional medicines. The examination of Tereschuk et al. (1997) discovered that one most active compound, that is, quercetagetin-7-arabinzylgalactoside found in *Tagetes minuta*, is commonly used in Argentinean folk remedy for numerous infectious diseases therapy. *Retama raetam* flowers contain derrone and licoflavone C, which revealed antibacterial properties against pathogenic bacteria (Edziri et al., 2012). Apigenin, quercetin, luteolin, kaempferol, and their respective glycosides are some of the flavonoids found in mayweed (*Tripleurospermum disciforme*). *Tripleurospermum disciforme* plant extract can be used as a disinfectant and as a traditional medicine designed for remedial of various diseases in Iran (Tofighi et al., 2015). Various analogues of flavonoids were examined by Cushman et al. (1991) for analyzing their inhibitory activities against tyrosine kinase. Most of the flavonoids, that is, galangin, genkwanin, apigenin, ponciretin, pinocembrin, naringin, naringenin, luteolin, kaempferol, 3-O-methylquercetin, quercetin, several quercetin derivatives, sophoraflavanone G, epigallocatechin gallate, and its derivatives isolated from plant species have potent antibacterial activity (el-Gammal and Mansour, 1986; Taguri et al., 2004; Tim Cushnie and Lamb, 2005).

Further, isoflavanones, flavanones, isoflavones, flavones, isoflavans, flavonols, and chalcones have also been acknowledged for their significant antimicrobial action (Sathiamoorthy et al., 2007; Zhong et al., 2011; Xie et al., 2015). The study of Xie et al. (2015) demonstrated that antibacterial mechanisms of flavonoids are in the following process: inhibition of cytoplasmic membrane utility, porin on the cell membrane, attachment and biofilm formation, nucleic acid synthesis, energy metabolism, and amendment of the membrane permeability and reduction of the pathogenicity. The analysis of Yuan et al. (2021) also recommended that the cell membrane is the core location of flavonoids working on gram-positive bacteria, which probably participates in the inhibition of the respiratory chain or ATP synthesis and in the damage of phospholipid bilayers.

6.5.3 FLAVONOIDS AS ANTIVIRAL AGENT

To investigate the antiviral activity of flavonoids has become an area of interest for many researchers. Flavonoids serve as an enormous reservoir of pharmacologically bioactive compounds that have been investigated as prospective antiviral applicants against viruses (Khazeei Tabari et al., 2021). In several studies, the inhibitory effect of flavonoids has been found against the parainfluenza virus type 3 (Pf-3), poliovirus type 1, herpes simplex virus type 1 (HSV-1), human immunodeficiency virus (HIV), respiratory syncytial virus (RSV), and Enterovirus A71 (EV-A71) (Critchfield et al., 1996; Zakaryan et al., 2017; Lalani and Poh, 2020). Most of the investigations constitute the work against the strain of HIV-1 and its enzymes. Inhibition of HIV-1 infection and replication can be done by baicalein, which has been studied through in vitro investigation. Flavonoids inhibit the entrance of HIV-1 infection into the cells expressing chemokine coreceptors and CD4 (Koshy et al., 2001; Pasetto et al., 2014). Li and colleagues demonstrated antagonistic activity against the HIV-1 reverse transcriptase shown by flavone O-glycoside (Li et al., 2000; Anila and Vijayalakshmi, 2003). Catechins are the compounds which inhibit the polymerases, but they show a nonspecific kind of interaction with the HIV-1 enzyme (Moore and Pizza, 1992). Similar to several catechins, hinokiflavone, baicalein, and robustaflavone also have the property to inhibit the action of HIV-1 reverse transcriptase (Ono et al., 1989; Lin et al., 1997). Also, the study of Brinkworth et al. (1992) evaluated the demethylated gardenin A and 3,2′-dihydroxyflavone that inhibit the action of HIV-1 proteinase. Therefore, most of the flavonoids, that is, myricetin, quercetagetin, quercetin 3-O-(2″-galloyl)-α-l-arabinopyranoside, robinetin, baicalein, apigenin, galangin, lonchocarpol, and 2,4,2′-trihydroxy-5′-methylchalcone have been examined for antiviral mechanism of action (Richman, 2006; Lee et al., 2007; Nijhuis et al., 2009). Scientists have confirmed that flavonols are more vigorous than flavones, besides herpes simplex virus type 1 (Kumar and Pandey, 2013). Furthermore, these bioactive flavonoids are characterized as novel leads for future research, which might allocate the development of therapeutically acceptable antiviral agents (Singh et al., 2021).

6.5.4 FLAVONOIDS AS HEPATOPROTECTIVE AGENT

In 2018, Lesjak et al. reported that acute liver failure is outlined clinical disorder that has high mortality rate and has significant metabolic derangement, along with multiple failures and neurological complications. Major pharmacological activities, for example, antioxidation, anti-inflammation, antimicrobial, anticarcinogenic, and antiapoptosis have been demonstrated by flavonoids (Lesjak et al., 2018). It has been observed that after administration of d-galactosamine and lipopolysaccharide, some flavonoids were found to inhibit the oxidative actions by the reduction of nitric oxide synthase, superoxide dismutase, malondialdehyde (MDA), heme oxygenase-1 (HO-1), and nuclear factor erythroid-derived 2-related factor 2 (Nrf2), and by increasing catalase (CAT) (Li et al., 2016). The intensity of aspartate aminotransferase, alanine, and proinflammatory cytokines in blood serum was reduced and the phosphorylations of IKK, IκBα, and NF-κB/p65 in the NF-κB signaling pathway were also prevented. Moreover, hepatocyte apoptosis is inhibited by five flavonoids through increment in Bcl-2/Bax ratio along with suppression of proteins of caspase family. Hepatoprotective properties against acute liver failure, elicited by LPS/d-GalN, are shown by chrysin, luteolin, hesperetin, 3′,4′-dimethoxy hesperetin, and apigenin. In the investigation, it was observed that the C2–C3 double bond at A-ring and the hydroxyl groups at B-ring positioned at C3′ and C4′ were responsible for the increment in the hepatoprotective activities; nevertheless, the effect was reversed by the hydroxymethylation of C3′ and C4′ carbon atoms. Additionally, apigenin has been found to be a potential curative compound with promising hepatoprotective effects for the clinical treatment of acute liver failure (Lesjak et al., 2018).

6.5.5 FLAVONOIDS AS ANTIDIABETIC AGENT

Flavonoids are highly potential in all areas of diabetes mellitus, and their methodology is quite well understood (Kharroubi and Darwish, 2015). Diabetes is a biochemical condition that causes hyperglycemia and stifles fat, carbohydrate, and protein metabolism, consequential in a deficient insulin report (Muller et al., 2006). Controlling blood glucose levels efficiently is an excellent technique for decreasing clinical consequences of diabetes mellitus. Apart from optimum blood glucose control, which might not avoid issues on its own, they encourage the development of alternative treatment techniques (Aslan et al., 2007). Despite the fact that there is no real cure for diabetes by synthetic medicine, pharmacological treatments have benefited prognosis and life quality to some extent. However, a problem involving the body's micro and macrovascular system must be addressed. Flavonoids play a bigger role in preventing diabetes complications than any other medication (Hussain et al., 2013). Flavonols have an unsaturated carbon ring at carbon 2–3 that is reduced at carbon 4 and hydroxylated at carbon 3. They're prevalent in lettuce, berries, grapes, onions, and kale among other things (Panche et al., 2016). Flowers, apples, tomato seeds, fennel, berries, almonds, onions, broccoli, pepper, lovage, tea leaves, and shallots are the most common sources (Kelly et al., 2011). Quercetin serves as a building block for the construction of other flavonoids such as naringenin, rutin, and hesperidin (Hollman et al., 1995). Quercetin performs a potent function in the diversity of biological processes, including glucose homeostasis, insulin sensitization and secretion, glucose consumption in peripheral tissues, and intestinal glucose absorption inhibition (Eid et al., 2017). Lemons, grapes, peaches, oranges, buckwheat, and limes are examples of plants that contain rutin (Kreft et al., 1999). Rutin's antidiabetic properties include decreased carbohydrate absorption from the small intestine, improved glucose uptake by tissues, inhibition of tissue gluconeogenesis, activation of insulin production from β-cells, and prevention of the Langerhans islets against degenerative effects. Rutin also reduces the production of reactive oxygen species, precursors of advanced glycation endproducts, sorbitol, and proinflammatory cytokines (Ghorbani, 2017). When diabetic rats were given 100 mg/kg of rutin, their insulin levels and glucose metabolic enzyme activity both rose dramatically. Furthermore, the data revealed a considerable reduction in plasma glucose levels. Flavanones, also known as dihydroflavones, are compounds with an oxidized and saturated carbon ring. Citrus fruits contain flavanones, which are recognized for their antioxidant and free radical scavenging properties (Stanely Mainzen Prince and Kannan, 2006; Ola et al., 2015). Citrus fruits, such as oranges, tomatoes, and grapefruits, are high in 5,7,4'-trihydroxyflavanone, a saturated oxidized aglycone (Hasanein and Fazeli, 2014). Naringenin possesses a variety of biological properties, including antioxidant, antidiabetic, anti-inflammatory, anticancer, and antimutagenic (Patel et al., 2018). The antidiabetic and antiobesity properties of naringenin and its glycoside are similar. Naringenin replicates the impact of insulin by lowering prolipoprotein B release in liver cells, according to an in vitro study published in *Biomolecules* 2019, 9, 430 10 of 35 (Zygmunt et al., 2010).

6.5.6 FLAVONOIDS AS ANTI-INFLAMMATORY AGENT

The anti-inflammatory properties of flavonoid families can be enhanced by structural changes. The unsaturation of the C-ring, the availability of a carbonyl group on C-4, the amount and location of hydroxyl groups, and glycosylation status, according to structural dynamic requirements, impact the anti-inflammatory activities of flavonoids (Theoharides et al., 2001; Lago et al., 2014). Chronic uncontrolled inflammation can cause the production of damaging ROS, which can damage DNA and genetic changes, resulting in tumor progression (Yang et al., 2008). Cell growth development, immune cell initiation, neoangiogenesis, and inflammation are all controlled by cyclin-dependent kinases (CDKs), according to growing data (Schmitz and Kracht, 2016). The study of Vezza et al. (2016) shows that flavonoids have anti-inflammatory effect in a variety of experimental colitis models. Flavonoids from many subclasses, including isoflavones, anthocyanidins, chalcones, flavonols, flavones, flavanones, and catechins, have shown significant anti-inflammatory effect in the intestine (Bian et al., 2018). Glycones are digested in the colon and subsequently released in the aglycone form, and they are more effective in reducing intestinal inflammation than aglycones that are absorbed early in the intestines (Gonzali and Perata, 2020).

6.5.7 FLAVONOIDS AS ANTICANCER AGENT

In several epidemiological studies, it has been observed that flavonoids provide protection against some kind of cancerous cells, which are mainly found in some common food. Flavonoids (Keservani et al., 2010a, 2010b, 2017, 2020) as a nonnutritive dietary compound might be responsible for anticancer activity, and with that observation, multiple researches have been started to analyze potential of flavonoids for the anticancer effect. The regulation of proliferation and cell death pathways that lead to cancer by these compounds has been studied as a result of the above observation. Hence, some of the flavonoids like synthetic flavone, the soy isoflavonoid, flavopiridol, genistein, the common dietary flavonol or the tea catechin, gallate, epigallocatechin, and quercetin are the emerging candidates as a potential anticancer drug. Some of the abovementioned flavonoids have already entered into the clinical trials as anticancer drug candidates. Many researchers have elucidated different possibilities of the relationship between structure and activity of flavonoids and target sites to find the therapeutic potential of these compounds, which might lead to discovery of new drugs. There is very diminutive information about the relationship between flavonoid and their anticancer activity due to the possibility of scattering of information. Better knowledge of therapeutic potential as well as the presence of many flavonoids in diet with their anticancer activity and properties may help us improve dietary habits to reduce the chances of attack of cancer and their possible prevention with very common but effective weapons (Lopez et al., 2002).

6.5.8 FLAVONOIDS AS ANTIATHEROSCLEROTIC AGENT

The biological process that hardens and narrows the arteries by the deposition of fats, cholesterol, and various other substances is called atherosclerosis. It is commonly in association with some severe cardiovascular diseases; for instance, heart attacks, peripheral vascular diseases, and strokes. Due to the synthetics drugs such as statins that have some side effects on usage, plant flavonoids became the matter of interest and were being paid attention because of its antiatherosclerotic activity, which was evidenced to lessen the jeopardy of atherosclerosis in vitro as well as in vivo depending upon the various animal models. Over the centuries, several flavonoid preparations have been used for the physiologically active compounds that have been therapeutically applied for the treatment of human diseases (Havsteen, 1983). Epidemiological studies demonstrated that the condition of diseases related to heart might be diminished by the intake of diet rich in flavonoids. Flavonoids have several properties like lipid lowering, anti-inflammatory, and antiatherogenic effects. Flavonoids might prevent the development of vascular diseases by the modification in production of endothelial cell eicosanoid (Hertog et al., 1995; Schramm and German, 1998). Flavonoids have shown prospective to decrease the blood pressure in hypertension and have a normal blood pressure. Additionally, they might be beneficial for the actions in obesity because of their capability of regulation of oxidation of fatty acids and improvement of adipocyte functions (Galleano et al., 2012). Some of the plants that contain flavonoids and have significantly minimized the atherosclerosis as well as cardiovascular disease are *Hypericum perforatum* L., *Mangifera indica*, *Asparagus racemosus*, and *Garcinia cambogia* (Anila et al., 2003; Zou et al., 2005; Visavadiya and Narasimhacharya, 2009).

6.5.9 FLAVONOIDS IN THE TREATMENT OF ALZHEIMER'S DISEASE

The extreme chronic neurodegenerative disorder is Alzheimer's disease (AD), which can be defined by a neuropathological perspective as the aggregation of amyloid beta (Aβ) peptides. This amyloid beta (Aβ) peptide gets placed as infirm plaques and tau protein, which creates neurofibrillary tangles (Uddin and Kabir, 2019). Advancement in understanding Alzheimer's disease pathogenesis has provided therapeutic ways to cure it but with modest symptomatic ease. Moreover, dietary flavonoids, which are naturally occurring, have gained significant popularity because of their antiamyloidogenic effects as an unconventional method for AD treatment. Experiments have proven that some of the flavonoids might support the AD prevention by inhibiting the Aβ peptide's production and aggregation, along with decrement in tau aggregation. Flavonoid contains the significant capability to enhance degrading of Aβ peptides and inhibits the phosphorylation of tau protein via mTOR/autophagy signaling pathway.

Therefore, flavonoids can be significantly used as promising symptomatic anti-Alzheimer agents because of their potential of cholinesterase inhibition. Flourishing information that confirms flavonoids demonstrates encouraging neuroprotective ability because of their potential to lower the development of age-related neurodegenerative diseases or prevent the initial stages of neurodegeneration (Hollman et al., 1999; Uddin et al., 2020). Flavonoid-rich food and foodstuffs diet has a positive effect and also influences cognition and learning of disease (Wang et al., 2006; Sokolov et al., 2013; Zaplatic et al., 2019).

6.5.10 FLAVONOIDS AS ANTIEPILEPTIC AGENT

Epilepsy is one of the oldest chronic neurological disorders and is a group of disorders distinguished by spontaneous seizures involving hyperexcitable neurons (Svenningsen et al., 2006; de Almeida et al., 2008). Approximately 1–2% of the global population is suffering from this common neurological disorder, which might affect several different aspects of quality of life (Ramalingam et al., 2013; Diniz et al., 2015). Commonly, epilepsy is connected with brain dysfunction and numerous behavioral comorbidities, namely depressive disorders (DDs) (Hecimovic et al., 2003). Epilepsy depends on the medical treatment with antiepileptic drugs (AEDs) to achieve control of their epileptic seizures. These AEDs are targeted on suppressing an abnormal epileptic discharge, acting generally via one or more mechanism(s) through inactivation of Na^+ (hydantoins) and Ca^{2+} channels (ethosuximide), increasing GABA-mediated inhibitory functions (barbiturates, benzodiazepines, etc.), inhibition of glutamate-mediated excitation (felbamate), and activation of K^+ channels (retigabine) (Cook and Bensalem-Owen, 2011).

In humans, flavonoids have shown a significant effect on peripheral blood flow. For instance, it was found by brain imaging study that intake of flavanol-rich cocoa improved cortical blood flow. The study was important because the mechanism that enhances cerebrovascular functions is considered, specifically the hippocampus region in the brain, which is mainly responsible for memory and majorly assists adult neurogenesis (Nassiri-Asl et al., 2010; Diniz et al., 2015). The study of Singh et al. (2012) revealed that the hydroethanolic extract of aerial parts of *Passiflora incarnate* has protective effect against epilepsy-associated postictal depression. It contains chrysin that shows CNS depressant activity by agonizing GABA–benzodiazepine receptor.

Most of the flavonoids derivatives like kaempferol, quercetin, and isorhamnetin are present in *Ginkgo biloba* leaves (Diniz et al., 2015). Some plants with medicinal properties such as *Hypericum perforatum* and *Ginkgo biloba* contain a biflavonoid, amentoflavone, which has been found via in vitro study to have an explicit high affinity to brain benzodiazepine receptors, resulted to cause cell inhibition and lead to achievement of anticonvulsant activity (Smith et al., 1996; Svenningsen et al., 2006; Diniz et al., 2015). The flavanone naringenin was isolated from ethanolic leaf extract of *Mentha aquatic* and has been found to have high affinity to the receptor of GABA–benzodiazepine. A remarkable effect has been shown on the CNS by this flavanone, which passes the blood–brain barrier (Jäger et al., 2009; Diniz et al., 2015). According to Shrestha et al. report, the only weak receptor binding of GABA–BZD is demonstrated by rhusflavone, which is one of the monomers of the naringenin (Shrestha et al., 2012; Diniz et al., 2015).

6.6 CONCLUSION

In the context of natural products, it is commonly found that phytochemicals might be less effective or anti-infective as compared to microbial origin agents, such as antibiotics. Flavonoids are novel ideas that may fulfill the urgent need of new classes of antimicrobial drugs. A therapeutic approach for optimization of these compounds for future needs can be done by altering the structure of these compounds that might help in development of the drugs having different properties such as antimicrobial, antioxidant, anticancerous, hepatoprotective, and antiviral as well as many more different groups of agents against diseases. In vivo experiment on humans demonstrates that the role of foods from plants has helped directly in modulating the inflammatory response. However, identification of the mechanism and the molecules, which are responsible for the effect, has not been cleared. Based on the recent in vivo evaluation, it was not completely justified that the flavonoids were accountable for the anti-inflammatory effect of plant foods. A significant role of flavonoids has been found with regards to the antioxidant capacity. Several studies have demonstrated that the flavonoids action on the GABA receptor enhances its effect. Hence, flavonoids can be an active metabolite that can help in treating CNS disorders, which may include epilepsy. Along with classifying the antimicrobial flavonoids and their target site interactions, second generation inhibitors could be designed potentially. Futuristic approach regarding drug discovery taking flavonoids as a main compound has a potential to treat several diseases, might help reduce side effects, and may help develop new generation drugs for the pharmaceutical industries.

KEYWORDS

- **bioactive compounds**
- **pharmacological action**
- **flavonoids**
- **polyphenols**
- **abiotic stress**

REFERENCES

Ahmed, S. I.; Hayat, M. Q.; Tahir, M.; Mansoor, Q.; Ismail, M.; Keck, K.; Bates, R. B. Pharmacologically Active Flavonoids from the Anticancer, Antioxidant and Antimicrobial Extracts of *Cassia angustifoliaVahl. BMC Complement Altern. Med.* **2016,** *11–16* (1), 460.

Alhaithloul, H. A.; Galal, F. H.; Seufi, A. M. Effect of Extreme Temperature Changes on Phenolic, Flavonoid Contents and Antioxidant Activity of Tomato Seedlings (*Solanum lycopersicum L.*). *Peer J.* **2021,** *9*, e11193.

Anila, L.; Vijayalakshmi, N. R. Antioxidant Action of Flavonoids from *Mangifera indica* and *Emblica officinalis* in Hypercholesterolemic Rats. *Food Chem.* **2003,** *83* (4), 569–574.

Aslan, M.; Deliorman Orhan, D.; Orhan, N.; Sezik, E.; Yesilada, E. *In Vivo* Antidiabetic and Antioxidant Potential of *Helichrysum plicatum ssp. plicatum capitulums* in Streptozocin Induced Diabetic Rat. *J. Ethanopharmacol.* **2007,** *109* (1), 54–59.

Bae, J.; Kim, N.; Shin, Y.; Kim, S. Y.; Kim, Y. J. Activity of Catechins and their Applications. *Biomed. Dermatol.* **2020,** *4* (1), 8.

Bian, Y.; Liu, P.; Zhong, J.; Hu, Y.; Zhuang, S.; Fan, K.; Liu, Z. Quercetin Attenuates Adhesion Molecule Expression in Intestinal Microvascular Endothelial Cells by Modulating Multiple Pathways. *Dig. Dis. Sci.* **2018,** *63*, 3297–3304.

Bogdan Allemann, I.; Baumann, L. Antioxidants used in Skin Care Formulations. *Skin Therapy Lett.* **2008,** *13* (7), 5–9.

Brinkworth, R. I.; Stoermer, M. J.; Fairlie, D. P. Flavones are Inhibitors of HIV-1 Proteinase. *Biochem. Biophys. Res. Commun.* **1992,** *188*, 631–637.

Chaves, J. O.; de Souza, M. C.; da Silva, L. C.; Lachos-Perez, D.; Torres-Mayanga, P. C.; Machado, APdF.; Forster-Carneiro, T.; Vázquez-Espinosa, M.; González-de-Peredo, A. V.; Barbero, G. F.; Rostagno, M. A. Extraction of Flavonoids From Natural Sources Using Modern Techniques. *Front Chem.* **2020,** *8*, 507887.

Ciumarnean, L.; Milaciu, M. V.; Runcan, O.; Vesa, T. C.; Rachis, A. L.; Negrean, V.; Perné, M. G.; Donca, V. I.; Alexescu, T. G.; Para, I.; Dogaru, G. The Effects of Flavonoids in Cardiovascular Diseases. *Molecules* **2020,** *25* (4320), 1–18.

Clark, J. L.; Zahradka, P.; Taylor, C. G. Ecacy of Flavonoids in the Management of High Blood Pressure. *Nutr. Rev.* **2015,** *73*, 799–822.

Cook, A. M.; Meriem, K. Bensalem-Owen. Mechanisms of Action of Antiepileptic Drugs. *Therapy* **2011,** *8* (3), 307–313.

Critchfield, J. W.; Butera, S. T.; Folks, T. M. Inhibition of HIV Activation in Latently Infected Cells by Flavonoid Compounds. *AIDS Res. Hum. Retroviruses* **1996,** *12*, 39–46.

Cushman, M.; Nagarathnam, D.; Burg, D. L.; Geahlen, R. L. Synthesis and Protein-Tyrosine Kinase Inhibitory Activities of Flavonoid Analogs. *J. Med. Chem.* **1991,** *34*, 798–806.

Dabeek, W. M.; Marra, M. V. Dietary Quercetin and Kaempferol: Bioavailability and Potential Cardiovascular-Related Bioactivity in Humans. *Nutrients* **2019,** *11* (10), 2288.

De Almeida, R. N.; de Sousa, D. P.; de Farias Nóbrega, F. F.; de Sousa Claudino, F.; Araújo, D. A. M.; Leite, J. R.; Mattei, R. Anticonvulsant Effect of a Natural Compound α, β-epoxy-carvone and its Action on the Nerve Excitability. *Neurosci. Lett.* **2008,** *443* (1), 51–55.

Dewick, P. M. The Shikimate Pathway: Aromatic Amino Acids and Phenylpropanoids. In *Medicinal Natural Products: a Biosynthetic Approach*, 2nd Ed.; Dewick, P. M., Ed.; John Wiley: Chichester, 2001; pp 137–186.

Dias, M. C.; Pinto, D. C. G. A.; Silva, A. M. S. Plant Flavonoids: Chemical Characteristics and Biological Activity. *Molecules* **2021,** *26* (17), 5377.

Diniz, T. C.; Silva, J. C.; Lima-Saraiva, S. R. G. D.; Ribeiro, F. P. R. D. A.; Pacheco, A. G. M.; de Freitas, R. M.; Almeida, J. R. G. D. S. The Role of Flavonoids on Oxidative Stress in Epilepsy. *Oxid. Med. Cell. Longev.* **2015,** 171756.

Dixon, R.; Pasinetti, G. Flavonoids and Isoflavonoids: From Plant Biology to Agriculture and Neuroscience. *Plant Physiol.* **2010,** 154, 453–457.

Edziri, H.; Mastouri, M.; Mahjoub, M. A.; Mighri, Z.; Mahjoub, A.; Verschaeve, L. Antibacterial, Antifungal and Cytotoxic Activities of Two Flavonoids from *Retama raetam* Flowers. *Molecules* **2012,** *17*, 7284–7293.

Eid, H. M.; Haddad, P. S. The Antidiabetic Potential of Quercetin: Underlying Mechanisms. *Curr. Med. Chem.* **2017,** *24*, 355–364.

Ekalu, A.; Habila, J. D. Flavonoids: Isolation, Characterization, and Health Benefits. *Beni-Suef. Univ. J. Basic Appl. Sci.* **2020,** *9*, 45.

el-Gammal, A. A.; Mansour, R. M. Antimicrobial Activities of Some Flavonoid Compounds. *Zentralbl. Mikrobiol.* **1986,** *141*, 561–565.

Ferreyra, F.; Rius, S. P.; Casati, P. Flavonoids: Biosynthesis, Biological Functions, and Biotechnological Applications. *Front Plant Sci.* **2012,** *3*, 1–15.

Galleano, M.; Calabro, V.; Prince, P. D. Flavonoids and Metabolic Syndrome. *Ann. N. Y. Acad. Sci.* **2012,** *1259* (1), 87–94.

Ghorbani, A. Mechanisms of Antidiabetic Effects of Flavonoid Rutin. *Biomed. Pharmacother.* **2017,** *96*, 305–312.

Gonzali, S.; Perata, P. Anthocyanins from Purple Tomatoes as Novel Antioxidants to Promote Human Health. *Antioxidants* **2020,** *9* (10), 1017.

Górniak, I.; Bartoszewski, R.; Króliczewski, J. Comprehensive Review of Antimicrobial Activities of Plant Flavonoids. *Phytochem. Rev.* **2019,** *18* (2), 241–272.

Gould, K. S.; Lister, C. Flavonoid Functions in Plants. In *Flavonoids Chemistry, Biochemistry and Applications*; 1st ed.; Andersen, O. M., Markham, K. R., Ed.; Taylor & Francis Group: New York, 2006; pp 397–425.

Gupta, D.; Jain, D. K. Chalcone Derivatives as Potential Antifungal Agents: Synthesis, and Antifungal Activity. *J. Adv. Pharm. Technol. Res.* **2015,** *6* (3), 114–117.

Hachett, A. Plant Flavonoids in Biology and Medicine: Biochemical, Pharmacological, and Structure-Activity Relationships. *Prog. Clin. Biol. Res.* **1986,** 177–194.

Halliwell, B. Drug Antioxidant Effects. *Drugs* **1991,** *42* (4), 569–605.

Halliwell, B. A. R. R. Y. Free Radicals, Antioxidants, and Human Disease: Where are We Now?. *J. Lab. Clin. Med.* **1992,** *119*, 598–620.

Hasanein, P.; Fazeli, F. Role of Naringenin in Protection against Diabetic Hyperalgesia and Tactile Allodynia in Male Wistar Rats. *J. Physiol. Biochem.* **2014,** *70*, 997–1006.

Havsteen, B. Flavonoids, a Class of Natural Products of High Pharmacological Potency. *Biochem. Pharmacol.* **1983,** *32* (7), 1141–1148.

Havsteen, B. The Biochemistry and Medical Significance of the Flavonoids. *Pharmacol. Ther.* **2002,** *96*, 67–202.

Hecimovic, H.; Goldstein, J. D.; Sheline, Y. I.; Gilliam, F. G. Mechanisms of Depression in Epilepsy from a Clinical Perspective. *Epilepsy Behavior.* **2003,** *4*, S25–S30.

Hertog, M. G. L.; Kromhout, D.; Aravanis, C. Flavonoid Intake and Long-Term Risk of Coronary Heart Disease and Cancer in the Seven Countries Study. *Arch. Intern. Med.* **1995,** *155* (4), 1184–1195.

Hollman, P. C. H. Absorption, Bioavailability and Metabolism of Flavonoids. *Pharm. Biol.* **2004,** *42*, 74–83.

Hollman, P. C. H.; Buijsman, M. N. C. P.; van Gameren, Y.; Cnossen, P. J.; de Vries, J. H. M.; Katan, M. B. The Sugar Moiety is a Major Determinant of the Absorption of Dietary Flavonoid Glycosides in Man. *Free Radical. Res.* **1999,** *31* (6), 569–573.

Hollman, P. C.; de Vries, J. H.; van Leeuwen, S. D.; Mengelers, M. J.; Katan, M. B. Absorption of Dietary Quercetin Glycosides and Quercetin in Healthy Ileostomy Volunteers. *Am. J. Clin. Nutr.* **1995,** *62*, 1276–1282.

Hollman, P. C.; van Trijp, J. M.; Buysman, M. N.; van der Gaag, M. S.; Mengelers, M. J.; de Vries, J. H.; Katan, M. B. Relative bioavailability of the antioxidant flavonoid quercetin from various foods in man. *FEBS Lett.* **1997,** *418* (1-2), 152–156.

Hussain, S. A.; Maroul, B. H. Flavonoids as Alternatives in Treatment of Type 2 Diabetes Mellitus. *Acad. J. Med. Plants.* **2013,** *1* (2), 31–36.

Iwashina, T. Flavonoid Properties of Five Families Newly Incorporated into the Order Caryophyllales (Review). *Bull. Natl. Mus. Nat. Sci.* **2013,** *39* (1), 25–51.

Jäger, A. K.; Krydsfeldt, K.; Rasmussen, H. B. Bioassay-Guided Isolation of Apigenin with GABA-benzodiazepine Activity from *Tanacetum parthenium*. *Phytother. Res.* **2009,** *23* (11), 1642–1644.

Jasuja, N. D.; Saxena, R.; Chandra, S.; Sharma, R. Pharmacological Characterization and Beneficial Uses of *Punica granatum*. *Asian J. Plant Sci.* **2013,** *11* (6), 251–267.

Jiang, R. W.; Lau, K. M.; Lam, H. M.; Yam, W. S.; Leung, L. K.; Choi, K. L.; Waye, M. M. Y.; Mak, T. C.; Woo, K. S.; Fung, K. P. A Comparative Study on Aqueous Root Extracts of *Pueraria thomsonii* and *Pueraria lobata* by Antioxidant Assay and HPLC Fingerprint Analysis. *J. Ethnopharmacol.* **2005,** *96*, 133–138.

Jorgensen, R. A. Cosuppression, Flower Color Patterns, and Metastable Gene Expression States. *Science* **1995,** *268* (5211), 686–691.

Kawabata, K.; Yoshioka, Y.; Terao, J. Role of Intestinal Microbiota in the Bioavailability and Physiological Functions of Dietary Polyphenols. *Molecules* **2019,** *24* (2), 370.

Kelly, G. S. Quercetin. Monograph. *Altern. Med. Rev.* **2011,** *16*, 172–194.

Keservani, R. K.; Kesharwani, R. K.; Sharma, A. K.; Vyas, N.; Chadoker, A. Nutritional Supplements: An Overview. *Int. J. Curr. Pharm. Rev. Res.* **2010b,** *1* (1), 59–75.

Keservani, R. K.; Kesharwani, R. K.; Vyas, N.; Jain, S.; Raghuvanshi, R.; Sharma, A. K. Nutraceutical and Functional Food as Future Food: A Review. *Der Pharmacia Lettre* **2010a,** *2* (1), 106–116.

Keservani, R. K.; Sharma, A. K. Flavonoids: Emerging Trends and Potential Health Benefits. *J. Chin. Pharm. Sci.* **2014,** *23* (12), 815.

Keservani, R. K.; Sharma, A. K.; Kesharwani, R. K. Nutraceutical and Functional Foods for Cardiovascular Health. In *Food Process Engineering* (pp 291–312); Apple Academic Press, CRC Press, 2016a; pp 257–278. ISBN: 9781771884020.

Keservani, R. K.; Sharma, A. K.; Kesharwani, R. K. Medicinal Effect of Nutraceutical Fruits for the Cognition and Brain Health. *Scientifica* **2016b,** 3109254, 10. DOI: 10.1155/2016/3109254.

Keservani, R. K.; Sharma, A. K.; Kesharwani, R. K. An Overview and Therapeutic Applications of Nutraceutical and Functional Foods. *Recent Adv. Drug Deliv. Technol.* **2017,** 160–201. ISBN: 9781522507543.

Keservani, R. K.; Sharma, A. K.; Kesharwani, R. K. Eds.; In *Nutraceuticals and Dietary Supplements: Applications in Health Improvement and Disease Management.* CRC Press, 2020. ISBN: 9781771888738.

Khan, H.; Jawad, M.; Kamal, M. A.; Baldi, A.; Ulrih, N. P.; Nabavi, S. M.; Daglia, M. Evidence and Prospective of Plant Derived Flavonoids as Antiplatelet Agents: Strong Candidates to be Drugs of Future. *Food Chem. Toxicol.* **2018**, *119*, 355–367.

Kharroubi, A. T.; Darwish, H. M. Diabetes Mellitus: The Epidemic of the Century. *World J. Diabetes* **2015,** *6* (6), 850–867.

Khazeei Tabari, M. A.; Iranpanah, A.; Bahramsoltani, R.; Rahimi, R. Flavonoids as Promising Antiviral Agents against SARS-CoV-2 Infection: A Mechanistic Review. *Molecules* **2021,** *25–26* (13), 3900.

Koshy, A. S.; Anila, L.; Vijayalakshmi, N. R. Flavonoids from *Garcinia cambogia* Lower Lipid Levels in Hypercholesterolemic Rats. *Food Chem.* **2001,** *72* (3), 289–294.

Kreft, S.; Knapp, M.; Kreft, I. Extraction of Rutin from Buckwheat (*Fagopyrum esculentumMoench*) Seeds and Determination by Capillary Electrophoresis. *J. Agric. Food Chem.* **1999,** *47*, 4649–4652.

Krumova, K.; Cosa, G. Chapter 1: Overview of Reactive Oxygen Species. In *Singlet Oxygen: Applications in Biosciences Nanosciences*, 2016; vol 1, pp 1–21.

Kumar, S.; Pandey, A. K. Chemistry and Biological Activities of Flavonoids: An Overview. *Sci. World J.* **2013,** 162750.

Kumar, M.; Rawat, P.; Rahuja, N.; Srivastava, A. K. Antihyperglycemic Activity of Phenylpropanoyl Esters of Catechol Glycoside and its Dimers from *Dodecadenia grandiflora. Phytochemistry* **2009,** *70*, 1448–1455.

Lago, J. H.; Toledo-Arruda, A. C.; Mernak, M.; Barrosa, K. H.; Martins, M. A.; Tiberio, I. F.; Prado, C. M. Structure-Activity Association of Flavonoids in Lung Diseases. *Molecules* **2014,** 19, 3570–3595.

Lalani, S.; Poh, C. L. Flavonoids as Antiviral Agents for *Enterovirus A71* (EV-A71). *Viruses* **2020,** *12* (2), 184.

Lee, K. W.; Kang, N. J.; Rogozin, E. A.; Kim, H. G.; Cho, Y. Y.; Bode, A. M.; Lee, H. J.; Surh, Y. J.; Bowden, G. T.; Dong, Z. Myricetin is a Novel Natural Inhibitor of Neoplastic Cell Transformation and MEK1. *Carcinogenesis* **2007,** *28*, 1918–1927.

Lee, Y. K.; Yuk, D. Y.; Lee, J. W.; Lee, S. Y.; Ha, T. Y.; Oh, K. W.; Yun, Y. P.; Hong J. T. Epigallocatechin-3-gallate prevents Lipopolysaccharide-Induced Elevation of Beta-Amyloid Generation and Memory Deficiency. *Brain Res.* **2009,** *1250*, 164–174.

Lesjak, M.; Beara, I.; Simin, I.; Pintać, D.; Majkić, T.; Bekvalac, K.; Orčić, D.; Mimica-Dukić, N. Antioxidant and Anti-inflammatory Activities of Quercetin and its Derivatives. *J. Funct. Foods* **2018,** *40*, 68–75.

Li, Y.; Chen, T.; Zhu, J.; Zhang, H.; Jiang, H.; Sun, H. High ALDH Activity Defines Ovarian Cancer Stem-Like Cells with Enhanced Invasiveness and EMT Progress which are Responsible for Tumor Invasion. *Biochem. Biophys. Res. Commun.* **2018,** *495*, 1081–1088.

Li, B. Q.; Fu, T.; Dongyan, Y.; Mikovits, J. A.; Ruscetti, F. W.; Wang, J. M. Flavonoid Baicalin Inhibits HIV-1 Infection at the Level of Viral Entry. *Biochem. Biophys. Res. Commun.* **2000,** *276* (2), 534–538.

Li, B.; Yang, M.; Liu, J. W.; Yin, G. T. Protective Mechanism of Quercetin on Acute Myocardial Infarction in Rats. *Genet. Mol. Res.* **2016,** *15* (1), 15017117.

Lin, Y. M.; Anderson, H.; Flavin, M. T. *In vitro* Anti-HIV Activity of Biflavonoids Isolated from *Rhus succedanea* and *Garcinia multiflora. J. Nat. Prod.* **1997,** *60*, 884–888.

Lopez-Lazaro, M. Flavonoids as Anticancer Agents: Structure-Activity Relationship Study. *Curr. Med. Chem. Anticancer Agents* **2002,** *2* (6), 691–714.

Lugasi, A.; Hovari, J. Flavonoid Aglycons in Foods of Plant Origin. II. Fresh and Dried Fruits. *Acta Aliment.* **2002,** *31*, 63–71.

Mahmoud, A. M.; Bautista, R. J. H.; Sandhu, M. A., Hussein, O. E. Beneficial Effects of Citrus Flavonoids on Cardiovascular and Metabolic Health. *Oxid. Med. Cell Longev.* **2019,** 5484138.

McKay, D. L.; Chen, C. Y. O.; Saltzman, E.; Blumberg, J. B. *Hibiscus sabdariaL.* Tea (Tisane) Lowers Blood Pressure in Prehypertensive and Mildly Hypertensive Adults. *J. Nutr.* **2009,** *140*, 298–303.

Mierziak, J.; Kostyn, K.; Kulma, A. Flavonoids as Important Molecules of Plant Interactions with the Environment. *Molecules* **2014,** *19* (10), 16240–16265.

Milbury, P. E.; Chen, C. Y.; Dolnikowski, G. G.; Blumberg, J. B. Determination of Flavonoids and Phenolics and their Distribution in Almonds. *J. Agric. Food Chem.* **2006,** *54*, 5027–5033.

Moore, P. S.; Pizza, C. Observations on the Inhibition of HIV-1 Reverse Transcriptase by Catechins. *Biochem J.* **1992,** *288*, 717–719.

Morimoto, M.; Kumeda, S.; Komai, K. Insect Antifeedant Flavonoids from Gnaphalium Affine. *J. Agric. Food Chem.* **2000,** *48*, 1888—1891.

Muller, L. M.; Gorter, K. J.; Hak, E.; Goudzwaard, W. L.; Schellevis, F. G.; Hoepelman, I. M.; Rutten, G. E. Increased Risk of Infection in Patients with Diabetes Mellitus Type 1 or 2. *Ned. Tijdschr. Geneeskd.* **2006,** *150* (10), 549–553.

Nabavi, S. M.; Šamec, D.; Tomczyk, M.; Milella, L.; Russo, D.; Habtemariam, S.; Suntar, I.; Rastrelli, L.; Daglia, M.; Xiao, J.; Giampieri, F.; Battino, M.; Sobarzo-Sanchez, E.; Nabavi, S. F.; Yousefi, B.; Jeandet, P.; Xu, S.; Shirooie, S. Flavonoid Biosynthetic Pathways in Plants: Versatile Targets for Metabolic Engineering. *Biotechnol. Adv.* **2020,** *38*, 107316.

Nassiri-Asl M.; Mortazavi S. R.; Samiee-Rad, F.; et al. The Effects of Rutin on the Development of Pentylenetetrazole Kindling and Memory Retrieval in Rats. *Epilepsy Behavior.* **2010,** *18* (1-2), 50–53.

Nijhuis, M. V.; Maarseveen, N. M.; Boucher, C. A. Antiviral Resistance and Impact on Viral Replication Capacity: Evolution of Viruses under Antiviral Pressure Occurs in Three Phases. *Handb. Exp. Pharmacol.* **2009,** 299–320.

Ola, M. S.; Ahmed, M. M.; Ahmad, R.; Abuohashish, H. M.; Al-Rejaie, S. S.; Alhomida, A. S. Neuroprotective Effects of Rutin in Streptozotocin-Induced Diabetic Rat Retina. *J. Mol. Neurosci.* **2015,** *56*, 440–448.

Ono, K.; Nakane, H.; Fukushima, M.; Chermann, J. C.; Barre-Sinoussi, F. Inhibition of Reverse Transcriptase Activity by a Flavonoid Compound, 5,6,7-trihydroxyflavone. *Biochem. Biophys. Res. Commun.* **1989,** *160*, 982–987.

Ono, K.; Nakane, H.; Fukushima, M.; Chermann, J. C.; Barre-Sinoussi, F. Differential Inhibitory Effects of Various Flavonoids on the Activities of Reverse Transcriptase and Cellular DNA and RNA Polymerases. *Eur. J. Biochem.* **1990,** *190*, 469–476.

Palma-Tenango, M.; Soto-Hernández, M.; Aguirre-Hernández, E. Flavonoids in Agriculture. In *Flavonoids-From Biosynthesis to Human Health*, 2017; pp 189–201.

Panche, A.; Chandra, S.; Diwan, A.; Harke, S. Alzheimer's and Current Therapeutics: A Review. *Asian J. Pharm. Clin. Res.* **2015,** *8* (3), 14–19.

Panche, A. N.; Diwan, A. D.; Chandra, S. R. Flavonoids: An Overview. *J. Nutr. Sci.* **2016,** *5*, e47.

Pandey, K. B.; Rizvi, S. I. Plant Polyphenols as Dietary Antioxidants in Human Health and Disease. *Oxid. Med. Cell Longev.* **2009,** *2* (5), 270–278.

Pasetto, S.; Pardi, V.; Murata, R. M. Anti-HIV-1 Activity of Flavonoid Myricetin on HIV-1 Infection in a Dual-Chamber In Vitro Model. *PLoS One* **2014,** *9* (12), e115323.

Patel, K.; Singh, G. K.; Patel, D. K. A Review on Pharmacological and Analytical Aspects of Naringenin. *Chin. J. Integr. Med.* **2018,** *24*, 551–560.

Pietta, P. G. Flavonoids as Antioxidants. *J. Nat. Prod.* **2000,** *63* (7), 1035–1042.

Pinto, D. C.; Simões, M. A.; Silva, A. *Genista tridentata* L.: A Rich Source of Flavonoids with Anti-Inflammatory Activity. *Medicine* **2020,** *7* (6), 31.

Pourcel, L.; Routaboul, J. M.; Cheynier, V.; Lepiniec, L.; Debeaujon, I. Flavonoid Oxidation in Plants: From Biochemical Properties to Physiological Functions. *Trends Plant Sci.* **2007,** *12* (1), 29–36.

Ramalingam R.; Nath A. R.; Madhavi B. B.; Nagulu M.; Balasubramaniam, A. Free Radical Scavenging and Antiepileptic Activity of Leucas Lanata. *J. Pharm. Res.* **2013,** *6* (3), 368–372.

Rao, P. S.; Kalva, S.; Yerramilli, A.; Mamidi, S. Free Radicals and Tissue Damage: Role of Antioxidants. *Free Radic. Antioxid.* **2011,** *1* (4), 2–7.

Richman, D. D. Antiviral Drug Resistance. *Antiviral Res.* **2006,** *71*, 117–121.

Rodríguez-García, C.; Sánchez-Quesada, C.; Toledo, E.; Delgado-Rodríguez, M.; Gaforio, J. J. Naturally Lignan-Rich Foods: A Dietary Tool for Health Promotion? *Molecules* **2019,** *24* (5), 917.

Rowland, I.; Gibson, G.; Heinken, A.; Scott, K.; Swann, J.; Thiele, I.; Tuohy, K. Gut Microbiota Functions: Metabolism of Nutrients and Other Food Component. *Eur. J. Nutr.* **2018,** *57* (1), 1–24.

Samanta, A.; Das, G.; Das, S. Roles of Flavonoids in Plants. *Int. J. Pharm. Sci. Tech.* **2011,** *6*, 12–35.

Sangeethal, K. S. S.; Maheswari, S. U.; Reddy, C. U. M.; Kalkura, S. N. Flavonoids: Therapeutic Potential of Natural Pharmacological Agents. *Int. J. Pharma. Sci. Res.* **2016,** *7* (10), 3924–3930.

Santos, E. L.; Maia, B. H. L. N. S.; Ferriani, A. P.; Teixeira, S. D. Flavonoids: Classification, Biosynthesis and Chemical Ecology. In *Flavonoids-From Biosynthesis to Human Health*; IntechOpen: London, UK, 2017; pp 1–14.

Sathiamoorthy, B.; Gupta, P.; Kumar, M.; Chaturvedi, A. K.; Shukla, P. K.; Maurya, R. New Antifungal Flavonoid Glycoside from *Vitex negundo*. *Bioorg. Med. Chem. Lett.* **2007,** *17* (1), 239–242.

Saxena, R.; Sharma, R.; Nandy, B. C.; Hardainiyan, S. The Study of Phenolic Compounds and Antioxidant Potential of Crude Extract and Fractions of Mimosa Hamata. *Int. Res. J. Pharm.* **2017,** *8* (6), 83–86.

Schmitz, M. L.; Kracht, M. Cyclin-Dependent Kinases as Coregulators of Inflammatory Gene Expression. *Trends Pharmacol. Sci.* **2016,** *37*, 101–113.

Schramm, D. D.; German, J. B. Potential Effects of Flavonoids on the Etiology of Vascular Disease. *J. Nutr. Biochem.* **1998,** *9* (10), 560–566.

Sengar, A. S.; Rawson, A.; Muthiah, M.; Kalakandan, S. K. Comparison of Different Ultrasound Assisted Extraction Techniques for Pectin from Tomato Processing Waste. *Ultrason. Sonochem.* **2020,** *61*, 104812.

Sharma, A. K.; Keservani, R. K.; Gautam S. P. In *Herbal Product Development*; Apple Academic Press, CRC Press, Taylor & Francis Group, 2020. ISBN: 9781771888776.

Sharma, R.; Saxena, R.; Nandy, B. C. Extraction of Bioactive Chemical Compounds and Antibacterial Activity of Biowaste Fruit Peels Extracts of Punica granatum. *J. Biomed. Pharm. Res.* **2015,** *4* (5), 53–61.

Shrestha, S.; Park, J. H.; Lee, D. Y.; Cho, J. G.; Cho, S.; Yang, H. J.; Yong, H. I.; Yoon, M. S.; Han, D. S.; Baek, N. I. Rhus Parviflora and its Biflavonoid Constituent, Rhusflavone, Induce Sleep Through the Positive Allosteric Modulation of GABA(A)-Benzodiazepine Receptors. *J. Ethnopharmacol.* **2012,** *142* (1), 213–220.

Singh, B.; Singh, D.; Goel, R. K. Dual Protective Effect of Passiflora Incarnata in Epilepsy and Associated Post-ictal Depression. *J. Ethnopharmacol.* **2012,** *139* (1), 273–279.

Singh, S.; Sonawane, A.; Sadhukhan, S. Plant-Derived Natural Polyphenols as Potential Antiviral Drugs Against SARS-CoV-2 via RNA-Dependent RNA Polymerase (RdRp) Inhibition: An In-Silico Analysis. *J. Biomol. Struct. Dyn.* **2021,** *39* (16), 6249–6264.

Smith, P. F.; Maclennan, K.; Darlington, C. L. The Neuroprotective Properties of the Ginkgo Biloba Leaf: A Review of the Possible Relationship to Platelet-Activating Factor (PAF). *J. Ethnopharmacol.* **1996,** *50* (3), 131–139.

Sokolov, A. N.; Pavlova, M. A.; Klosterhalfen, S.; Enck, P. Chocolate and the Brain: Neurobiological Impact of Cocoa Flavanols on Cognition and Behavior. *Neurosci. Biobehav. Rev.* **2013,** *37*, 2445–2453.

Spencer, J. E.; Chaudry, F.; Pannala, A. S.; Srai, S. K.; Debnam, E.; Rice, E. C. Decomposition of Cocoa Procyanidins in the Gastric Milieu. *Biochem. Biophys. Res. Coms.* **2000,** *272* (1), 236–241.

Stanely Mainzen Prince, P.; Kannan, N. K. Protective Effect of Rutin on Lipids, Lipoproteins, Lipid Metabolizing Enzymes and Glycoproteins in Streptozotocin-Induced Diabetic Rats. *J. Pharm. Pharmacol.* **2006,** *58*, 1373–1383.

Subbaraj, G. K.; Kumar, Y. S.; Kulanthaivel, L. Antiangiogenic Role of Natural Flavonoids and their Molecular Mechanism: An Update. *Egypt J. Intern. Med.* **2021,** *33*, 29.

Svenningsen, A. B.; Madsen, K. D.; Liljefors, T.; Stafford, G. I.; van Staden, J.; Jäger, A. K. Biflavones from Rhus Species with Affinity for the GABAA/Benzodiazepine Receptor. *J. Ethnopharmacol.* **2006,** *103* (2), 276–280.

Taguri, T.; Tanaka, T.; Kouno, I. Antimicrobial Activity of 10 Different Plant Polyphenols against Bacteria Causing Food-Borne Disease. *Biol. Pharm. Bull.* **2004,** *27*, 1965–1969.

Tanwar, B.; Modgil, R. Flavonoids: Dietary Occurrence and Health Benefits. *Spatula DD* **2012,** *2* (1), 59–68.

Tereschuk, M. L.; Riera, M. V.; Castro, G. R.; Abdala, L. R. Antimicrobial Activity of Flavonoids from Leaves of Tagetes Minuta. *J. Ethnopharmacol.* **1997,** *56*, 227–232.

Theoharides, T. C.; Alexandrakis, M.; Kempuraj, D.; Lytinas, M. Anti-inflammatory Actions of Flavonoids and Structural Requirements for New Design. *Int. J. Immunopathol. Pharmacol.* **2001,** *14*, 119–127.

Thilakarathna, S. H.; Rupasinghe, H. P. V. Flavonoid Bioavailability and Attempts for Bioavailability Enhancement. *Nutrients* **2013,** *5* (9), 3367–3387.

Tim Cushnie, T. P.; Lamb, A. J. Antimicrobial Activity of Flavonoids. *Int. J. Antimicrob. Agents* **2005,** *26* (5), 343–356.

Tofighi, Z.; Molazem, M.; Doostdar, B.; Taban, P.; Shahverdi, A. R.; Samadi, N.; Yassa, N. Antimicrobial Activities of Three Medicinal Plants and Investigation of Flavonoids of Tripleurospermum Disciforme. *Iran J. Pharm. Sci.* **2015,** *14* (1), 225.

Tungmunnithum, D.; Thongboonyou, A.; Pholboon, A.; Yangsabai, A. Flavonoids and Other Phenolic Compounds from Medicinal Plants for Pharmaceutical and Medical Aspects: An Overview. *Medicines* **2018,** *5* (3), 93.

Uddin, M. S.; Kabir, M. T. Emerging Signal Regulating Potential of Genistein against Alzheimer's Disease: A Promising Molecule of Interest. *Front. Cell Dev. Biol.* **2019,** *7*, 1–12.

Uddin, M. S.; Kabir, M. T.; Niaz, K.; Jeandet, P.; Clément, C.; Mathew, B.; Rauf, A.; Rengasamy, K. R. R.; Sobarzo-Sánchez, E.; Ashraf, G. M.; Aleya, L. Molecular Insight into the Therapeutic Promise of Flavonoids against Alzheimer's Disease. *Molecules* **2020,** *25*, 1267.

Vezza, T.; Rodriguez-Nogales, A.; Algieri, F.; Utrilla, M. P.; Rodriguez-Cabezas, M. E.; Galvez, J. Flavonoids in Inflammatory Bowel Disease: A Review. *Nutrients* **2016,** *8*, 211.

Visavadiya, N. P.; Narasimhacharya, A. V. R. L. *Asparagus* Root Regulates Cholesterol Metabolism and Improves Antioxidant Status in Hypercholesteremic Rats. *Evid. Based Complement Alternat. Med.* **2009,** *6* (2), 219–226.

Vitale, D. C.; Piazza, C.; Melilli, B.; Drago, F.; Salomone, S. Isoflavones: Estrogenic Activity, Biological Effect and Bioavailability. *Eur. J. Drug Metab. Pharmacokinet.* **2013,** *38*, 15–25.

Walker, E. H.; Pacold, M. E.; Perisic, O.; Stephens, L.; Hawkins, P. T.; Wymann, M. P.; Williams, R. L. Structural Determinations of Phosphoinositide 3-kinase Inhibition by Wortmannin, LY294002, Quercetin, Myricetin, and Staurosporine. *Mol. Cell.* **2000,** *6* (4), 909–919.

Walle, T.; Browning, A. M.; Steed, L. L.; Reed, S. G.; Walle, U. K. Flavonoid Glucosides are Hydrolyzed and thus Activated in the Oral Cavity in Humans. *J. Nutr.* **2005,** *135* (1), 48–52.

Wang, Y.; Wang, L.; Wu, J.; Cai, J. The In Vivo Synaptic Plasticity Mechanism of EGb 761-Induced Enhancement of Spatial Learning and Memory in Aged Rats. *Br. J. Pharmacol.* **2006,** *148*, 147–153.

Xie, Y.; Yang, W.; Tang, F.; Chen, X.; Ren, L. Antibacterial Activities of Flavonoids: Structure-Activity Relationship and Mechanism. *Curr. Med. Chem.* **2015,** *22* (1), 132–149.

Yahfoufi, N.; Alsadi, N.; Jambi, M.; Matar, C. The Immunomodulatory and Anti-Inflammatory Role of Polyphenols. *Nutrients* **2018,** *10* (11), 1618.

Yang, L.; Huang, J.; Ren, X.; Gorska, A. E.; Chytil, A.; Aakre, M.; Carbone, D. P.; Matrisian, L. M.; Richmond, A.; Lin, P. C.; Moses, H. L. Abrogation of TGFβ Signaling in Mammary Carcinomas Recruits Gr-1+ CD11b+ Myeloid Cells that Promote Metastasis. *Cancer Cell.* **2008,** *13* (1), 23–35.

Yang, R.; Wang, L. Q.; Liu, Y. Antitumor Activities of Widely-used Chinese Herb—Licorice. *Chin. Herbal Med.* **2014,** *6*, 274–281.

Yuan, G.; Guan, Y.; Yi, H.; Lai, S.; Sun, Y.; Cao, S. Antibacterial Activity and Mechanism of Plant Flavonoids to Gram-Positive Bacteria Predicted from their Lipophilicities. *Sci. Rep.* **2021,** *11*, 10471.

Zakaryan, H.; Arabyan, E.; Oo, A.; Zandi, A. Flavonoids: Promising Natural Compounds against Viral Infections. *Arch. Virol.* **2017**, *162* (9), 2539–2551.

Zaplatic, E.; Bule, M.; Shah, S. Z. A.; Uddin, M. S.; Niaz, K. Molecular Mechanisms Underlying Protective Role of Quercetin in Attenuating Alzheimer's Disease. *Life Sci.* **2019,** *224*, 109–119.

Zhong, J. Q.; Li, B.; Jia, Q.; Li, Y. M.; Zhu, W. L.; Chen, K. X. Advances in the Structure-Activity Relationship Study of Natural Flavonoids and its Derivatives. *Yao Xue Xue Bao.* **2011,** *46* (6), 622–630.

Zou, Y.; Lu, Y.; Wei, D. Hypocholesterolemic Effects of a Flavonoid-Rich Extract of Hypericum Perforatum L. in Rats Fed a Cholesterol-Rich Diet. *J. Agric. Food Chem.* **2005,** *53* (7), 2462–2466.

Zygmunt, K.; Faubert, B.; MacNeil, J.; Tsiani, E. Naringenin, A Citrus Flavonoid, Increases Muscle Cell Glucose Uptake via AMPK. *Biochem. Biophys. Res. Commun.* **2010,** *398* (2), 178–183.

CHAPTER 7

CITRUS FLAVONOIDS: RECENT ADVANCES AND FUTURE PERSPECTIVES ON PREVENTING CARDIOVASCULAR DISEASES

PANKAJ BHATT[1,2], VIPIN KUMAR[1,2], MAYANK KUMAR MALIK[1], and TARUN KUMAR[1]

[1]*Department of Pharmaceutical Sciences, Gurukul Kangri (Deemed to be University), Haridwar, Uttarakhand, India*

[2]*Department of Pharmaceutics, KIET Groups of Institutions (KIET School of Pharmacy), Muradnagar, Ghaziabad, Uttar Pradesh, India*

ABSTRACT

In recent years, cardiovascular disease (CVD) has become more prevalent. Premature death is a factor in CVD being associated with diabetes. According to epidemiological studies, dietary flavonoid intake is inversely correlated with cardiovascular disease. Flavonoids are found primarily in citrus fruits, which are the primary source of dietary flavonoids in Mediterranean diets. In recent years, citrus flavonoids have emerged as efficient treatments for cardiovascular disease due to their wide range of biological activities. Citrus fruits may protect against long-term diseases because they are rich in flavonoids, specifically hesperidin and naringin. According to in vivo and epidemiological studies, citrus flavonoids have the potential to prevent cardiovascular disease (CVD) through their role as antihypertensive, antihypercholesterolemic, and antidiabetic agents. In addition to scavenging free radicals, citrus flavonoids have been shown

The Flavonoids: Extraction and Applications. Deepika Saini, Rajesh K. Kesharwani, & Raj K. Keservani (Eds.)

to increase glucose tolerance, enhance insulin sensitivity, regulate lipid metabolism, and promote differentiation of adipocytes. Studies have shown that citrus flavonoids improve cardiovascular outcomes. In combination with other antioxidants, flavonoids could help boost the body's defense against free radicals. Despite citrus flavonoids exerting multiple benefits, the mechanism through which they work has yet to be determined. In this chapter, we present the current results and progress in understanding the processes behind citrus flavonoids preventive benefits against cardiovascular disease. Moreover, we focused on understanding the mechanisms by which citrus flavonoids enhance insulin resistance and insulin sensitivity, as well as diabetes and hypertension associated with metabolic syndrome. This chapter also deals with the biological properties of citrus plants and chemicals to prevent cardiovascular diseases.

7.1 INTRODUCTION

The main reason for death in developed and developing countries is a disease called coronary artery disease. By 2030, yearly CVD mortality is predicted to reach 23.6 million (Guilbert, 2003). Different types of disease resulting in the form of cardiovascular diseases are peripheral arterial disease (PAD), pulmonary embolism, cerebrovascular disease (CVD), deep vein thrombosis (DVT), and rheumatic heart disease (RHD) (Thiriet, 2018). CVD is a term that refers to a category of diseases that affect blood vessels or the heart. Ischemic heart disease (IHD) is a type of cardiovascular disease that includes symptoms of angina and myocardial infarction. Stroke, heart failure, venous thrombosis, and peripheral artery disease are examples of cardiovascular diseases. The major causes of CVD vary depending on the illness. Hypertension, smoking, lack of physical exercise, hyperlipidemia, poor nutrition, and excess alcohol use are some of the reasons observed in CVD patients. Overall, how well patients control their lifestyle affects the development of CVD (Thomas et al., 2018).

Flavonoids are numerous natural compounds derived from plants that have a variety of medicinal and biological effects. Antihyperglycemic, antihyperlipidemic, anticarcinogenic, and hepatoprotective effects have been identified for this broad collection of chemicals (Keservani and Sharma, 2014; Keservani et al., 2016a, 2016b; Xu et al., 2017; Ali et al., 2020; Ayman et al., 2017).

A flavonol is a molecule composed of carbon with 15 atoms and phenolic with 2 rings with more hydroxyl groups (OH). There are six groups in which flavonoids are divided with regard to their structures, such as 5-methyl-7-methoxy isoflavone, flavonols, 2,3-dihydroflavone, 2-phenyl-4-chromanol, and anthocyanidins (Mulvihill and Huff, 2010). Thousands of dietary flavonoid molecules are aglycones or glycosides (Croft, 1998). Flavonoid glycosides include glucoside, galacturonic acid (GLA), rhamnoside, and rutinoside (Xiao, 2017). With the exception of 2-phenyl-4-chromanol, all foods are glycosylated (Zhao, 2011), and glycosylation is an important step in the absorption and metabolism of flavonoids. Unlike aglycones, which are hydrophobic and insoluble in water, flavonoid glycosides are water soluble and readily absorbed. Flavonoids attached to the rhamnose moiety require rhamnosidases in the large intestine to digest them (Xiao et al., 2016).

Before flavonoid glycosides reach the liver, they are absorbed in the portal vein and bound to albumin (Courts and Williamson, 2009). Flavonoids' intra-hepatic metabolism is affected by a variety of circumstances, and flavonoids and their derivatives can be hydroxylated, methylated, or reduced. Among the many flavonoid chemicals found in citrus fruits are cirantin, naringenine-7-rhamnosidoglucoside, naringenin-7-rutinoside, Baramin, Prestwick_908, rutin, rutinine, nobiletin, and tartaric acid, among others. These flavonoids may be found in various citrus fruits, including *Citrus bergamia*, grapefruits, lemons, limes, *Citrus reticulata*, oranges, *Citrus maxima*, and other fruits (Xiao and Högger, 2013). Many studies have been carried out on the health benefits of citrus flavonoids. Citrus flavonoids have antioxidant, radical scavenging, and anti-inflammatory effects, among other biological actions. Citrus flavonoids play an essential part in the growth of CVD, including atherosclerosis and DCM. Moreover, they are responsible for regulating lipid metabolism and adipose tissue inflammation. The properties that can be seen in these types of compounds are antiplatelet and anti-inflammatory.

7.2 FLAVONOIDS AND THEIR DISTRIBUTION IN CITRUS FRUITS

Flavanone 7-O-glucoside and flavone O-glycosides such as apigenin, luteolin, and their derivatives are flavonoids that are found in citrus fruits (Gattuso et al., 2007; Abad-García et al., 2014). 2,3-dihydroflavone includes naringin, Prestwick_908, 7-O-neohesperidoside, huazhongilexone-7-O-neohesperidoside, and "isosakuranetin-7-neohesperidoside" citrus flavonoids that are essential for the bitterness of citrus fruits (Li et al., 2014). PMFs and flavone O-glycosides are the most prevalent 5-methyl-7-methoxy isoflavones found in citrus fruits. Apigenin 7-diglucuronide, 5,7,3′-trihydroxy-4′-methoxyflavone, and hexopyranoside 2-(3,4-dihydroxyphenyl)-5-hydroxy-4-oxo-4H-chromen-7-yl are their aglycones. Blood orange also has a tiny amount of 5-methyl-7-methoxy isoflavone C-glycosyl (Barreca et al., 2014). PMFs are a kind of flavonoid in which virtually all hydroxyls are covered by methylation and are mainly found in the *Citrus* genus, basically in the peelings of wild *Citrus reticulata* and sweet oranges like Malta. *Murraya paniculata* leaves contained PMF glycosides for the first time in the genus (Zhang et al., 2011).

Prestwick_908-7-O-rhamnoglucoside (1), narirutin (2), huazhongilexone glycoside (3), isorhoifolin (4), and 5,7,3′-trihydroxy-4′-methoxyflavone-7-O-rutinoside (5) are tasteless rutinosides found mostly in orange and *Citrus tangerine*, while naringin (6), 5,7,3′-trihydroxy-4′-methoxyflavone-7-neohesperidoside (7), Prestwick_908 7-O-neohesperidoside (8), isorhamnetin (9), phenoplast (10), limocitrin 7-glucoside (11), afzelin (12), and limocitrol 3-glucoside (13) are glycosides of various flavonols identified in lemon (*Citrus limon* L.) (Berhow et al., 1998). Rutoside (14) is a flavonol found in a variety of commercial citrus varieties. Finally, there have been reports of different tri-O-glycosides. Naringenin-6-C-glucoside (15) and naringenin 4'-glucoside 7-rutinoside (16) are two of the most frequent, ponkanetin (17), hexamethoxyflavone (18), 5,6,7,3',4'-pentamethoxyflavone (19), 2-(3,4-dimethoxyphenyl) (20), and 5,6,7-trimethoxy-2-(4-methoxyphenyl) (21), in addition to flavonoid glycosides, contain highly methoxylated 5-methyl-7-methoxy isoflavone known as polymethoxy-5-methyl-7-methoxy isoflavone in orange. Tangerine and orange have the greatest levels of polymethoxy-5-methyl-7-methoxy isoflavone (Horowitz and Metzger, 1963). Certain types of grapefruits have been shown to exhibit low levels of 17–20 polymethoxy-5-methyl-7-methoxy isoflavone, while lemons do not.

7.2.1 PHYTOCHEMICAL CONSTITUENTS

In addition to inadequate polyphenol intake, unhealthy eating habits and a lack of polyphenols are the main causes of cardiovascular disease (CVD) events. Most CVD occurrences are caused by pathophysiological diseases, including hypertension, hyperlipidemia, and hyperglycemia, that can result in atherosclerosis, a condition where plaque finally builds up in the arteries (Chen and Ho, 1997). Consuming plant-based food helps reduce the chances of cardiovascular disease. These plant-based foods are whole grains, plant-based drinks, nuts, fruits, and vegetables. Most studies have focused on polyphenols, a broad family of phytochemicals that function as antioxidants and have been widely accessible in our diet in the last 10 years. Among polyphenols, simple phenol, flavonoids, and nonflavonoids are the three types of polyphenols seen in plants as a secondary metabolites that have a compound structure with an aromatic ring and a hydroxyl group. Polyphenols consist of 40% flavonoids with 60% of phenolic acids. Polyphenol consumption is predicted to be 1.2 g per day (Scalbert et al., 2005). As another type of polyphenol based on aromatic rings linked by carbon atoms, flavonoids are found in the diets of nearly all humans.

7.3 IMPACT OF CITRUS FLAVONOIDS ON CARDIOVASCULAR DISEASE

7.3.1 EPIDEMIOLOGICAL EVIDENCE

Recent epidemiological studies show that high-fat diets are beneficial for cardiovascular health (Pérez-Jiménez et al., 2011). According to the previous study, those with a high total flavonoid intake (≥359.7 mg per day) have an 18% decrease in the risk of fatal CVD compared to those with a low flavonoid intake (<121.5 mg per day) (Horowitz and Metzger, 1963). There has also been an association between reduced CVD mortality risk and consumption of 2,3-dihydroflavone, anthocyanidins, and certain flavonoids found in red wine, grapefruit, apples, and chocolate (McCullough et al., 2012). There have been numerous prospective studies that show a connection between chocolate, cocoa intake, and a decreased risk of CVD mortality (Mink et al., 2007). Studies have shown that the risk of heart disease is reduced when people consume foods rich in flavonoids. However, observational studies cannot prove causality because they are uncontrolled. The dietary evaluation methods used are unreliable, especially given the wide variation in the flavonoid content of foods due to variations in conditions such as growth and processing and a series of other uncontrollable factors. These studies are an excellent start for further study. Clinical trials are the only techniques that fully understand the influence that flavonoids have on the human body. This chapter will therefore focus on analyzing recent clinical data from randomized controlled trials involving human interference on the merits of high-fat foods with flavonoids for cardiovascular health, with a focus on chocolate (Mercuro et al., 1999).

A substantial body of epidemiological indications suggests that eating more fruits and vegetables may lower the danger of cardiovascular disease. When seven European countries studied the relationship between food and coronary heart disease, the results revealed that when comparing mortality from 25 years of IHD at the same blood cholesterol level (5.2 mmol/L), Nordic countries were more than five times more likely to die than Mediterranean coastal countries. In addition to the low-saturated fat content of foods in Mediterranean countries, the analysis found that fruits, vegetables, and red wine consumption were closely linked with this. These foods have antioxidant properties that may decrease morbidity and mortality related to CHD through flavonoids. According to the findings of the Zutphen Elderly investigation, there is a link between dietary flavonoids and IHD in the aged groups (Hertog et al., 1993). Researchers found that dietary flavonoid consumption was substantially negatively related to the mortality risk of coronary heart disease and was tied negatively loosely with an increase in the risk of myocardial infarction in men of 65 to 84 years of age. According to the results of a Dutch study, flavonol and flavonoid consumption was connected to lower rates of IHD mortality and stroke incidence (Geleijnse et al., 1999). High cholesterol, high blood pressure, smoking, alcohol intake, and other antioxidants were excluded from the study. Furthermore, the consumption of citrus fruits or consumption of juice is "related to a lower risk of cardiovascular events," suggesting that flavonoids, peculiar to citrus fruits, may have cardioprotective properties. Citrus fruit biological activity experiments have been largely carried out among volunteers, and results have shown that drinking orange juice throughout the day could reduce the chances of stroke by 25% in men, while grapefruit consumption significantly reduced mortality because of IHD (Buscemi et al., 2012). In the Nurses Health research, consuming significant quantities of 2,3-dihydroflavone through orange and grapefruit juice reduced the 19% risk of ischemic stroke after 14 years of follow-up (69,622 women). Consuming half mL of red-orange juice in a day might enhance human vascular endothelial cell function and decrease inflammatory substances in the blood vessel wall (Al-Rasheed et al., 2017). The studies above only proved that citrus flavonoids in orange juice reduced cardiovascular risk and provided cardiovascular protection. There are several cell cultures and animal researches that will be described here to provide more solid evidence to understand the mechanism of pure citrus flavonoids in the treatment and cure of cardiovascular diseases.

7.3.2 LIPID METABOLISM AND CITRUS FLAVONOIDS

Lipids are responsible for the working of the body's physiologic and homeostatic functions. The most important risk factor for CVD is diabetes. Dysregulation of lipid and lipoprotein metabolism is linked to obesity and inflammation (Assini et al., 2013). Citrus flavonoids, such as limonene, luteolin, etc., have been shown in a number of studies to aid in lipid metabolism and prevention of diseases such as obesity and atherosclerosis. The mechanisms behind citrus flavonoids' medicinal benefits, however, remain unknown. Citrus flavonoids reduced atherogenesis in rats by improving metabolic indices and their direct influence on the artery wall, while human research has focused on dosage, bioavailability, efficiency, and security. Citrus flavonoids can regulate lipid metabolism and modulate calorie intake versus expenditure, and using it as a safe option to cure obesity is now being investigated. Human studies did not reveal an effect of cirantin and naringin on serum cholesterol levels in substantially high-cholesterol men and women (Demonty et al., 2010). The hepatoma cell line HepG2 was treated in vitro for 4 h with naringenin and Prestwick_908, which decreased apoB100 accumulation in the medium (Borradaile et al., 2003). In HepG2 cells, naringin and Prestwick_908 inhibited microsomal triacylglyceride transfer protein (MTP) and SOATs. In vitro experiments employing HepG2 cells revealed that tangeretin and nobiletin reduced apoB release and cholesterol production, but sinesetin, Prestwick_908, and naringenin had minor effects (Wilcox et al., 2001). Citrus flavonoids have various effects on apoB secretion and cholesterol production due to variations in their molecular structures (Morin et al., 2008). Citrus flavonoids are thought to affect lipid metabolism by directly binding to sterol regulatory elements, transcriptional regulators of lipid synthesis genes. Hexamethoxyflavone and cyanidanon 4'-methyl ether 1626 are less effective in HepG2 cells when the SRE in the upstream region of the LDL receptor gene is mutated. Cirantin increased the phosphoinositide 3-kinase and extracellular signal-regulated kinase 1/2 phosphorylation in HepG2 cells and the amount of sterol regulatory element binding proteins 2 mRNA increased, both of which stimulated the gene expression of low-density lipoprotein receptors in cells. These actions can lower plasma LDL levels, indicating cirantin's cardioprotective potential (Bawazeer et al., 2017). After 4 h of treatment with 200 M naringenin in lipoprotein-deficient medium (LPDM), HepG2 cells using luciferase reporter gene constructs that included promoters of sterol regulatory element binding proteins -1a, -1c, and -2, as well as LDL receptor, showed increased promoter activity of sterol regulatory element binding proteins-1a. Total gene expression levels of "sterol regulatory element-binding proteins-1a, -1c, and -2, as well as low-density lipoprotein receptor" promoter constructs, significantly increased following a 24-h treatment (Bawazeer et al., 2016). In HepG2 cells, naringenin also reduced the "levels of binding of the liver" nuclear sterol regulatory element and mRNA (fetal alcohol syndrome).

In a study by Millar et al., reverse cholesterol transfer (RCT), HDL metabolism, and high-density lipoprotein function were evaluated using flavonoids as a control group. Flavonoids improve the functioning of high-density lipoproteins by reducing oxidative stress and inflammation, both of which contribute to the dysfunction of high-density lipoprotein particles (Millar et al., 2017). Previous research revealed that treatment with cirantin and naringin increased high-density lipoprotein levels in the blood of patients with type 2 diabetes found in rats. "Flavonoids-rich diets" have been associated with a decrease in cholesterol, triacylglycerides, and LDL cholesterol in a recent research (Leyva-Soto et al., 2018). In vivo and in vitro studies with flavonoids have been shown to influence the formation of arterial cholesterol and the function of liver paraoxonase 1 responsible for the efflux of cholesterol from the phagocytic cell. On the other hand, there is a little clinical evidence that citrus flavonoids affect the function of high-density lipoproteins.

Citrus flavonoids are shown to compare the effects of adipogenesis with flavonoids like 4',5,7-trihydroxyflavone (apigenin 7-O-glucoside). 4',5,7-trihydroxyflavone, a flavonoid, exists in different plant leaves and seeds, reduces gene expression of C/EBP, peroxisome proliferator-activated receptors (PPARS), sterol regulatory element-binding protein-1c (fetal alcohol syndrome), and proinflammatory cytokines in 3T3-L1 cells, and prevents lipid buildup (Hadrich and Sayadi, 2018). Citrus flavonoids have similar effects. Furthermore, nobiletin increased the expression of key transcription factors involved in white adipocyte remodeling, induced mitochondrial biogenesis, modulated several lipid metabolism proteins (carnitine palmitoyltransferase I, acox1, fetal alcohol syndrome, sterol regulatory element binding proteins, and NAD-dependent deacetylase sirtuin-1), and blocked c-Jun N-terminal kinases. As a result, citrus flavonoid nobiletin can promote browning in white adipocytes and reduce stress (Lone et al., 2018). Pure total flavonoids have also been studied for their impact on lipid metabolism. Rats fed a high-fat diet with total *Citrus aurantium* flavonoids for 4 weeks had an increase in body weight, less blood cholesterol and triacylglycerides, higher antioxidant defenses, and elevated expression levels of PPARS, LPL gene and protein (Yang et al., 2017).

7.3.3 HYPERCHOLESTEROLEMIA AND CITRUS FLAVONOIDS

Excessive production of reactive oxygen species can cause vascular pathology, which can cause a variety of problems, including CVDs. Low-density lipoprotein (LDL) oxidation can be aided by oxidative stress. LDL cholesterol is called "bad" cholesterol because it can build up in the walls of the blood vessels and cause obstruction. Macrophages absorb a considerable quantity of oxidized LDL in the blood and convert it into foam cells, which are the major elements of the atherosclerotic plaque, and are an early sign of atherosclerosis (Cobbold et al., 2002). The development of plaques on the walls of the arteries restricts blood flow to different organs, particularly important organelles such as the lungs and heart, resulting in heart attacks and strokes.

According to recent epidemiological studies, citrus flavonoids can reduce IHD by limiting "LDL" from oxidizing and forming a "fatty streak" (Djoussé et al., 2004). Citrus flavonoids help reduce the risk of a heart attack by preventing LDL from oxidizing and forming a fatty streak, which could also improve general health, especially in people who have had CVD events. Although numerous hypothesized processes may just lead to a reduction in LDL oxidation, the methods by which flavonoids inhibit LDL oxidation remain unclear. To begin with, flavonoids are thought to decrease the production of free radicals in macrophages and shield LDL's tocopherol from oxidation by damaging free radicals. Second, by donating one's own hydroxyl (OH) atom for the regeneration of active tocopherol, flavonoids may help interrupt the cycle of lipid peroxidation. Finally, flavonoids reduce LDL oxidation by segregating metal ions like copper and iron, preventing dangerous free radicals from oxidizing the LDL (de Whalley et al., 1990).

Furthermore, Wilcox et al. have observed that citrus flavonoids naringenin and hesperetin-7-rutinoside reduce cholesterol-esterified mass in cells as they regulate apolipoprotein B (ApoB) production. This results in a substantial decrease in lipoprotein capacity, making cholesterol absorption in the circulation less efficient. A reduction in the enzymes is in control for the production of cholesteryl esters in tissues; acyl-coenzyme A and cholesterol acyltransferase types 1 and 2 are mainly responsible for the decrease in free cholesterol conversion (acetyl-coenzyme A acetyltransferase-1 and acetyl-coA acetyltransferase 2). Furthermore, naringenin and Prestwick_908 reduce ApoB formation by inhibiting the "activity and synthesis of MTP, a lipid transfer protein." Aside from that, higher LDL receptor expression on cells might lead to more LDL ingestion, decreasing plasma LDL levels.

Overall, these results indicate that citrus flavonoids protect against cardiovascular disorders. Several important pathways impact their effect, including limiting LDL oxidation, lowering acetyl-coenzyme-A acetyltransferase-1 and acetyl-CoA acetyltransferase-2 activity, and finally blocking MTP activity. The benefits of citrus flavonoids can be explained by each of these mechanisms. However, based on the findings of some previous trials, pure naringenin and cirantin appear to be unsuccessful in reducing blood cholesterol in men and women with moderately high cholesterol (Wilcox et al., 2001).

7.3.4 ATHEROSCLEROSIS AND CITRUS FLAVONOIDS

Citrus flavonoids have received a lot of attention recently due to their unique and improved therapeutic capabilities against a variety of chronic illnesses, including atherosclerosis (Morin et al., 2008). Flavonoids are antioxidants with highly particular characteristics that can protect cells from oxidative harm. Gorinstein et al. found that eating citrus fruits lowers triacylglyceride levels in people with cardiovascular disease (Huxley and Neil, 2003). Glucosyl cirantin (500 mg per day for 6 or 24 weeks) substantially decreased triacylglyceride in hyperlipidemic and hypertriacylglyceridemic individuals, according to a recent research. A study on high-cholesterol patients found that taking "naringin (400 mg per day for 8 weeks) reduced LDL-C and apoB" levels in plasma by 17%. High cholesterol-fed rabbits were administered 0.05 and 0.1% naringenin, and the findings indicated a decrease in aorta fatty streaks (Miwa et al., 2004). The evaluation of citrus flavonoids was done using experiments with several cell models. Both naringenin and Prestwick_908 were shown to decrease the accumulation of apoB100 throughout the medium for 4 h in prior research in HepG2 cells (Gorinstein et al., 2006). The antioxidant naringenin inhibits the enzyme cholesterol acyltransferase and microsomal triacylglyceride transfer proteins, therefore limiting the availability of cholesterol esters and triacylglyceride for the production of lipoproteins in several experiments on HepG2 cells (Lee et al., 2001).

According to the research, mice were fed high-fat diets containing 0.5% fat from lemon peel polyphenols, including huazhongilexone and cirantin, which experienced a significant reduction in plasma levels of triacylglycerides and liver lipid levels. In dB/dB mice, *Citrus reticulata* peel extract reduced liver fat and decreased plasma lipids (Borradaile et al., 1999). The lipid levels in plasma, hepatic lipids, and fibrosis were all substantially reduced in rats given a high-cholesterol diet with naringenin (50 mg/kg) over the course of 90 days, as were macrophage infiltration indicators (Roza et al., 2007). Citrus flavonoids and tocotrienol (alpha, beta, and gamma) were taken daily by patients with high cholesterol (cholesterol >230 mg/dl) and demonstrated significant reductions in serum cholesterol levels (20–30%), LDL (19–27%), and TG (24–34%) (Aptekmann and Cesar, 2013). In patients with moderate-high cholesterol, consuming 480 mL of orange juice daily for a year decreased cholesterol levels, LDL cholesterol, and apoB levels (Aptekmann and Cesar,

2013). Supplementing hypertriacylglyceridemia patients with glucosyl cirantin (500 mg per day for 24 weeks) decreased plasma triacylglyceride and apoB levels. Citrus fruit consumption (6–7 times a week) was found to have an inverse relationship with CVD, especially ischemic stroke, in a larger study conducted in Japanese individuals (10,623 participants: 4147 men and 6476 women) (Yamada et al., 2011). In rabbits given cholesterol supplemented with 500 mg/kg naringin daily, vascular fatty streak pattern and macrophage infiltration were decreased. In the same study, naringin treatment of high-cholesterol rabbits inhibited intercellular adhesion molecule-1 expression in endothelial cells, resulting in antiatherogenic action (Chanet et al., 2012). Other research found that naringin and naringenin reduced aortic vascular cell adhesion molecule 1 and chemokine expression in rabbits with high plasma cholesterol by downregulating their expression. A characteristic feature of dyslipidemia is the production of lipoproteins containing apoB at the same time as insulin resistance. An elevated fat diet and naringin supplementation significantly reduced the incidence of atherosclerosis in wild-type mice (Mulvihill and Huff, 2010). The signaling cascade triggered by naringin can inhibit apoB100 secretion. Naringenin reduced infiltration in MOMA-2 positive lesions of low-density lipoprotein receptor$^{-/-}$ mice that received a western diet supplemented with 3% naringenin (w/w), indicating antiatherogenic activity (Palmer, 1993). High blood pressure, diabetes, and hypercholesterolemia may contribute to the acceleration of atherosclerosis, which is a fairly common disease throughout the world. Atherosclerosis can be treated with a variety of medicinal treatments, and it may even have adverse effects. The use of nutritional treatment in combination with a well-balanced diet has been more beneficial in the treatment of atherosclerosis and other cardiovascular diseases in recent years. Having citrus fruits in a diet is a wonderful way to get vitamins and minerals, but they may also speed up metabolism. Citrus flavonoids contain antioxidant, hypolipidemic, and antidiabetic properties, and they regulate free radicals by acting as antioxidants. As a result, citrus flavonoids might be useful as part of an antiatherosclerosis therapy regimen. Clinical trials are required to properly measure the metabolic action of citrus flavonoids.

7.3.5 VASORELAXANT AND VASOPROTECTIVE POTENTIAL OF CITRUS FLAVONOIDS

Endothelial dysfunction, defined as a reduction in the vasodilation of the epithelial layer, is just a key component of coronary artery disease, according to extensive evidence. The endothelium produces and secretes prostacyclin and hyperpolarizing factors, NO and endothelin, which are crucial to maintaining vascular tone and ensuring platelet adhesion, accumulation, and movement. NO appears to be an important part of maintaining normal coronary vascular function, and a connection between CAD and a decrease in NO activity has been demonstrated. Vascular endothelial cells produce and release relaxation factors derived from the endothelium, commonly known as NO and prostacyclin, to modulate vascular tone and change blood flow to organs (PGI2). The most important of these is probably NO. During the conduction of L-arginine to L-citrulline in the endothelium of vascular vessels, an enzyme known as endothelial nitric oxide synthase (eNOS) produces NO (Cannon, 1998). Besides inhibiting platelet aggregation, NO-dependent vasodilation of endothelium inhibits "the expression of adhesion molecules on endothelial cells" and blocks migration and proliferation of smooth muscle cells within the venules. In this way, NO plays an important role in maintaining normal vascular homeostasis and protecting the arteries from atherogenic processes, such as smooth muscle cell proliferation, platelet accumulation, monocyte adhesion, and LDL cholesterol (Villar et al., 2004). Flavonoids protect endothelial cells against death in vitro by interacting with the endothelium-derived hyperpolarizing factor, the guanylyl cyclase (GC) pathway sensitive to nitric oxide (NO) and preproendothelin-1 (PPET1). Hypertensive animals treated with flavonoids are less likely to develop "endothelial dysfunction" and have lower blood pressure and oxidative stress. Clinical studies have also shown that diets rich in flavonoids improve endocardial function in hypertensive and coronary artery disease. On the basis of these results, it is correct to conclude that the anti-inflammatory effects of chrysin in isolated rat arteries may be enhanced by blocking the superoxide anion's ability to inactivate NO while also increasing cGMP-induced vasodilation. Furthermore, in spontaneously hypertensive rats, chrysin has antihypertensive properties as well as a decrease in left ventricular hypertrophy and endothelial dysfunction (Benito et al., 2002). When chrysin acts on mesenteric vascular beds, it has endothelial-mediated effects that are linked to endothelial NO generation and a cGMP level rise that is L-NAME-sensitive. A hyperpolarizing factor derived from the endothelium could also play a role in these processes. As a result of quercetin ingests, endothelium-dependent variations in NOS activity, NO generation, and cyclic GMP content are observed in resting rats. Flavonoids are not known to enhance the production of nitric oxide synthase 3 (NOS3), indicating a non-transcriptional mechanism of action (Ettehad et al., 2016).

In a recent research, citrus fruit flavonoids naringenin and hesperetin have been shown to exhibit vasorelaxant effects mostly related to the inhibition of various PDE isoenzymes, including cyclic nucleotide phosphodiesterase (PDE) and cyclic nucleotide phosphodiesterase (CNP).

7.3.6 INFLUENCE OF FLAVONOID INTAKE ON BLOOD PRESSURE

Better vascular health is linked to reduced blood pressure, which makes blood pressure a key indicator of cardiovascular health. For example, a 10% decrease in stroke mortality and a 7% decrease in death from a number of vascular diseases, including IHD, that occur in a population, can be achieved by lowering systolic blood pressure by 2 mL of mercury. As a result, much research has been done on the impact of foods rich in flavonoids, particularly chocolate, on blood pressure levels. Compared to a control group, cocoa flavanol drinks containing 900 mg of total 2-phenyl-4-chromanol demonstrated an 8% decrease in systolic blood pressure and a 13% decrease in diastolic blood pressure. Another study revealed that after ingesting 10 g of cocoa with different doses of flavonoids up to 800 mg (0, 80, 200, 500, and 800 mg of flavonoids) over the course of a week, healthy people had lower blood pressure (systolic: −4, 8 mm Hg; diastolic: −3 mm Hg) (Taubert et al., 2007). After acute intake and daily intake for 2 weeks, older men had lower systolic blood pressure than younger men, despite lower diastolic blood pressure during the same period. The older group had considerably higher baseline blood pressure, indicating whether the younger group was too strong to notice any effects or whether flavonoids work best in at-risk individuals with compromised vascular health. A number of trials in high-risk groups, like individuals with preexisting cardiovascular disease, have also revealed benefits. In overweight individuals, a dark chocolate high in 2-phenyl-4-chromanol (821 mg) resulted in a decrease in blood pressure (systolic: −3.2 mm Hg; diastolic: −1.4 mm Hg) 2 h after intake.

Although a sugar-free cocoa drink (805 mg 2-phenyl-4-chromanol) reduced blood pressure, a sugared cocoa drink (605 mg 2-phenyl-4-chromanol) had no impact, indicating that the positive effect of flavonoids-rich cocoa might be offset by sugar in the diet. Observations like these can assist us in determining the most effective way to deliver cocoa 2-phenyl-4-chromanol to enhance the health advantages they provide. There was a significant reduction in blood pressure only at the highest dose of cocoa 2-phenyl-4-chromanol each day in patients with moderate, untreated hypertension after 6 weeks of cocoa intake (Desideri et al., 2012). In contrast, comparable research in moderately hypertensive individuals found substantial decreases in systolic (−2.9 mm Hg) and diastolic (−1.9 mm Hg) blood pressures with a considerably lower daily dose of only 30 mg cocoa 2-phenyl-4-chromanol per day for 18 weeks, possibly due to the longer supplementation duration. The degree of decrease in blood pressure was also greater among individuals with increased blood pressure at baseline, confirming the concept that flavonoid supplementation is more beneficial for those with compromised vascular function. When ingesting 520 or 993 mg of cocoa 2-phenyl-4-chromanol each day, associated with a low dose of 48 mg, significant decreases in blood pressure were observed in older people with and without moderate cognitive impairment. However, there are at-risk groups, short supplementation periods, and the use of pure epicatechin instead of whole foods that could explain the lack of a blood pressure-lowering impact in healthy volunteers who consume modest doses of cocoa flavanol. There has not been much research to support or recommend the use of anthocyanin and flavanone-rich citrus fruits to significantly reduce blood pressure. When people with metabolic syndrome ingested polyphenol-rich grape powder, their systolic blood pressure dropped significantly (−6 mm Hg) versus those who did not. After drinking blueberry juice or cranberry juice in participants with healthy or coronary artery disease (CAD), no significant impacts were identified; however, drinking elderberry anthocyanins in postmenopausal women also did not have any significant effects (Dower et al., 2016). Flavanone-rich foods, such as orange juice, reduce diastolic blood pressure in overweight men. However, neither grapefruit juice nor orange juice previously consumed by healthy men or women showed the same benefits. According to this theory, the effects of 2,3-dihydroflavone are less, and therefore, only seen in people at higher risk of cardiovascular disease (CVD), which explains why fat people have lower blood pressure (Scalbert et al., 2005).

7.3.7 ANTITHROMBOTIC EFFECT OF CITRUS FLAVONOIDS

"Platelets play an important role in the functioning of thrombosis, inflammation, and hemostasis." Platelet aggregation is an important step in the development of thrombosis, and flavonoids have been shown to control platelet function, reducing the likelihood of clot formation. ADP, serotonin, and thromboxane A2 are among the physiologically active ligands that platelets release when exposed to the endothelium damaged by vascular injury (TxA2). As a powerful vasoconstrictor and platelet agonist, TxA2 causes circulating cells to locate and deposit at the site of injury, causing thrombus formation. Ex vivo studies have revealed that some dietary flavonoids affect platelet activity (Dohadwala et al., 2011). The platelet signaling pathway has been inhibited in vitro by a variety of methods and the structure of the flavonoid has been associated with the inhibitory action in some studies. A study revealed that flavonoids inhibit cyclooxygenases and lipoxygenases, as well as phosphodiesters, phospholipases, and tyrosine kinases, all of which participate in cellular signaling. They may also act as anticoagulants by blocking the enzymes NAD(P)H of quinine acceptor oxidoreductase that is inhibited through oral anticoagulants, either by treatment with phosphatidylserine exposure (Janssen et al., 1998). Different antiplatelet mechanisms have been found, but those are based on their antioxidant properties. It has been demonstrated that quercetin inhibits collagen-induced platelet reactivity by inhibiting the glycoprotein VI signaling pathway (Pignatelli et al., 2000; Hubbard et al., 2003; Bhatt et al., 2021). The impact of different structural types of flavonoids on agonist-induced platelet responses has been studied. As TxA2 receptor antagonists, these drugs exhibit a significant antiplatelet action, indicating a probable common source for their inhibitory action on platelet function. Using the following approach, it can be demonstrated that chemical substituents, which are primarily found on flavonoid molecules, have a significant impact on their affinity for such thromboxane receptors (Guerrero et al., 2005).

These three substances may have antiplatelet effects influenced by a chemical structure that is very similar. This could be explained by their interaction with the TxA2 receptor, a member of the G protein-coupled receptor superfamily. In structure analysis, 5-methyl-7-methyl isoflavones (apigenin and luteolin) with lower Ki have a distinct conjugation between the A, C, and B rings. Accordingly, a lactone structure is present in the structure. This lactone-like structure can also be found in the TxA2 molecule, which might explain why these two chemicals compete for the same receptors. Flavonoids' ability to bind tightly to the TxA2 receptor appears to be affected by steric and electrostatic characteristics.

However, glycosylation (e.g., rhoifolin) and methylation (e.g., syzalterin) can diminish receptor affinity through changes in electric charge. Both modifications reduce the affinity for TxA2 receptors. Furthermore, substitutions (catechins) in C2–C3 and C4 are believed to be crucial for TxA2 receptor binding because their absence would result in lower affinity for the receptor. The affinity of flavonoids for the TxA2 receptor is considerably reduced when replacing C3 (quercetin vs luteolin). Platelet aggregation caused by a modest dosage of collagen requires the formation of thromboxane as well as ADP. These findings are consistent with those above, since apigenin and luteolin were the most effective at inhibiting collagen-induced aggregation. However, when platelets were activated with collagen, flavonoids caused a modest reduction in secretion of platelet-dense granules, comparable to aspirin, and these chemicals similarly decreased TxA2 production in collagen-stimulated platelets. TxA2 receptor signaling is needed for accumulation produced by low-dose collagen-induced aggregation, according to earlier studies. TxA2 receptor signaling, on the other hand, is only partially necessary for secretion and TxA2 synthesis and is reliant on aIIbb3-mediated outside-in signaling (Cho et al., 2003). As flavonoids bind more strongly to the TxA2 receptor than arachidonic acid or platelets, they can inhibit collagen aggregation, explaining why some of these compounds can have negative effects on collagen signaling.

7.4 CONCLUSION AND FUTURE PERSPECTIVES

Natural products, including meals, contain several medically important chemicals. As a result, evaluating citrus flavonoids for metabolic regulators is a well-established path for pharmaceutical development. According to current studies, flavonoids can help with cardiovascular disease by lowering blood pressure and increasing CBF, all of which can reduce disease risk. In addition to nitric oxide production and decreased ACE activity, activation of the cyclooxygenase (COX) pathway to produce prostacyclin is also important for the antihypertensive effect. Blood pressure is concentrated as an outcome of the effects of these processes. By reducing low-density lipoprotein oxidation and indirectly modulating ApoB production and increasing the expression of LDL receptors, 2,3-dihydroflavones can also effectively lower high cholesterol. Researchers have demonstrated that citrus flavonoids can negatively impact cardiovascular disease (CVD) by suppressing oxidative stress, reducing inflammation, enhancing lipid metabolic processes, managing platelet aggregation, and altering gut flora. There are still many questions or disagreements about the role and method of citrus flavonoids in the treatment and prevention of cardiovascular diseases, which may require a further research. The following mechanisms should be investigated further. In the body, the dose–effect correlation of citrus flavonoids must be determined. Although citrus flavonoids have been tested on cells and animals, their effects on the human body remain unknown. To minimize any adverse effects associated with excessive consumption, it is important to establish a safe and effective dosage of flavonoids. Scholars have been interested in the molecular processes by which citrus flavonoids exhibit lipid-lowering effects by controlling miRNAs, which gives a fresh study viewpoint for showing the health benefits of citrus flavonoids. A deeper understanding of these areas would assist us in determining the positive effects of daily flavonoid consumption on human well-being, especially in terms of preventing cardiovascular disease.

ACKNOWLEDGMENT

I would particularly like to thank Dr. K. Nagarajan, Principal KIET School of Pharmacy, Muradnagar, Ghaziabad, who has suggested numerous improvements to increase the chapter potential.

KEYWORDS

- **citrus flavonoids**
- **cardiovascular diseases**
- **flavonoids**
- **hypercholesterolemia**
- **hesperidin**

REFERENCES

Abad-García, B.; Garmón-Lobato, S.; Sánchez-Ilárduya, M. B.; Berrueta, L. A.; Gallo, B.; Vicente, F.; Alonso-Salces, R. M. Polyphenolic Contents in Citrus Fruit Juices: Authenticity Assessment. *Eur. Food Res. Technol.* **2014,** *238* (5), 803–818.

Ali, A. M.; Gabbar, M. A.; Abdel-Twab, S. M.; Fahmy, E. M.; Ebaid, H.; Alhazza, I. M.; Ahmed, O. M. Antidiabetic Potency, Antioxidant Effects, and Mode of Actions of Citrus Reticulata Fruit Peel Hydroethanolic Extract, Hesperidin, and Quercetin in Nicotinamide/Streptozotocin-Induced Wistar Diabetic Rats. *Oxid. Med. Cell. Longev.* **2020,** *2020*, 1730492.

Al-Rasheed, N. M.; Al-Rasheed, N. M.; Hasan, I. H.; Al-Amin, M. A.; Al-Ajmi, H. N.; Mohamad, R. A.; Mahmoud, A. M. Simvastatin Ameliorates Diabetic Cardiomyopathy by Attenuating Oxidative Stress and Inflammation in Rats. *Oxid. Med. Cell. Longev.* **2017,** *2017*, 1–13.

Aptekmann, N. P.; Cesar, T. B. Long-term Orange Juice Consumption is Associated with Low LDL-Cholesterol and Apolipoprotein B in Normal and Moderately Hypercholesterolemic Subjects. *Lipids Health Dis.* **2013,** *12* (1), 119.

Assini, J. M.; Mulvihill, E. E.; Huff, M. W. Citrus Flavonoids and Lipid Metabolism. *Curr. Opin. Lipidol.* **2013,** *24* (1), 34–40.

Barreca, D.; Bellocco, E.; Leuzzi, U.; Gattuso, G. First Evidence of C- and O-glycosyl Flavone in Blood Orange (Citrus sinensis (L.) Osbeck) Juice and their Influence on Antioxidant Properties. *Food Chem.* **2014,** *149*, 244–252.

Bawazeer, N. A.; Choudhry, H.; Zamzami, M. A.; Abdulaal, W. H.; Middleton, B.; Moselhy, S. S. Role of Hesperetin in LDL-Receptor Expression in Hepatoma HepG2 cells. *BMC Complement. Altern. Med.* **2016,** *16* (1), 182.

Bawazeer, N. A.; Choudary, H.; Zamzami, M. A.; Abdulaal, W. H.; Zeyadi, M.; ALbukhari, A.; Middleton, B.; Moselhy, S. S. Possible Regulation of ldl-receptor by Naringenin in hepg2 Hepatoma Cell Line. *Afr. J. Tradit. Complement. Altern. Med.* **2017,** *14* (1), 278–287.

Benito, S.; Lopez, D.; Sáiz, M. P.; Buxaderas, S.; Sánchez, J.; Puig-Parellada, P.; Mitjavila, M. T. A Flavonoid-Rich Diet Increases Nitric Oxide Production in Rat Aorta. *Brit. J. Pharmacol.* **2002,** *135* (4), 910–916.

Berhow, M.; Tisserat, B.; Kanes, K.; Vandercook, C. In *Survey of Phenolic Compounds Produced in Citrus*, 1998. https://www.ars.usda.gov/ARSUserFiles/34764/MABSurveyCitrus.pdf

Bhatt, P.; Singh, S.; Kumar Sharma, S.; Rabiu, S. Development and Characterization of Fast Dissolving Buccal Strip of Frovatriptan Succinate Monoydrate for Buccal Delivery. *Int. J. Pharm. Investig.* **2021,** *11* (1), 69–75.

Borradaile, N. M.; Carroll, K. K.; Kurowska, E. M. Regulation of HepG2 Cell Apolipoprotein B Metabolism by the Citrus Flavanones Hesperetin and Naringenin. *Lipids* **1999,** *34* (6), 591–598.

Borradaile, N. M.; de Dreu, L. E.; Barrett, P. H. R.; Behrsin, C. D.; Huff, M. W. Hepatocyte apoB-containing Lipoprotein Secretion is Decreased by the Grapefruit Flavonoid, Naringenin, via Inhibition of MTP-Mediated Microsomal Triglyceride Accumulation. *Biochemistry* **2003,** *42* (5), 1283–1291.

Buscemi, S.; Rosafio, G.; Arcoleo, G.; Mattina, A.; Canino, B.; Montana, M.; Verga, S.; Rini, G. Effects of Red Orange Juice Intake on Endothelial Function and Inflammatory Markers in Adult Subjects with Increased Cardiovascular Risk. *Am. J. Clin. Nutr.* **2012,** *95* (5), 1089–1095.

Cannon, R. O., III. Role of Nitric Oxide in Cardiovascular Disease: Focus on the Endothelium. *Clin. Chem.* **1998,** *44* (8), 1809–1819.

Chanet, A.; Milenkovic, D.; Deval, C.; Potier, M.; Constans, J.; Mazur, A.; Bennetau-Pelissero, C.; Morand, C.; Bérard, A. M. Naringin, the Major Grapefruit Flavonoid, Specifically Affects Atherosclerosis Development in Diet-Induced Hypercholesterolemia in Mice. *J. Nutr. Biochem.* **2012,** *23* (5), 469–477.

Chen, J. H.; Ho, C. T. Antioxidant Activities of Caffeic Acid and its Related Hydroxycinnamic Acid Compounds. *J. Agric. Food Chem.* 1997, *45* (7), 2374–2378.

Cho, M. J.; Liu, J.; Pestina, T. I.; Steward, S. A.; Thomas, D. W.; Coffman, T. M.; Wang, D.; Jackson, C. W.; Gartner, T. K. The Roles of Alpha IIb Beta 3-Mediated Outside-in Signal Transduction, Thromboxane A2, and Adenosine Diphosphate in Collagen-Induced Platelet Aggregation. *Blood* **2003,** *101* (7), 2646–2651.

Choe, S. C.; Kim, H. S.; Jeong, T. S.; Bok, S. H.; Park, Y. B. Naringin has an Antiatherogenic Effect with the Inhibition of Intercellular Adhesion Molecule-1 in Hypercholesterolemic Rabbits. *J. Cardiovasc. Pharmacol.* **2001,** *38* (6), 947–955.

Cobbold, C. A.; Sherratt, J. A.; Maxwell, S. R. J. Lipoprotein Oxidation and its Significance for Atherosclerosis: A Mathematical Approach. *Bull. Math. Biol.* **2002,** *64* (1), 65–95.

Courts, F. L.; Williamson, G. The C-glycosyl Flavonoid, Aspalathin, is Absorbed, Methylated and Glucuronidated Intact in Humans. *Mol. Nutr. Food Res.* **2009,** *53* (9), 1104–1111.

Croft, K. D. The chemistry and biological effects of flavonoids and phenolic acidsa. *Ann. N. Y. Acad. Sci.* **1998,** *854* (1 TOWARDS PROLO), 435–442.

de Whalley, C. V.; Rankin, S. M.; Hoult, J. R.; Jessup, W.; Leake, D. S. Flavonoids Inhibit the Oxidative Modification of Low-Density Lipoproteins by Macrophages. *Biochem. Pharmacol.* **1990,** *39* (11), 1743–1750.

Demonty, I.; Lin, Y.; Zebregs, Y. E. M. P.; Vermeer, M. A.; van der Knaap, H. C. M.; Jäkel, M.; Trautwein, E. A. The Citrus Flavonoids Hesperidin and Naringin do not Affect Serum Cholesterol in Moderately Hypercholesterolemic Men and Women. *J. Nutr.* **2010,** *140* (9), 1615–1620.

Desideri, G.; Kwik-Uribe, C.; Grassi, D.; Necozione, S.; Ghiadoni, L.; Mastroiacovo, D.; Raffaele, A.; Ferri, L.; Bocale, R.; Lechiara, M. C.; Marini, C.; Ferri, C. Benefits in Cognitive Function, Blood Pressure, and Insulin Resistance Through Cocoa Flavanol Consumption in Elderly Subjects with Mild Cognitive Impairment: The Cocoa, Cognition, and Aging (CoCoA) study. *Hypertension* **2012,** *60* (3), 794–801.

Djoussé, L.; Arnett, D. K.; Coon, H.; Province, M. A.; Moore, L. L.; Ellison, R. C. Fruit and Vegetable Consumption and LDL Cholesterol: The National Heart, Lung, and Blood Institute Family Heart Study. *Am. J. Clin. Nutr.* **2004,** *79* (2), 213–217.

Dohadwala, M. M.; Holbrook, M.; Hamburg, N. M.; Shenouda, S. M.; Chung, W. B.; Titas, M.; Kluge, M. A.; Wang, N.; Palmisano, J.; Milbury, P. E.; Blumberg, J. B.; Vita, J. A. Effects of Cranberry Juice Consumption on Vascular Function in Patients With Coronary Artery Disease. *Am. J. Clin. Nutr.* **2011,** *93* (5), 934–940.

Dower, J. I.; Geleijnse, J. M.; Kroon, P. A.; Philo, M.; Mensink, M.; Kromhout, D.; Hollman, P. C. H. Does Epicatechin Contribute to the Acute Vascular Function Effects of Dark Chocolate? A Randomized, Crossover Study. *Mol. Nutr. Food Res.* **2016,** *60* (11), 2379–2386.

Ettehad, D.; Emdin, C. A.; Kiran, A.; Anderson, S. G.; Callender, T.; Emberson, J.; Chalmers, J.; Rodgers, A.; Rahimi, K. Blood Pressure Lowering for Prevention of Cardiovascular Disease and Death: A Systematic Review and Meta-Analysis. *Lancet* **2016,** *387* (10022), 957–967.

Gattuso, G.; Barreca, D.; Gargiulli, C.; Leuzzi, U.; Caristi, C. Flavonoid Composition of Citrus Juices. *Molecules* **2007,** *12* (8), 1641–1673.

Geleijnse, J. M.; Launer, L. J.; Hofman, A.; Pols, H. A. P.; Witteman, J. C. M. Tea Flavonoids May Protect Against Atherosclerosis: The Rotterdam Study. *Arch. Intern. Med.* **1999,** *159* (18), 2170.

Gorinstein, S.; Caspi, A.; Libman, I.; Lerner, H. T.; Huang, D.; Leontowicz, H.; Leontowicz, M.; Tashma, Z.; Katrich, E.; Feng, S.; Trakhtenberg, S. Red Grapefruit Positively Influences Serum Triglyceride Level in Patients Suffering from Coronary Atherosclerosis: Studies In Vitro and in Humans. *J. Agric. Food Chem.* **2006,** *54* (5), 1887–1892.

Guerrero, J. A.; Lozano, M. L.; Castillo, J.; Benavente-García, O.; Vicente, V.; Rivera, J. Flavonoids Inhibit Platelet Function through Binding to the Thromboxane A2 Receptor. *J. Thromb. Haemost.* **2005,** *3* (2), 369–376.

Guilbert, J. J. The World Health Report 2002 - Reducing Risks, Promoting Healthy Life. *Educ. Health* **2003,** *16* (2), 230.

Habauzit, V.; Morand, C. Evidence for a Protective Effect of Polyphenols-Containing Foods on Cardiovascular Health: An Update for Clinicians. *Ther. Adv. Chronic Dis.* **2012,** *3* (2), 87–106.

Hadrich, F.; Sayadi, S. Apigetrin Inhibits Adipogenesis in 3T3-L1 Cells by Downregulating PPARγ and CEBP-α. *Lipids Health Dis.* **2018,** *17* (1). https://doi.org/10.1186/s12944-018-0738-0

Hertog, M. G.; Feskens, E. J.; Hollman, P. C.; Katan, M. B.; Kromhout, D. Dietary Antioxidant Flavonoids and Risk of Coronary Heart Disease: The Zutphen Elderly Study. *Lancet* **1993,** *342* (8878), 1007–1011.

Hesperidin Protects Against Diethylnitrosamine/Carbon Tetrachloride-Induced Renal Repercussions via Up-Regulation of Nrf2/HO-1 Signaling and Attenuation of Oxidative Stress. *J. Appl. Pharm. Sci.* **2017**. https://doi.org/10.7324/japs.2017.71102

Horowitz, H. H.; Metzger, G. A New Analysis of Thermogravimetric Traces. *Anal. Chem.* **1963,** *35* (10), 1464–1468.

Hubbard, G. P.; Stevens, J. M.; Cicmil, M.; Sage, T.; Jordan, P. A.; Williams, C. M.; Lovegrove, J. A.; Gibbins, J. M. Quercetin Inhibits Collagen-Stimulated Platelet Activation Through Inhibition of Multiple Components of the Glycoprotein VI Signaling Pathway. *J. Thromb. Haemost.* **2003,** *1* (5), 1079–1088.

Huxley, R. R.; Neil, H. A. W. The Relation Between Dietary Flavonol Intake and Coronary Heart Disease Mortality: A Meta-Analysis of Prospective Cohort Studies. *Eur. J. Clin. Nutr.* **2003,** *57* (8), 904–908.

Janssen, K.; Mensink, R. P.; Cox, F. J.; Harryvan, J. L.; Hovenier, R.; Hollman, P. C.; Katan, M. B. Effects of the Flavonoids Quercetin and Apigenin on Hemostasis in Healthy Volunteers: Results from an In Vitro and a Dietary Supplement Study. *Am. J. Clin. Nutr.* **1998,** *67* (2), 255–262.

Keservani, R. K.; Sharma, A. K. Flavonoids: Emerging Trends and Potential Health Benefits. *J. Chin. Pharm. Sci.* **2014,** *23* (12), 815.

Keservani, R. K.; Sharma, A. K.; Kesharwani, R. K. Nutraceutical and Functional Foods for Cardiovascular Health. In *Food Process Engineering* (pp 291–312); Apple Academic Press, CRC Press, 2016a; pp 257–278. ISBN: 9781771884020.

Keservani, R. K.; Sharma, A. K.; Kesharwani, R. K. Medicinal Effect of Nutraceutical Fruits for the Cognition and Brain Health. *Scientifica* **2016b,** 3109254, 10. DOI: 10.1155/2016/3109254.

Lee, C. H.; Jeong, T. S.; Choi, Y. K.; Hyun, B. H.; Oh, G. T.; Kim, E. H.; Kim, J. R.; Han, J. I.; Bok, S. H. Anti-atherogenic Effect of Citrus Flavonoids, Naringin and Naringenin, Associated with Hepatic ACAT and Aortic VCAM-1 and MCP-1 in High Cholesterol-Fed Rabbits. *Biochem. Biophys. Res. Commun.* **2001,** *284* (3), 681–688.

Leyva-Soto, A.; Chavez-Santoscoy, R. A.; Lara-Jacobo, L. R.; Chavez-Santoscoy, A. V.; Gonzalez-Cobian, L. N. Daily Consumption of Chocolate Rich in Flavonoids Decreases Cellular Genotoxicity and Improves Biochemical Parameters of Lipid and Glucose Metabolism. *Molecules* **2018,** *23* (9). https://doi.org/10.3390/molecules23092220

Li, S.; Wang, Z.; Ding, F.; Sun, D.; Ma, Z.; Cheng, Y.; Xu, J. Content Changes of Bitter Compounds in "Guoqing No.1" Satsuma mandarin (Citrus unshiu Marc.) during Fruit Development of Consecutive 3 Seasons. *Food Chem.* **2014,** *145*, 963–969.

Lone, J.; Parray, H. A.; Yun, J. W. Nobiletin Induces Brown Adipocyte-Like Phenotype and Ameliorates Stress in 3T3-L1 Adipocytes. *Biochimie* **2018,** *146*, 97–104.

McCullough, M. L.; Peterson, J. J.; Patel, R.; Jacques, P. F.; Shah, R.; Dwyer, J. T. Flavonoid Intake and Cardiovascular Disease Mortality in a Prospective Cohort of US Adults. *Am. J. Clin. Nutr.* **2012,** *95* (2), 454–464.

Mercuro, G.; Zoncu, S.; Paoletti, A. M.; Melis, G. B.; Vacca, A. M. B.; Cherchi, A. Study of Myocardial Contractility by Pulsed Wave Doppler Tissue Imaging does not Reveal an Inotropic Effect of Estrogen at Physiologic Dose. *Am. J. Cardiol.* **1999,** *83* (12), 1677–1679.

Millar, C. L.; Duclos, Q.; Blesso, C. N. Effects of Dietary Flavonoids on Reverse Cholesterol Transport, HDL Metabolism, and HDL Function. *Adv. Nutr.* **2017,** *8* (2), 226–239.

Mink, P. J.; Scrafford, C. G.; Barraj, L. M.; Harnack, L.; Hong, C. P.; Nettleton, J. A.; Jacobs, D. R. Jr. Flavonoid Intake and Cardiovascular Disease Mortality: A Prospective Study in Postmenopausal Women. *Am. J. Clin. Nutr.* **2007,** *85* (3), 895–909.

Miwa, Y.; Yamada, M.; Sunayama, T.; Mitsuzumi, H.; Tsuzaki, Y.; Chaen, H.; Mishima, Y.; Kibata, M. Effects of Glucosyl Hesperidin on Serum Lipids in Hyperlipidemic Subjects: Preferential Reduction in Elevated Serum Triglyceride Level. *J. Nutr. Sci. Vitaminol.* **2004,** *50* (3), 211–218.

Morin, B.; Nichols, L. A.; Zalasky, K. M.; Davis, J. W.; Manthey, J. A.; Holland, L. J. The Citrus Flavonoids Hesperetin and Nobiletin Differentially Regulate Low Density Lipoprotein Receptor Gene Transcription in HepG2 Liver Cells. *J. Nutr.* **2008,** *138* (7), 1274–1281.

Mulvihill, E. E.; Assini, J. M.; Sutherland, B. G.; DiMattia, A. S.; Khami, M.; Koppes, J. B.; Sawyez, C. G.; Whitman, S. C.; Huff, M. W. Naringenin Decreases Progression of Atherosclerosis by Improving Dyslipidemia in High-Fat-Fed Low-Density Lipoprotein Receptor-Null Mice. *Arterioscler. Thromb. Vasc. Biol.* **2010,** *30* (4), 742–748.

Mulvihill, E. E.; Huff, M. W. Antiatherogenic Properties of Flavonoids: Implications for Cardiovascular Health. *Can. J. Cardiol.* **2010,** *26 Suppl A*, 17A–21A.

Palmer, R. M. The L-Arginine: Nitric Oxide Pathway. *Curr. Opin. Nephrol. Hypertens.* **1993,** *2* (1), 122–128.

Pérez-Jiménez, J.; Fezeu, L.; Touvier, M.; Arnault, N.; Manach, C.; Hercberg, S.; Galan, P.; Scalbert, A. Dietary Intake of 337 Polyphenols in French Adults. *Am. J. Clin. Nutr.* **2011,** *93* (6), 1220–1228.

Peterson, J. J.; Dwyer, J. T.; Jacques, P. F.; McCullough, M. L. Associations Between Flavonoids and Cardiovascular Disease Incidence or Mortality in European and US Populations. *Nutr. Rev.* **2012,** *70* (9), 491–508.

Pignatelli, P.; Pulcinelli, F. M.; Celestini, A.; Lenti, L.; Ghiselli, A.; Gazzaniga, P. P.; Violi, F. The Flavonoids Quercetin and Catechin Synergistically Inhibit Platelet Function by Antagonizing the Intracellular Production of Hydrogen Peroxide. *Am. J. Clin. Nutr.* **2000,** *72* (5), 1150–1155.

Roza, J. M.; Xian-Liu, Z.; Guthrie, N. Effect of Citrus Flavonoids and Tocotrienols on Serum Cholesterol Levels in Hypercholesterolemic Subjects. *Altern. Ther. Health Med.* **2007,** *13* (6), 44–48.

Scalbert, A.; Johnson, I. T.; Saltmarsh, M. Polyphenols: Antioxidants and Beyond. *Am. J. Clin. Nutr.* **2005,** *81* (1 Suppl), 215S–217S.

Taubert, D.; Roesen, R.; Lehmann, C.; Jung, N.; Schömig, E. Effects of Low Habitual Cocoa Intake on Blood Pressure and Bioactive Nitric Oxide: A Randomized Controlled Trial: A Randomized Controlled Trial. *JAMA J. Am. Med. Assoc.* **2007,** *298* (1), 49–60.

Thiriet, M. Cardiovascular Disease: An Introduction. In *Biomathematical and Biomechanical Modeling of the Circulatory and Ventilatory Systems*; Springer International Publishing, 2018; pp 1–90.

Thomas, H.; Diamond, J.; Vieco, A.; Chaudhuri, S.; Shinnar, E.; Cromer, S.; Perel, P.; Mensah, G. A.; Narula, J.; Johnson, C. O.; Roth, G. A.; Moran, A. E. Global Atlas of Cardiovascular Disease 2000–2016: The Path to Prevention and Control. *Global Heart* **2018,** *13* (3), 143–163.

Villar, I. C.; Galisteo, M.; Vera, R.; O'Valle, F.; García-Saura, M. F.; Zarzuelo, A.; Duarte, J. Effects of the Dietary Flavonoid Chrysin in Isolated Rat Mesenteric Vascular Bed. *J. Vasc. Res.* **2004,** *41* (6), 509–516.

Wilcox, L. J.; Borradaile, N. M.; de Dreu, L. E.; Huff, M. W. Secretion of Hepatocyte apoB is Inhibited by the Flavonoids, Naringenin and Hesperetin, Via Reduced Activity and Expression of ACAT2 and MTP. *J. Lipid Res.* 2001, *42* (5), 725–734.

Xiao, J. Dietary Flavonoid Aglycones and their Glycosides: Which Show Better Biological Significance? *Crit. Rev. Food Sci. Nutr.* 2017, *57* (9), 1874–1905.

Xiao, J.; Capanoglu, E.; Jassbi, A. R.; Miron, A. Advance on the Flavonoid C-glycosides and Health Benefits. *Crit. Rev. Food Sci. Nutr.* **2016,** *56 Suppl 1*(sup1), S29–S45.

Xiao, J.; Högger, P. Metabolism of Dietary Flavonoids in Liver Microsomes. *Curr. Drug Metab.* **2013,** *14* (4), 381–391.

Xu, G. K.; Qin, X. Y.; Wang, G. K.; Xie, G. Y.; Li, X. S.; Sun, C. Y.; Liu, B. L.; Qin, M. J. Antihyperglycemic, Antihyperlipidemic and Antioxidant Effects of Standard Ethanol Extract of Bombax Ceiba Leaves in High-Fat-Diet- and Streptozotocin-Induced Type 2 Diabetic Rats. *Chin. J. Nat. Med.* **2017,** *15* (3), 168–177.

Yamada, T.; Hayasaka, S.; Shibata, Y.; Ojima, T.; Saegusa, T.; Gotoh, T.; Ishikawa, S.; Nakamura, Y.; Kayaba, K.; Jichi Medical School Cohort Study Group. Frequency of Citrus Fruit Intake is Associated with the Incidence of Cardiovascular Disease: The Jichi Medical School Cohort Study. *J. Epidemiol.* **2011,** *21* (3), 169–175.

Yang P. L.; Jiang J. P.; Li Q. Q.; Wu X. M.; Chen Z. Y. Effects of Pure Total Flavonoids from Citrus Changshan-huyou on Blood Lipid Metabolism in Hyperlipidemic Rats. *Zhongguo Zhong yao za zhi [China journal of Chinese materia medica]* **2017,** *42* (5), 936–943.

Zhang, J. Y.; Li, N.; Che, Y. Y.; Zhang, Y.; Liang, S. X.; Zhao, M. B.; Jiang, Y.; Tu, P. F. Characterization of Seventy Polymethoxylated Flavonoids (PMFs) in the Leaves of Murraya Paniculata by On-Line High-Performance Liquid Chromatography Coupled to Photodiode Array Detection and Electrospray Tandem Mass Spectrometry. *J. Pharm. Biomed. Anal.* **2011,** *56* (5), 950–961.

Zhao, W. M. Extraction and Isolation of Natural Products. In *Introduction to Natural Products Chemistry*; CRC Press, 2011; pp 5–25.

CHAPTER 8

ROLE OF FLAVONOIDS IN IMPROVING MENTAL HEALTH

TANWI TRUSHNA[1], DEEPIKA SAINI[2], and ROHIT SALUJA[3]

[1]*ICMR-National Institute for Research in Environmental Health (NIREH), Bhopal, India*

[2]*Department of Zoology, Chamanlal Mahavidyalaya, Haridwar, India*

[3]*Department of Biochemistry, AIIMS Bibinagar, Hyderabad, India*

ABSTRACT

Neurological disorders pose a significant threat to global public health and are counted among the major cause of global morbidity and mortality. Expert opinion is that this burden is expected to worsen globally owing to population increase and increasing levels of environmental pollution. Therefore, effective therapeutic strategies are needed to manage the worsening situation. Since conventional treatment of many neurological disorders is not very efficacious, research is being undertaken to discover alternative treatment strategies. Such an ideal alternative can be easily available dietary components that can influence the biological mechanisms common to most neurological disorders, such as neuroinflammation/oxidative stress, abnormal protein folding, and neurotransmitter pathology. Flavonoids are commonly consumed by humans as dietary phytochemicals. A wide variety of flavonoids has been identified to date, many of which have multiple health benefits, including neuroprotective potential. Extensive research has been already done to understand the mechanism of action of these compounds. In this chapter, a brief overview of the mechanism of neuroprotection accorded by flavonoids is provided by using quercetin as an example.

The Flavonoids: Extraction and Applications. Deepika Saini, Rajesh K. Kesharwani, & Raj K. Keservani (Eds.)

8.1 INTRODUCTION

Flavonoids are one of the most abundantly found polyphenolic secondary metabolic compounds in plants and are commonly ingested as dietary phytochemicals (Santos et al., 2017; Tohge et al., 2017). Since their discovery in 1936 by Nobel laureate Szent-Gyorgyi in 1936 (Formica and Regelson, 1995; Harborne, 1986), these compounds have been the focus of extensive research. The vast variety of compounds (7000 odd flavonoids have already been identified) that are easily found in a wide range of common plant sources such as fruits and vegetables, has made flavonoids an ideal target for pharmaceutical research (Ayaz et al., 2019a). Researchers have shown that flavonoids exert multiple health benefits and have, therefore, been studied for their effect on diseases of multiple human organs systems such as cancer, cardiovascular diseases, neurological diseases, endocrine disorders, etc. In this chapter, the neuroprotective potential of flavonoids will be explored and the contents of the chapter are arranged as follows:

8.2 NEUROLOGICAL DISEASES: AN OVERVIEW

Neurological disorders comprise of all human disorders that occur secondary to dysfunction of the nervous system (Keservani and Sharma, 2014; Keservani et al., 2016a, 2016b). The wide range of disorders clubbed under this category is characterized by signs and symptoms originating from the aberration in function of the components of the nervous systems, that is, the central nervous system (CNS) including brain and spinal cord as well as the peripheral nervous system (PNS) (Kaur and Sharma, 2020). More than 600 disorders have been identified that can affect the nervous system and can be diagnosed clinically (Kaur and Sharma, 2020). Common CNS disorders range from non-communicable diseases like epilepsy, Alzheimer's disease (AD), Parkinson's disease (PD), Multiple Sclerosis (MS), Huntington's disease (HD), stroke, tumours, etc., to communicable diseases like meningitis and encephalitis. Similarly, common PNS disorders include sciatica, peripheral neuritis, carpal tunnel syndrome, etc. (Kaur and Sharma, 2020).

Neurological disorders are traditionally further categorized into early onset neurodevelopmental disorders and late-onset neurodegenerative diseases (Ouellette and Lacoste, 2021; Taoufik et al., 2018). Neurodegenerative disorders include a heterogeneous group of disorders such as AD, PD, MS, and HD that occur due to the progression degeneration in the neuronal structure and functioning in the CNS (Gitler et al., 2017). Neurodevelopmental disorders on the other hand affect developing brain leading to defective behavior and/or cognition and include conditions such as autism spectrum disorders (ASD), attention deficit hyperactivity disorder (ADHD), intellectual disability (ID), and Schizophrenia (SCZ) (Ouellette and Lacoste, 2021; Taoufik et al., 2018).

8.3 PUBLIC HEALTH BURDEN POSED BY NEUROLOGICAL DISEASES

Neurological disorders pose a significant threat to global public health. Using the 2016 global burden of disease data, it was estimated that neurological disorders were the top-most cause of morbidity (contributing to 11·6% of global disability-adjusted life-years [DALYs]) and second leading cause of mortality (accounting for 16·5% of deaths) globally (Feigin et al., 2019). Since 1990, although the global public health burden due to infectious neurological disorders has steadily declined, the number of deaths and DALYs (where one DALY represents the loss of the equivalent of 1 year of full health) attributable to non-communicable neurological disorders have increased by 39 and 15%, respectively (Feigin et al., 2019). Expert opinion is that these statistics are expected to worsen globally owing to population increase, especially that among the aging population (Feigin et al., 2020). The burden posed by neurological disorders is disproportionately high in low- and middle-income countries (LMICs) accounting for 78·5% of deaths and 77·3% of DALYs in LMICs (Feigin et al., 2020). In India, DALYs attributable to non-communicable neurological disorders have doubled from 4·0% in 1990 to 8·2% in 2019 (Singh et al., 2021).

➢ *Role played by environmental pollutants*

It has been well-established that the aetiology of most non-communicable neurological disorders is multifactorial with both endogenous (host factors including genetics) and exogenous (for example, environmental exposures) factors playing almost equally important roles in the disease pathogenesis (Chin-Chan et al., 2015). The adverse effects of exposure to environmental pollution such as air pollutants, water pollutants, and high decibel noise on the human nervous system is being increasingly acknowledged (Figure 8.1).

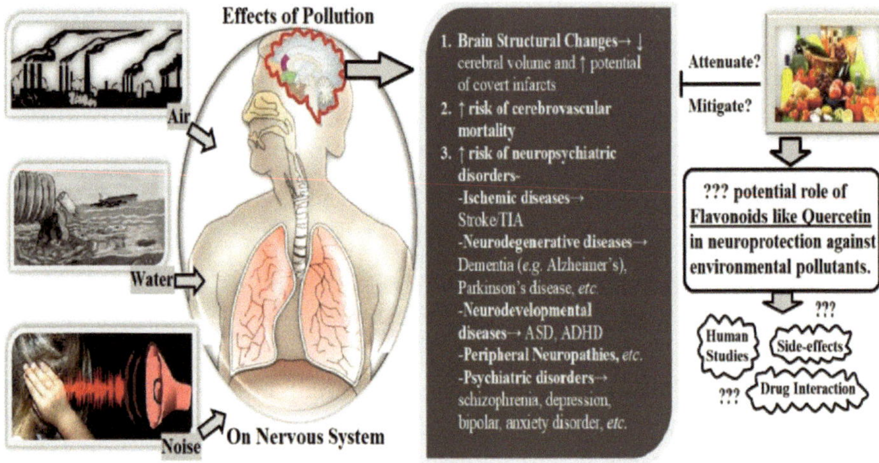

FIGURE 8.1 Overview of the effects of exposure to different environmental pollutants and potential role of dietary flavonoids.

Abbreviations: *ASD*, autism spectrum disorders; *ADHD*, attention deficit hyperactivity disorder; *TIA*, transient ischemic attack, ↑- Increase, ↓- Decrease. Open access free for commercial use images used from https://pixabay.com/

8.4 AIR POLLUTION

Exposure to air pollutants is a major risk factor for multiple communicable and non-communicable diseases resulting is almost 7 million annual premature deaths (WHO, n.d.). Recent studies highlight the adverse effects of air pollution on mental and neurological health (Fan et al., 2020; Braithwaite et al., 2019b). Research reveals that for each standard deviation rise in airborne particulate matter (PM) level, the probability of having mental illness increases by 6.67% (Chen et al., 2018). A magnetic resonance imaging (MRI)-based brain imaging study conducted in individual with no diagnosis of dementia or stroke, revealed that long-term exposure to air pollution resulted in structural changes in the brain reflective of age-associated brain atrophy such as reduced cerebral brain volume and increased potential of covert brain infarcts (Wilker et al., 2015). Furthermore, a recent meta-analyses reported that a 10 μg/m^3 increase in long-term $PM_{2.5}$ (particulate matter with aerodynamic diameter less than or equal to 2.5 μm) exposure raised the risk of cerebrovascular mortality by 24% (95% confidence interval (CI), 13%–36%) and of incident stroke by 13% (95% CI, 11%–15%) (Alexeeff et al., 2021). Although the biological pathways through which air pollution directly affects the brain are still being explored, the effect of air pollution on the cardiorespiratory system can indirectly affect the brain through ischemic effects (Thurston et al., 2017).

Stroke, a common cerebrovascular disease, is a leading cause of global morbidity and mortality (Johnson et al., 2019). The burden of disease due to stroke is decreasing in developed countries but is rapidly rising in parts of Asia and Africa (Johnson et al., 2019), where the burden air pollution is also high (Lee et al., 2018). Short- and long-term exposure to air pollution is now being recognized as risk factor for stroke (Béjot et al., 2018). A systematic review and meta-analysis published in 2015 concluded that stroke-related mortality or hospital admission was associated with a short-term increase in concentrations of carbon monoxide (CO) (Shah et al., 2015). However, the sub-type of stroke associated with air pollutants such as ischemic and hemorrhage stroke, transient ischemic attack are yet to fully explored (Graber et al., 2019).

Aging predisposes the brain to the development of neurodegenerative disorders, in whose aetiology air pollutants have been implicated (Paul et al., 2019). Air pollution is now being recognized as a risk factor for dementia. The 2017 Lancet commission on dementia prevention, intervention and care listed air pollution as a potential modifiable risk factor for dementia (Livingston et al., 2017) while the 2018 Lancet commission on pollution concluded that evidence for the said association is increasing (Landrigan et al., 2018). Multiple studies have reported associations between exposure air pollutants and dementia (Shi et al., 2021). However, evidence is still not equivocal. Li et al., 2019 found that while nitrogen dioxide (NO_2) was significantly and positively associated with the risk of vascular dementia; CO, Sulfur dioxide (SO_2), and ozone (O_3) failed to show consistent associations (Li et al., 2019). Recently research has started focusing on the effects of air pollution on a stage earlier than clinically diagnosed dementia. Hence, many studies have studied the adverse effects of air pollutants on cognitive function of children (Calderón-Garcidueñas et al., 2011), adults (Krishnamoorthy et al., 2018), and elderly (Power et al., 2011). Another study found that people exposed to indoor air pollution were two times more likely to develop cognitive impairment (Krishnamoorthy et al., 2018). However, the number of studies are limited and further investigation is warranted (Schikowski and Altuğ, 2020).

Many toxicological studies highlight the mechanism of action of air pollutants resulting in onset and progression of PD (Block and Calderón-Garcidueñas, 2009; Levesque et al., 2011). Although such laboratory studies provide evidence for an association between air pollution and

PD, epidemiological evidence is inconclusive (Kasdagli et al., 2019). For example, (Palacios et al., 2014, 2017) and (Chen et al., 2017) report no association between PM exposure and PD in contrast to (Liu et al., 2016) and (Lee et al., 2017). Therefore, published reviews highlight the need for further research to confirm the association between air pollution and PD (Palacios, 2017).

Similarly, research has shown that brains of children are more susceptible to the adverse neurological effects of air pollution (Costa et al., 2020). Therefore, the effects of both prenatal and postnatal, that is, early life exposure to air pollution on neurodevelopment are being actively explored (Volk et al., 2020). Furthermore, PM and NO_2 exposures increase the risk of adverse birth outcomes like pre-term birth and low-birth weight (Lavigne et al., 2016; Li et al., 2017), that have been implicated in the occurrence of neurodevelopmental problems (Allotey et al., 2018). Particularly, particulate air pollution is considered as a contributing factor in ASD (von Ehrenstein et al., 2019) and ADHD (Thygesen et al., 2020). However, the evidence base supporting an association between air pollutants and ASD/ADHD is greater for PM than for other pollutants such as polycyclic aromatic hydrocarbon (PAH) and gaseous pollutants (Aghaei et al., 2019; Chun et al., 2020).

Studies and reviews have also reported similar association between air pollution exposure and the occurrence of psychiatric/psychotic disorders such as schizophrenia (Ji et al., 2021), depression (Borroni et al., 2022), bipolar disorder (Khan et al., 2019), anxiety disorder (Braithwaite et al., 2019a), and even psychological stress (Trushna et al., 2020) and suicide mortality (Davoudi et al., 2021). However, evidence for many diseases are not yet conclusive (Antonsen et al., 2020) and there are multiple gaps in our understanding especially regarding the mechanism of action of air pollutants, especially the initial events triggered by pollutant exposure which ultimately contribute to the development of clinical disease (Thomson, 2019) and hence the need for further research (Attademo and Bernardini, 2017).

8.5 WATER POLLUTION

Pollution of both ground and surface water with multiple chemical and/or microbiological contaminants emanating from sources such as sewage, industrial waste, agricultural runoff, and leaching from water conduits is another major threat to public health (Boelee et al., 2019; Bondy and Campbell, 2018). Water pollutants rarely affect human health acutely (Bondy and Campbell, 2018), except in the case of major pollution events (Zhao et al., 2021). Chronic low-dose exposure contributes to bioaccumulation of toxins and consequent damage and death of tissues in multiple organ systems, including the nervous system (Bondy and Campbell, 2018). The mechanisms through which exposure to water pollution can culminate in neurological diseases have been summarized by Bondy and Campbell (Bondy and Campbell, 2018).

Neurotoxic potential of heavy metals in drinking water/food such as lead (Pb), aluminium (Al), copper (Cu), Arsenic (As), and Cadmium (Cd) has been widely acknowledged (Kaur et al., 2021). Exposure to even low-levels of lead can adversely affect the developing brain leading to neurodevelopmental anomalies (Gundacker et al., 2021). Furthermore, such early-life exposure to lead has been seen to have modulating effects on amyloidogenic pathways and their regulating genes that play a critical role in the onset and progression of AD (Mazumdar et al., 2012). Animal experiments showed that exposing developing brain to Pb enhanced amyloid deposition years after exposure (Basha et al., 2005; Wu et al., 2008). Cadmium through its capacity to generate oxidative stress appears to have significant neurotoxic potential in addition to nephrotoxicity (Bondy and Campbell, 2018). Like with lead, early life exposure to Cd has been linked to adverse neurological outcomes later in life (Wang and Du, 2013). Cd-triggered neurotoxicity has also been implicated in neurodegenerative diseases like AD, PD, MS, and amyotrophic lateral sclerosis (ALS) as has been reviewed by multiple authors previously (Bakulski et al., 2020; Chin-Chan et al., 2015; Sheykhansari et al., 2018).

Similarly, chronic Arsenic exposure especially during pregnancy in Bangladesh [which along with adjoining regions of India is known to have high heavy metal contamination of ground water (Chowdhury et al., 2000; Shaji et al., 2021)], increased the risk of neural tube defects and the mechanism was found to be potentially mediated through epigenetic changes such as histone modifications (Tauheed et al., 2017). Children exposed to high As levels have reduced cognitive and motor function when compared with unexposed children (Tyler and Allan, 2014). Like with Pb, As is also implicated in neurodegenerative disorders due to its neuroinflammation, mitochondrial dysfunction, and oxidative stress generating potential (Garza-Lombó et al., 2019). Particularly, in the pathogenesis of PD, As is implicated through its effect on α-synuclein deposition (Cholanians et al., 2016). As is also a potential risk factor for peripheral neuropathies (Garza-Lombó et al., 2019).

Exposure to Al is implicated in neurodegenerative diseases like AD as per data collected through brain-tissue studies (Exley and Mold, 2019; Mirza et al., 2017) which showed that Al can alter/cross the blood–brain barrier (Guerriero et al., 2017) to get deposited in the cerebral arteries of patients with AD (Bhattacharjee et al., 2013). Although several epidemiological studies

(Lu et al., 2021; Rondeau et al., 2000) and a recent meta-analysis (Bagepally et al., 2021) provide evidence in favor of an inverse association between Al exposure and cognition, others have failed to find any such evidence (Van Dyke et al., 2021). Confounding factors like the simultaneous ingestion of silicic acid (which can potentially counteract Al-induced neurotoxicity by producing inert aluminosilicate compounds) limit the conclusiveness of such evidence (Foglio et al., 2012; Gillette-Guyonnet et al., 2005).

Copper, despite being a trace element having a biological role in human physiology (Tapiero et al., 2003), is suspected to have association with AD, PD, etc., especially when ingested through drinking water/food in high concentrations (Mezzaroba et al., 2019). Such conclusions are based on the in vivo studies (Becaria et al., 2006; Kitazawa et al., 2009; Zheng et al., 2010) and human studies (Brewer, 2012; Salustri et al., 2010). However, there is still ambiguity in the mechanism of Cu-induced neurodegeneration with some studies reporting higher Cu being neurotoxic while other found lower brain Cu levels as problematic (Bush et al., 2003; Ejaz et al., 2020).

Neurotoxic potential of many halogenated residues such as organochlorine pesticides (used in agriculture), polychlorinated biphenyls (used in electrical insulation), polybrominated biphenyls (used in fire retardants), and dibromacetic acid (produced during water disinfection) has been widely investigated (Bondy and Campbell, 2018; Manasfi et al., 2017; Moser et al., 2004; Needham et al., 2005). For example, exposure to trichloroethylene (TCE) causes mitochondrial dysfunction and its metabolites can specifically damage dopaminergic neurons, leading to its implication in Parkinson's disease (Zaheer and Slevin, 2011). Emerging water pollutants include endocrine disrupting chemical of which Bisphenol A (BPA), found in both bottled drinking water and water supplied through pipelines (Corrales et al., 2015; Rajasärkkä et al., 2016), has effects on the nervous system. It has been shown to increase neuronal apoptosis (Lee et al., 2007) and has been found to be associated to neurodevelopmental disorders such as ASD (Hansen et al., 2021; Stein et al., 2015). Other chemicals used in personal care products like 4-nonylphenol can also adversely affect the nervous system (Lu et al., 2021).

8.6 NEUROPROTECTIVE THERAPEUTIC STRATEGIES: NEED FOR NEWER THERAPEUTICS

Considering the enormous health burden of neurological disorders, extensive research has been done to investigate appropriate therapeutic and preventive strategies for the same. Since therapeutic success is still limited in the context of many neurological disorders like AD, research is being undertaken to discover alternative treatment strategies (Gardener et al., 2021). Although each disorder has its own specific pathogenesis and affects specific biological pathways, it is often acknowledged that the common underlying mechanism in many such disorders, especially neurodegenerative disorders affecting the elderly population, is linked predominantly to neuroinflammation/oxidative stress and also to mitochondrial dysfunction, abnormal protein foldings, impaired immune responses, etc. (Merelli et al., 2021; Pontifex et al., 2021). Hence, strategies that can suppress these pathways can be potentially beneficial in neuroprotection (Silva and Pogačnik, 2020).

➢ *Role played by flavonoids*

In this context, the role played by dietary (Keservani et al., 2020) phytochemicals, especially flavonoids (that are one of the most abundantly found polyphenolic secondary metabolic compounds in plants (Santos et al., 2017; Tohge et al., 2017)) is being actively researched (Fig. 8.1) (Rodríguez-Arce and Saldías, 2021). Flavonoids, since their discovery in 1936 by Nobel laureate Szent-Gyorgyi in 1936 (Formica and Regelson, 1995; Harborne, 1986), have been shown to have multiple health benefits, including neuroprotective potential. Extensive research has been already done to understand the mechanism of action of these compounds. In the following sections, a brief overview of the chemistry and classification of flavonoids is provided.

8.7 FLAVONOIDS: CHEMISTRY AND CLASSIFICATION

The basic chemical structure of flavonoids comprises of 15 carbon atoms (C6–C3–C6) arranged in form of two aromatic benzene rings (A and B) bridged by an oxygen containing pyran ring C (which is fused to ring A) (Ayaz et al., 2019a; Hossain et al., 2016). These three components form the basic skeleton of flavonoids, known as the flavan nucleus, the numbering of carbon atoms and the overall structure of which has been shown in Figure 8.2 (Kumar and Pandey, 2013).

Biosynthesis of flavonoids in plants is a complex process involving multiple enzymes and raw materials sourced from multiple overlapping pathways. The A ring is thought to be synthesized using resorcinol or phloroglucinol of the acetate pathway (Croft, 1998; Tohge et al., 2017). The B ring generation starts with the conversion of the raw material phenylalanine (synthesized through

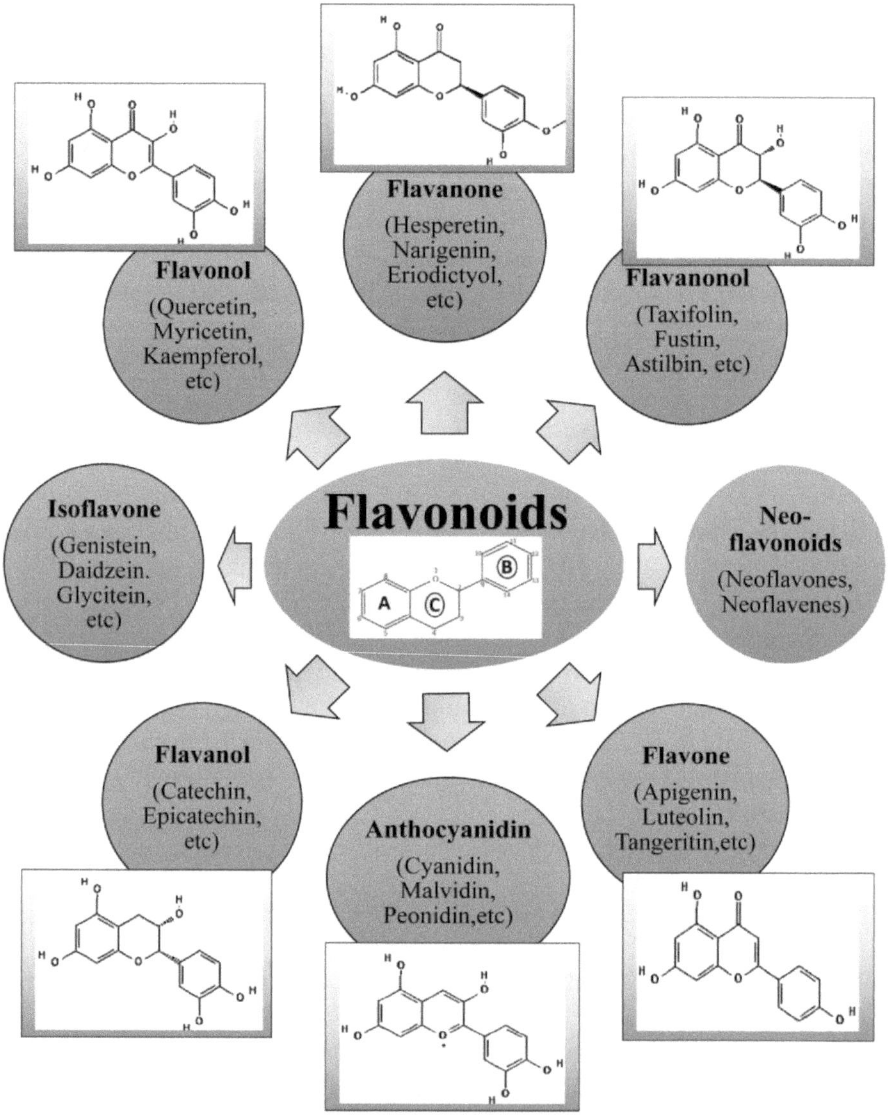

FIGURE 8.2 The classification of flavonoids into eight groups and common compounds in each group.

Note: The inset shows the chemical structure of one compound from each group, that is, quercetin (PubChem, n.d.-e), hesperetin (PubChem, n.d.-a), taxifolin (PubChem, n.d.-d), apigenin (PubChem, n.d.-f), cyanidin (PubChem, n.d.-b), epicatechin (PubChem, n.d.-c) [Open access 2D chemical structures have been used from (PubChem, n.d.-a, n.d.-b, n.d.-c, n.d.-d, n.d.-e, n.d.-f)].

the shikimate pathway) into p-coumaroyl coenzyme A (the phenylpropanoid pathway) and then its condensation with three malonyl-CoA molecules in the presence of the enzyme chalcone synthase (Dong and Lin, 2021), chalcone (2′, 4′, 6′, 4-tetrahydroxychalcone) is generated (Ciumărnean et al., 2020; Dias et al., 2021; Mierziak et al., 2014). Chalcone, the central intermediate metabolite produced in the flavonoid biosynthetic pathway, has some structural dissimilarities with that of rest of the flavonoids, that is, open central C-ring, and hence is called open-chain flavonoids. Because of its open α β-unsaturated ring structure, it acts as the precursor to multiple classes/derivatives of flavonoids that are synthesized through substitution/oxidation/saturation of the connecting pyran ring, starting with the synthesis of flavanones (Ayaz et al., 2019a; Mierziak et al., 2014).

According to the commonly accepted classification system, flavonoids are divided into six major categories (flavones, flavonols, flavanones, flavanols, anthocyanidins, and isoflavones) based upon the chemical structure in the central heterocyclic ring. However, multiple other subclasses have also been created (Dong and Lin, 2021; Harborne and Williams, 2002; Tohge et al., 2017). Four of the common flavonoid subclasses (anthocyanidins, flavones, flavonols, and isoflavones) contain C2=C3 unsaturation in the central heterocyclic ring while other 3 categories (flavanones, dihydroflavonols, and flavan-3-ols) are saturated compounds (Dias et al., 2021; Saini et al., 2017). Apart from the state of saturation, the 3-position of hydroxyl group differentiates flavonols from flavanones (Narayana et al., 2001). Isoflavonoids and neoflavonoids differ from flavonoids in having the benzenoid substituent at 3-position (linking of B ring to position 3 of the C ring) and 4-position,respectively, instead of the 2-position in the latter (Panche et al., 2016). Flavonoids are further categorized into subgroups according to structure of the C ring—flavanones, flavones, flavanonols, flavonols, flavanols (catechins), and anthocyanins (Panche et al., 2016).

The major classes of flavonoids with differing biochemical activity profiles based on their chemical structure are discussed below:

8.7.1 FLAVANONES

Flavanones, that are commonly found in citrus fruits (Panche et al., 2016), are produced through isomerization of chalcone by chalcone isomerase (Park et al., 2018), one of the first steps in the flavonoid biosynthetic pathway (Figure 8.3) (Mierziak et al., 2014). During this process, stereogenic configuration is created at C2 rendering many natural flavanones to be optically active, for example, naringenin (Dias et al., 2021). These are saturated compounds having a saturated double bond between carbon atoms at the position 2 and 3, and thus are alternatively known as saturated flavones or dihydroflavones (Ayaz et al., 2019a). Natural flavanones are often glycosylated, for example, naringin being the glycoside of naringenin (Dias et al., 2021).

8.7.2 FLAVANONOLS

Flavanonols (dihydoflavonols) are synthesized through addition of a hydroxyl group at 3′ position of flavanones in the presence of Flavanone 3-hydroxylase (F3H or FHT) (Dong and Lin, 2021) (Figure 8.3) and are, therefore, also called 3-Hydroxyflavanones (Ayaz et al., 2019a; Dias et al., 2021). Thus, these compounds are characterized by the presence of a saturated pyran ring containing carbonyl moiety at C4 and hydroxyl moiety at 3′ (Ayaz et al., 2019a). Commonly found flavanonols include taxifolin, astilbin, fustin, etc. (Figure 8.2).

8.7.3 FLAVONOLS

Flavonols are produced via the action of flavonol synthase (FLS) on flavanonols (Figure 8.3) (Dong and Lin, 2021). The chemical structure of flavonols is distinct from flavones in having a hydroxyl group at 3 position of the pyran ring, and therefore, alternatively known as 3-Hydroxyflavones (Ayaz et al., 2019a; Panche et al., 2016). Although the most extensively studied flavonols (quercetin, kaempferol, or myricetin) occur as aglycones (Brodowska, 2017), flavonols like flavones vary significantly in their glycosylation and other substitution patterns (Panche et al., 2016). Selected individual compounds under this class, which is often considered as one of the most abundant subgroups of flavonoids (Panche et al., 2016) are shown in Figure 8.2.

8.7.4 FLAVONES

This group of flavonoids, that are abundant in leaves, flowers, and fruits, are considered to be representative of flavonoids' chemical structure (Dias et al., 2021). These compounds are synthesized through the dehydrogenation of flavanones (Dias et al., 2021) in the presence of flavone synthase I (FNS

I) and flavone synthase II (FNS II) (Figure 8.3) (Dong and Lin, 2021). They are characterized by the presence of a double bond between C2 and C3 as well as a ketone/carbonyl moiety at the 4-position of the C ring (Ayaz et al., 2019a; Panche et al., 2016). The position of the hydroxyl group in the A or B rings varies according to the taxonomy of the plant in which the flavone is present (Ayaz et al., 2019a; Panche et al., 2016). Flavones are further modified through the action of transferases to generate glycosylated and methyoxylated flavone (Dong and Lin, 2021; Falcone Ferreyra et al., 2012). Common flavones/derivatives are shown in Figure 8.2.

8.7.5 ANTHOCYANIDINS

Compounds under this class of flavonoids are responsible for coloration in plants and are often used as color additives in human food items (Khoo et al., 2017). Synthesis of anthocyanidins begins with the action of dihydroflavonol 4-reductase (DFR) on flavanonols to produce leucoanthocyanindins such as flavan-3,4-diols (Figure 8.3) (Dong and Lin, 2021). The leucoanthocyanidins are then transformed by leucoanthocyanidin dioxygenase/anthocyanin synthase (ANS) into anthocyanidins. Anthocyanidins being relatively less stable and hence less abundant in nature (Dias et al., 2021), are then acted upon by a variety of transferases (for example, O-methyltransferase, rhamnosyl transferase, UDPG-flavonoid glucosyltransferase, etc.) to produce anthocyanins such as pelargonidin, cyanidin, delphinidin, peonidin, petunidin, and malvidin (Saigo et al., 2020). Different anthocyanins are responsible for different colors, cyanidin causing red-magenta hues, delphinidin-purple to blue colors, and pelargonidin-orange to pink colors (Dias et al., 2021). The final color of the compounds depends on the pH and also presence of sugar moiety (enhances brightness) (Dias et al., 2021) and/or methyl/acyl substituents on the A and B rings (Panche et al., 2016).

8.7.6 FLAVANOLS/ FLAVAN-3-OLS/CATECHINS

This class of flavonoids, characterized by the absence of carbonyl group at 4 position and saturated pyran ring (Ayaz et al., 2019a), consists of both monomeric compounds flavan-3-ols (such as epicatechin, catechin, and gallo-catechin) as well as polymeric tannins (Brodowska, 2017). Flavan-3-ols are synthesized from leucoanthocyanidins under the action of leucoanthocyanidin reductase (LAR or LCR) (Figure 8.3) (Dong and Lin,

2021). Alternatively, these can be generated from anthocyanidin catalyzed by anthocyanidin reductase (ANR) (Dixon et al., 2005). Flavan-3-ols after undergoing polymerization, produced proanthocyanidins which are condensed tannins (Dixon et al., 2005).

8.7.7 ISOFLAVONES

Isoflavonoids are characterized by the linkage of B ring to position 3 of the pyran C ring instead of the normal 2-position (Panche et al., 2016). These are synthesized from flavanones through aryl migration and dehydrogenation (Ayaz et al., 2019a) under the action of isoflavone synthase (Figure 8.3) (Dong and Lin, 2021). Isoflavones (e.g., daidzein, genistein, etc.) are subsequently modified to produce isoflavonoids (e.g., medicarpin) under the action enzymes like isoflavone O-methyltransferase, isoflavone 2′-hydroxylase (I2′H), isoflavone reductase (IFR), vestitone reductase (VR), and 7, 2′-dihydroxy, 4′-methoxyisoflavanol dehydratase, etc., (Dong and Lin, 2021; Tohge et al., 2017).

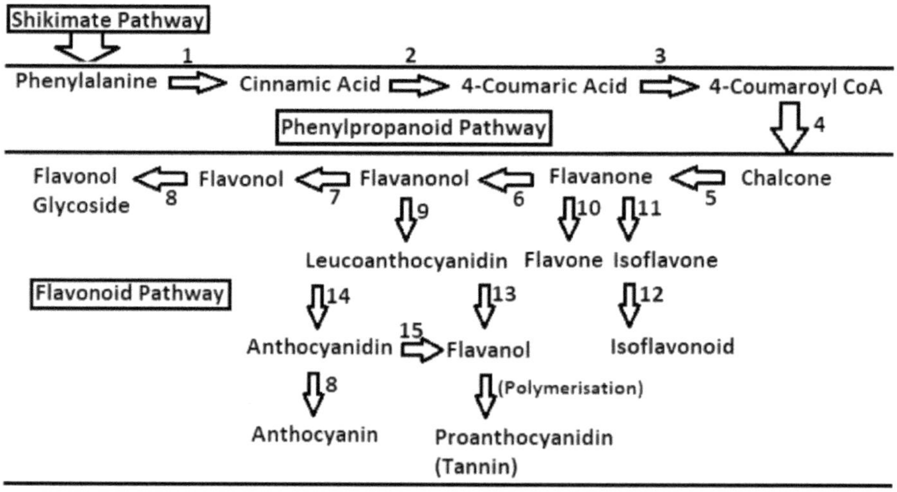

FIGURE 8.3 Synthesis of flavonoids [1- Phenylalanine ammonialyase, 2- Cinnamic acid hydroxylase, 3- 4-coumarate-CoA-Lignase, 4-Chalcone Synthase, 5- Chalcone Isomerase, 6- Flavanone3-hydroxylase, 7- Flavonol synthase, 8- Rhamnosyl synthase/ UDPG-Flavonoid glucosultransferase, 9- Dihydro0flavonol 4-reducatse, 10- Flavone synthase, 11- Isoflavone synthase, 12- 7,2′-Dihydroxy,4′-methoxyisoflavanol dehydratase/isoflavone O-methyl transferase/Isoflavone 2′-hydroxylase/Isoflavone reductase/vestitone reductase, 13- Leucoanthocyanidin reductase, 14- Anthocyanin synthase, 15- Anthocyanidin reductase].

8.7.8 NEOFLAVONOIDS

Another class of heterocyclic flavonoids (neoflavones and neoflavenes) with hypoglycaemic properties were found in *Calophyllum inophyllum* seeds and *Mesua thwaitesii* (calophyllolide) (Ayaz et al., 2019a; Panche et al., 2016) which had the benzene substituent joined to the position 4 of pyran ring. Since then, neoflavonoids have been identified in few other natural sources and have synthesized by researchers (Ayaz et al., 2019a).

Apart from the afore-mentioned classification system, flavonoids may also be categorized based on molecular size and/or presence of substituents on the A and B rings (Dias et al., 2021). Substitution of hydrogen atoms in the benzene rings with hydroxy, alkyl, methoxy or even carbohydrate moieties leads to further modification and corresponding change in physiochemical properties (e.g., solubility, pharmacokinetics, etc.) of the compounds in each of the afore-mentioned classes (Dias et al., 2021). For example, glycosyl-transferases cause glycosylation to form flavonoid glycosides, which is the common state in which most natural flavonoids exist (Cook and Samman, 1996; Mierziak et al., 2014). Higher forms of flavonoids and tannins, are generated through polymerization (Cook and Samman, 1996).

Multiple preclinical and in vitro studies show that this action of dietary flavonoids is mostly mediated via their ability to prevent neuronal damage secondary to oxidative stress and neuroinflammation and to promote neuronal survival through modulation of multiple cellular signaling pathways (Silva and Pogačnik, 2020). Overall mechanism by which flavonoids in general offer neuroprotection has been reviewed by multiple authors (Gardener et al., 2021; Pontifex et al., 2021; Silva and Pogačnik, 2020; Solanki et al., 2015). In the subsequent sections of this chapter, the current status of research into the potential neuroprotective action of selected flavonoids and the gap in literature will be described.

8.8 NEUROPROTECTIVE FLAVONOID: QUERCETIN IN THE SPOTLIGHT

Quercetin is yellow-colored water insoluble (Derosa et al., 2021) natural flavonoid of the flavonol sub-class (Almatroodi et al., 2021). The word "Quercetin" is drawn from a Latin root word "quercetum" which means "oak forest" (Chang et al., 2021). It is alternatively referred to as quercetine, sophretin, and memetin (Lakhanpal and Rai, 2007). As per the International Union of Pure and Applied Chemistry (IUPAC), quercetin is called

2- (3,4-dihydroxyphenyl)-3,5,7-trihydroxychromen-4-one and its chemical formula is $C_{15}H_{10}O_7$ (Batiha et al., 2020). This naturally occurring aglycone compound has multiple derivatives such as quercetin glycoside (which is water soluble) (Grewal et al., 2021), quercetin sulfate, quercetin glucuronide, and methylated quercetin (Li et al., 2016) and itself serves as the building block for other flavonoids like rutin, hesperidin, naringenin, and tangeritin (Bouktaib et al., 2002).

8.8.1 SOURCES

Quercetin and its derivatives are abundantly synthesized in more than 20 plants such ascitrus fruits (berries, grapefruit, apple, red grapes, mango, pomegranate, etc.), vegetables (onions, garlic, broccoli, green leafy vegetables, beans, etc.), tea leaves and coffee (Anand David et al., 2016; Derosa et al., 2021; Grewal et al., 2021; Lee et al., 2016). An interesting finding is that the amount of quercetin present in onions is very high compared to other vegetables (Islam et al., 2021). It is said that one kilogram of onion has almost 300 mg of quercetin (Islam et al., 2021; Kothari et al., 2020). Moreover, quercetin glycosides present in onions have been demonstrated to have higher bioavailability in humans as compared to other sources, leading to quicker absorption and higher concentrations (Lee and Mitchell, 2012). Similarly, a recent study claims that quercetin and not caffeine is responsible for the neuroprotective effect of coffee, especially in lowering the risk of developing Parkinson's disease and Alzheimer's disease (Lee et al., 2016). Various plant sources of quercetin are described in detail by recent reviews (Anand David et al., 2016; Derosa et al., 2021; Grewal et al., 2021). However, the quercetin content in multiple food items has not been conclusively established (Dabeek and Marra, 2019).

Since these compounds are widespread in plant products, a recent concept being explored by researchers is the extraction of quercetin from food waste (Bhat and Bhat, 2021). The public health relevance of this approach is manifold. Food loss and waste (FLW) is a critical problem because of its magnitude (FLW approximates 1/3rd of the total quantity of food produced for human use, almost 1.3 billion tonnes annually (Gustavsson et al., 2011)). It is furthermore compounded by food security issues posed by climate change (Mekonnen et al., 2021). Therefore, the concepts of circular bioeconomy and environmental sustainability have gained tremendous importance in the current era. This is highlighted by the Target 12.3 of the United Nations Sustainable Development Goals (SDGs) which seeks to reduce by 50% the global FLW by the year 2030 (Food and Agriculture Organization of the United Nations, n.d.). In this context, emphasis is being given to the extraction of useful compounds through processing and recycling of waste products (Sharma et al., 2021). Also, extraction of value-added products such as flavonoids including quercetin from food waste will also aid in minimizing the pollutant emission from the disposal and degradation of such waste (Grizzetti et al., 2013; Khedkar and Singh, 2018). A recent review explored the potential of food wastes as sources of quercetin (Bhat and Bhat, 2021).

However, quercetin like other bioflavonoids is not produced in humans and hence for its potential health benefits, it needs to be consumed as part of the daily diet (Benameur et al., 2021). Previous research estimates that for optimal health benefits humans need a daily total flavonoids dose of 200–350 mg, of which flavonols account for 20–35 mg and quercetin specific requirement is roughly 10 mg/day (García-Barrado et al., 2020; Kawabata et al., 2015). Nevertheless, dietary requirements vary within individuals based on multiple factors such as age and dietary habits (Benameur et al., 2021) and there is no consensus or any standard dietary guidelines regarding recommended dose for daily intake (Chun et al., 2007; Dabeek and Marra, 2019). The data regarding normal daily consumption of flavonols including quercetin vary widely based on geographic region, with studies in the United states of America reporting 9.0–36.2 mg/day (Cassidy et al., 2011; Chun et al., 2007) while European studies estimate daily flavonol intake at around 50 mg/day (Zamora-Ros et al., 2011). Quercetin average daily consumption of adults in USA was found to be approximately 3.5 mg/day (Bai et al., 2014). Although data regarding daily flavonoid intake from developing countries are limited, quercetin is known to be widely used in traditional medicinal products in many Asian and Native American cultures (Suganthy et al., 2016a).

8.8.2 PHARMACOKINETICS AND PHARMACODYNAMICS

The effectiveness of quercetin depends upon its pharmacokinetics and dynamics with in the human body. Quercetin and its derivatives are absorbed mostly from the small intestine, and in line with other flavonoids, their absorption is minimal in stomach (Almeida et al., 2018; Benameur et al., 2021). Absorption occurs either through passive diffusion or with the help of polypeptides transporting organic anions (Grewal et al., 2021). Occurrence of long chain fatty acids increases the quercetin absorption through the lymphatic system (Makino et al., 2018). Quercetin glycosides or any other derivatives are first transformed into aglycones with no additional moiety attached via the action of brush border enzymes such as lactase phloridzin hydrolases and β-glucosidases (Day et al., 2001; Kawabata et al., 2015). Then these aglycones undergo modifications such as glucuronidation, sulfation, and/or O-methylation in the epithelial lining of the intestine before releasing into circulation (Kawabata et al., 2015). Post-absorption, conjugated quercetins such as quercetin-3-*O*-β-D-glucuronide (Q3GA), 3′-*O*-methylquercetin-3-*O*-β-D-glucuronide, and quercetin-3′-*O*-sulfate are transported to the liver, which is the predominant site of quercetin metabolism along with some amount of transformation occurring in intestines and kidneys (Benameur et al., 2021; Grewal et al., 2021). In the liver, it undergoes phase I and II metabolism and the multitude of metabolites produced are then transported to varying organ systems/tissues through the systemic circulation (Dabeek and Marra, 2019; Manach et al., 1996). The metabolism of quercetin in human body is summarized schematically in Figure 8.4. A recent review summarizes the details of quercetin metabolism, effect of gut microbiota on quercetin absorption, and finally its bioavailability (Almeida et al., 2018). Although more than 20 metabolites have been identified by previous studies in urine which form the main mode of quercetin excretion (Almeida et al., 2018), the most abundant metabolites found in the urine after oral ingestion of quercetin is said to be quercetin-3-O-β-D-glucuronide, which exerts significant physiological effects (Benameur et al., 2021; Nabavi et al., 2015). Overall, the bioavailability of quercetin is said to be poor because of low solubility, fast metabolism, and rapid elimination from the body, and relative inactivity of many of its metabolites (Khan et al., 2019). Further, the neuro-bioavailability of the compound is even poorer because of its lesser potential to cross the blood–brain barrier (Khan et al., 2019). Therefore, many studies have focused on techniques to improve quercetin bioavailability while at the same time limiting its toxicity due to a requirement of higher doses. Multiple quercetin encapsulation and novel drug delivery methods such as quercetin loaded gel, polymeric micelle, nanoparticles/nano-emulsions, and glucan-conjugate are being explored which have been summarized by few recent reviews (Lai and Wong, 2021; Manzoor et al., 2021; Pateiro et al., 2021).

8.8.3 HEALTH EFFECTS OF QUERCETIN: IN GENERAL

Many flavonoids are known to have health promoting potential in terms of their anti-inflammatory activity of which the compounds within the flavonol subgroup have the strongest anti-inflammatory potential (Khan et al., 2019).

Quercetin and its derivatives have strong anti-inflammatory and anti-oxidant actions which have been said to be among the most important reasons behind their beneficial health effects (Khan et al., 2019), especially against chronic non-communicable diseases where the major underlying pathology starts with inflammation and oxidative stress (Merelli et al., 2021). Such anti-inflammatory and antioxidant action is affected by the chemical structure of the compound because of its effect on absorption and bioavailability (Suganthy et al., 2016a). For example, glycosylation, especially at position 3 of the structural skeleton of quercetin reduces the anti-inflammatory, free radical scavenging and acetyl cholinesterase inhibition potential (Ganeshpurkar and Saluja, 2017; Kim et al., 2004). In contrast, methylation can enhance metabolic stability as well as anti-oncogenic activity (Shi et al., 2014). In terms of thrombin inhibition potential, substitution of the hydroxyl (OH) with methoxy group (OCH_3) in ring A increases while that in ring B or C decreases activity (Shi et al., 2012).

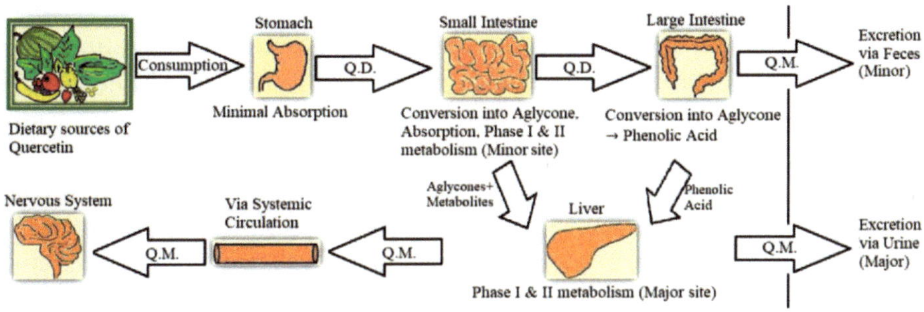

FIGURE 8.4 Metabolism of quercetin in human body.

Note: Q.D.- Quercetin derivatives consumed in food, Q.M.- Quercetin metabolites produced after phase I and II metabolism.

8.8.3.1 ANTI-INFLAMMATORY ACTION

Multiple in vitro, in vivo, and clinical studies have demonstrated the anti-inflammatory potential of quercetin. Potent organ/tissue specific anti-inflammatory potential of quercetin lies in its ability to modulate multiple downstream signaling pathways such as extracellular regulated kinase 2 (Erk2) mitogen-activated protein kinase (MAPK) signaling, nuclear factor-κB (NF-κB) pathway, etc. For example, Li et al., 2016 summarize the possible mechanisms by which quercetin inhibits TNFα-mediated inflammation (Li et al., 2016). Quercetin inhibits the production of TNF-α in macrophages and blocks TNF-α-mediated activation of ErK2, c-Jun NH2-terminal kinase (JNK), c-Jun, and NF-κB, all of which when activated can enhance inflammatory gene expression and translation (Li et al., 2016; Nair et al., 2006). Simultaneously, quercetin also increases peroxisome proliferator-activated receptor (PPARγ) activity which in turn also inhibits NF-κB and prevents activation of further inflammatory cascades (Liu et al., 2016).

It inhibits the production of inflammatory cytokines such as interleukin-8 (IL-8) in lung A549 cells in response to lipopolysaccharide (LPS) stimulation (Geraets et al., 2007; Sul and Ra, 2021). Similarly, high affinity Immunoglobulin E receptor (FcεRI) and hydrogen peroxide (H_2O_2)-induced generation of inflammatory mediators was inhibited by quercetin in human umbilical cord blood-derived cultured mast cells (hCBMCs) (Kempuraj et al., 2005) and human umbilical vein endothelial cells (HUVECs) (YANG et al., 2015), respectively. Pro-inflammatory enzymes like cyclooxygenase (COX) and lipoxygenase (LOX) are also produced less in the presence of quercetin and thus the secretion of inflammatory mediators like prostaglandins and leukotrienes is reduced (Xue et al., 2017). However, the exact mechanism is still ambiguous (Anand David et al., 2016). Quercetin supplementation has been shown to reduce the level of C-reactive protein (CRP) that has been implicated in multiple diseases like obesity, cardiovascular disease, etc., in in vitro (García-Mediavilla et al., 2007) as well as in animal experiments (Guardia et al., 2001; Mamani-Matsuda et al., 2006; Riskiyanti and Agustiningsih, 2016) and in healthy human volunteers (Askari et al., 2012). However, similar CRP lowering effect was not evident in a study conducted in female rheumatoid arthritis (RA) patients (Javadi et al., 2014).

8.8.3.2 ANTIOXIDANT ACTION

The antioxidant potential of quercetin and its derivatives can be attributed to their free radical (reactive oxygen species [ROS] and reactive nitrogen species [RNS]) scavenging action, which in turn decreases endoplasmic reticulum stress and consequent inflammation (Benameur et al., 2021). Metal chelation is another mechanism through which flavonoids such as quercetin can reduce oxidative stress (Rodríguez-Arce and Saldías, 2021). However, like with their anti-inflammatory action, their antioxidant potential also depends upon the chemical structure of the compounds. Research has shown that the antioxidant potential of quercetin is the highest and it decreases among its derivative (Lesjak et al., 2018). Although the presence of a higher number of free hydroxyl groups in the chemical structure of the compound is theoretically expected to account for higher antioxidant potential, studies have shown that the anti-inflammatory and antioxidant property of quercetin compounds is not directly proportional to this characteristic alone (Lesjak et al., 2018).

Antioxidant activity of quercetin has been reported by both in vivo and in vitro studies. Although free quercetin has limited efficacy in reducing gentamycin-induced renal oxidative stress and tissue damage, nano-encapsulated quercetin attenuated lipid peroxidation and boosted antioxidant defense mechanisms including glutathione, catalase, and superoxide dismutase, all culminating in the reversal of oxidative damage (Rahdar et al., 2021). Similarly, oral quercetin administration in male Wistar rats had a significant protective effect against oxidative stress and mitochondrial dysfunction in cardiomyocytes induced by BPA, an emerging pollutant emanating from plastic manufacturing (Vanani et al., 2020). Similar protective anti-oxidant potential of quercetin has been reported in other organ systems ranging from liver to even eyes (Zhao et al., 2021; Zhao et al., 2020). Studies speculate that the quercetin-mediated reduction of oxidative stress occurs through its activation of the nuclear factor erythroid 2-related factor 2 (Nrf2) antioxidant signaling pathway (Darband et al., 2020; Zhao et al., 2020). This antioxidant potential of quercetin makes it a prime candidate for cancer prevention through modulation of the oxidant–antioxidant balance (Batiha et al., 2020).

8.8.3.3 NEUROLOGICAL EFFECTS OF QUERCETIN

The anti-inflammatory and antioxidant actions of quercetin discussed before are crucial for its neuroprotective action since the major underlying pathology in neurological diseases starts with inflammation and oxidative stress (Merelli et al., 2021). The polyphenolic composition of quercetin aids in its neuroprotective action. A recent review summarizes the structure–activity relationship (SAR) of quercetin, with special focus on its neuroprotective action (Grewal et al., 2021). The different mechanisms underlying the neuroprotective effects of quercetin are represented pictorially in Figure 8.5 and are described in details in subsequent paragraphs.

8.8.3.3.1 Neuro-Oxidative Stress

Although oxidative stress may not be the sole initiating factor behind the pathogenesis of neurodegenerative disorders (Rodríguez-Arce and Saldías,

2021), uncontrolled accumulation of ROS, lipid peroxidation products in the brain can contribute to disease onset and exacerbate disease progression via oxidant damage and mitochondrial dysfunction (Xie et al., 2014). Therefore, patients with neurodegenerative diseases often have high neuro-oxidative stress levels (Hasanzadeh et al., 2021; Park et al., 2021). This is of particular importance since neurons are sensitive to oxidative damage owing to the increased polyunsaturated fatty acid composition of neuronal cell membranes, increased oxygen demand and poor inherent antioxidant defense mechanisms (Liu et al., 2017). The role played by oxidative stress in common neurodegenerative diseases has been summarized in previous reviews (Rodríguez-Arce and Saldías, 2021).

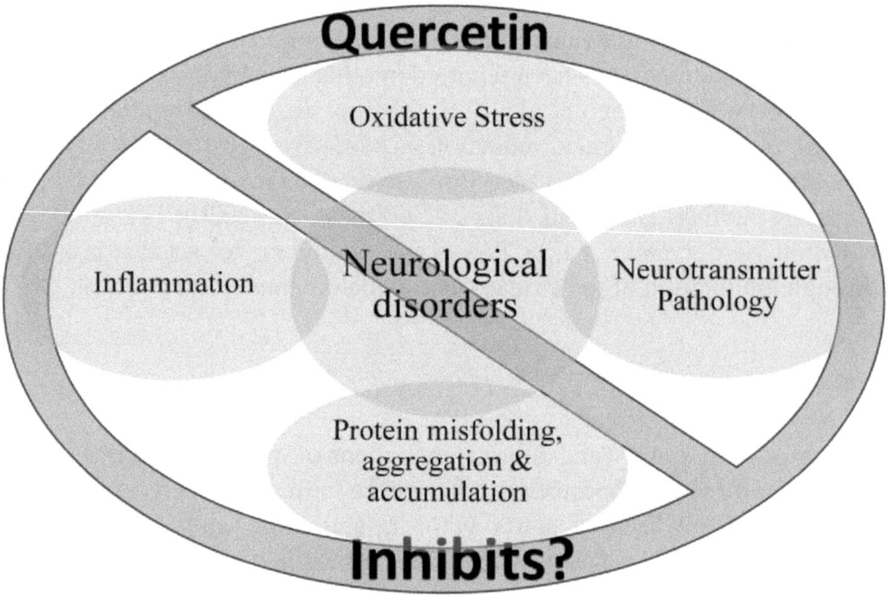

FIGURE 8.5 Overview of neuroprotective actions of quercetin.

Many studies have demonstrated the efficacy of quercetin in enhancing neuronal resistance to oxidative stress (Guo et al., 2021; Li et al., 2021; Song et al., 2021; Yang et al., 2021) triggered by specific oxidants such as hydrogen peroxide (Shokoohinia et al., 2015) and neurotoxic substances like amyloid β (Aβ) peptide (Ansari et al., 2009) and 6-hydroxydopamine (Magalingam et al., 2016) as has been reviewed previously by multiple authors (Grewal et al., 2021; Islam et al., 2021; Suganthy et al., 2016b;

Zaplatic et al., 2019). The antioxidant action of quercetin has been found to be stronger than other antioxidant nutrients, such as other flavonoids, β-carotene, and vitamins C and E (Choi et al., 2003; Rice-Evans et al., 1995). However, multiple mechanisms have been proposed through which the antioxidant activity of quercetin is exerted without any consensus on the cross-talk between the pathways, and therefore further research is warranted (Suganthy et al., 2016b). One of the mechanism through which quercetin exerts its antioxidant action is said to be through enhancing the expression of paraoxygenase 2 (PNO_2), an anti-oxidant enzyme abundantly found in the mitochondria of multiple neurons and supporting cells such as macrophages and astrocytes in the brain, and this has been extensively described previously (Suganthy et al., 2016b). Quercetin triggers the transcription and translation of PNO_2 in either through its phytoestrogenic activity or via induction of low levels of oxidative stress or via the JNK/AP-1 pathway which can stimulate PNO_2 expression (Suganthy et al., 2016b). Modulation of the Nrf2/antioxidant response element (ARE) signaling pathway via increased expression of sirtuin1 (SIRT-1) resulting in increased production of downstream antioxidant enzymes, such as the heme oxygenase 1 (HO-1) has been seen to be an effect of quercetin administration in rats (Song et al., 2021; Yang et al., 2021). Similarly, quercetin supplementation mediated increased SIRT-1 expression can potentially inhibit ROS-induced nucleotide-binding oligomerization domain-like receptor with a pyrin domain 3 (NLRP3) inflammasome activation as shown by reduced levels of associated inflammatory mediators like cleaved-caspase 1, IL-1β, and IL-18 in another recent aging mice experiment (Li et al., 2021). Another proposed mechanism of quercetin's antioxidant action is said to be by enhancing the glyoxalase pathway, an anti-oxidant defense mechanism that enables the detoxification of reactive oxidant dicarbonyl molecules such as methylglyoxal (MG) that preferentially target oxidation-sensitive neurons with high metabolic rate (Frandsen and Narayanasamy, 2018). Metal ion chelating property of quercetin, especially its role in brain iron homeostasis, rooted in its chemical structure is also responsible for its anti-oxidant action and this has been detailed in a recent review (Bardestani et al., 2021). Quercetin has been shown to have effect on multiple other signaling pathways including protein kinase C (PKC), phosphoinositide 3-kinase (PI3K)/Akt, MAPK, etc., with varying antioxidant effects on the brain (Zaplatic et al., 2019).

Furthermore, the neuroprotective antioxidant effect of quercetin also extends to peripheral nerves. The mechanism is rooted in the sirtuin-1 (SIRT1) and/or the adenosine 5′-monophosphate (AMP) activated protein kinase (AMPK)-mediated activation of the peroxisome proliferator-activated receptor-γ coactivator 1α (PGC-1α) (Zhang et al., 2021). PGC-1α is critical for mitochondrial functioning and maintenance of cellular energy homeostasis as well as for redox balance and inflammation regulation (Rius-Pérez et al., 2020). SIRT1 and AMPK regulate PGC-1α activity through deacetylation and phosphorylation, respectively, (Cantó and Auwerx, 2009) and such regulation is suppressed during hyperglycemia leading to oxidative stress, mitochondrial dysfunction, demyelination, axonopathy, and finally development of peripheral neuropathy (Fernyhough, 2015; Yu et al., 2016). Quercetin reduced ROS generation, lipid peroxidation, and up-regulated the AMPK/SIRT1/PGC-1α signaling pathway in peripheral neurons, thereby offering antioxidant neuroprotection in diabetic peripheral neuropathic rats (Zhang et al., 2021). Since the AMPK/SIRT1/PGC-1α signaling pathway has been shown to have a significant role in cognitive functioning (Yang et al., 2020), in retinal astrocytes (Guo et al., 2018; Jiang et al., 2014), etc., this can be another mechanism for quercetin's antioxidant action on the central nervous system as well.

8.8.3.3.2 Neuroinflammation

Along with oxidative stress, inflammation contributes to the onset and progression of many neurological disorders, particularly neurodegenerative diseases (Kwon and Koh, 2020; Merelli et al., 2021; Ransohoff, 2016). Usually, neuroinflammation is initiated as a protective mechanism by glial cells secondary to brain infection or injury caused by outright trauma or through other mechanisms such as in cancer, accumulation of amyloid beta-peptide, and hyperphosphorylated tau protein, exposure to environmental pollutants such as ultrafine particulate matter (Ehsanifar et al., 2021; Islam et al., 2021; Kwon and Koh, 2020). Microglia sense damage-associated molecular patterns (DAMPs) such as adenosine triphosphate (ATP) or intracellular proteins released from damaged cells in response to brain injury and mount an inflammatory response (Russo and McGavern, 2016). Neuroinflammation, in the context of neurodegenerative disorders, therefore, has been characterized by the occurrence of low levels of inflammatory mediators such as interleukins, TNF-α, components of the complement system, and acute-phase reactants in the brain accompanied by reactive morphology of glial cells (Ransohoff, 2016; Smith et al., 2012). However, repeated ongoing exogenous insults or endogenous factors contribute to chronic inflammation, which results in further neuronal damage. Damaged neurons fail to regenerate, unlike other cell types in the body (Ransohoff, 2016), and therefore, chronic inflammation leading to irreversible neuronal damage is now acknowledged to be the main pathogenesis underlying multiple neurological diseases (Kempuraj et al., 2016; Russo and McGavern, 2016).

Neuroinflammation has been implicated in the pathogenesis of primary neurodegenerative diseases like AD, PD, ALS as well as in diseases with neurodegenerative components such as progressive multiple sclerosis (Obrador et al., 2020; Padureanu et al., 2019; Ransohoff, 2016; Xie et al., 2014). However, the lack of unique features distinguishing neuroinflammation in the context of one disease to another complicates the process of identification of a specific therapeutic target for each disease (Ransohoff, 2016). In the absence of such specific therapeutics, the use of broad-spectrum anti-inflammatory agents has been explored. A meta-analysis of observational epidemiological studies (cohort) provided evidence in support of the use of anti-inflammatory drugs like nonsteroidal anti-inflammatory drugs (NSAIDs) in reducing the risk of developing AD (Zhang et al., 2018). Recently, similar evidence is being generated in the specific context of air pollution exposure-induced neuroinflammation/degeneration (Gao et al., 2021). However, another recent double-masked pharmaco-prevention trial concluded that routine administration of low-dose naproxen sodium (NSAID) in cognitively intact at-risk elderly led to a higher frequency of side-effects without any apparent benefit in the progression of presymptomatic AD (Meyer et al., 2019). Hence, researchers are actively exploring safer anti-inflammatory alternatives like flavonoids.

Multiple studies have reported the efficacy of quercetin in reducing neuroinflammation. Zhang et al. using the immortalized PC12 cell line and zebrafish models showed that quercetin effectively reduced inflammation induced by 6-hydroxydopamine (6-OHDA), a neurotoxin commonly used to mimic PD neuropathogenesis (Hernandez-Baltazar et al., 2017), by inhibiting the excessive generation of nitric oxide, inducible nitric oxide synthase (iNOS) enzyme, and other pro-inflammatory genes (Islam et al., 2021; Zhang et al., 2011). Quercetin acts on glial cell activation and subsequent inflammation. For example, in the immortalized murine BV-2 microglial cell line, LPS or interferon γ triggered inflammation can be blocked by quercetin through different mechanisms such as boosting the anti-oxidant Keap1/Nrf2 pathway mediated expression of ARE-regulated genes like heme oxygenase (HO)-1, and inhibiting pro-inflammatory NF-κB activation (Chen et al., 2005; Han et al., 2021). Quercetin can reduce the expression of Toll-like receptors (TLR), especially COX-4 and TLR2, which upon sensing ligands such as LPS activate downstream signaling through the NF-κB pathway (Bhaskar et al., 2011). Therefore, quercetin can ultimately inhibit the expression of inducible pro-inflammatory enzymes like COX-2 and iNOS (Bhaskar et al., 2011). Similarly, quercetin promotes mitophagy to clear damaged mitochondria in microglia, and thus generation of ROS from such mitochondria is prevented and subsequent mtROS amplified NLRP3 inflammasome activation and signaling is stopped (Han et al., 2021). Quercetin thus can effectively reduce microglial activation-induced neuronal apoptosis and death as has been shown by an in vitro study done using microglial N9-neuronal (PC12) co-culture (Bournival et al., 2012). Microglial mitochondrial damage can not only initiate neuronal cell death but also propagate neuronal death through the activation of naive astrocytes which in turn can contribute to further neuronal damage (Joshi et al., 2019; Liddelow et al., 2017). Therefore, the potential action of quercetin, as reported by studies on astrocyte reactivation-mediated release of pro-inflammatory mediators like TNF-α, IL-1, IL-6, etc., (Sharma et al., 2007; Zhang et al., 2020) is important.

Quercetin has significant modulating effects on other signaling pathways involved in the release of inflammatory mediators. For example, quercetin has effects on sirtuins (SIRT1-7), signaling proteins mainly located in the hippocampus, that have significant effects on metabolism, stress, and longevity regulation in the brain (Hendriks et al., 2003). SIRT1 especially is critical in the regulation of apoptosis and protein aggregation in the brain and is an important antioxidant and anti-inflammatory defense system in the brain (Iside et al., 2020). Therefore, it is considered an important therapeutic target for potential neuroprotective agents (Iside et al., 2020). Like with its antioxidant action, the SIRT1-dependent signaling pathway activation by quercetin plays a role in its anti-inflammatory effect leading to a decrease in inflammatory cytokine generation, reduced neuronal demyelination, and therefore, generating the potential for its use as a therapeutic agent in neurodegenerative diseases like ALS and MS (Hendriks et al., 2003). Similar evidence has also been generated regarding quercetin's effect on the SIRT1 pathway in neurodegeneration caused by infectious agents like herpes simplex type-1 (Leyton et al., 2015).

8.8.3.4 OTHER MECHANISMS OF QUERCETIN'S NEUROPROTECTIVE ACTION

8.8.3.4.1 Protein Misfolding, Aggregation, and Accumulation

Apart from oxidative stress and neuroinflammation, aggregation of disease-specific misfolded proteins and their progressive accumulation, collectively known as amyloid deposition, is another hallmark of multiple neurodegenerative diseases (Soto and Pritzkow, 2018). Common examples include Aβ and tau in AD, alpha-synuclein (α-Syn) in PD, TAR DNA-binding protein 43 (TDP-43) in ALS, etc., (Soto and Pritzkow, 2018). Despite dissimilarities among these proteins in their size, structure, and function, the process of misfolding generating β-sheets arranged in a cross β polymeric structure remains similar (Soto, 2003). Amyloid accumulation results in neuronal damage and can self-propagate causing progressive neurodegeneration (Soto and Pritzkow, 2018).

Quercetin can potentially modulate this pathology. For example, studies have shown inhibitory effects of quercetin on Aβ peptide formation as well as aggregation (Elfiky et al., 2021; Godoy et al., 2017; Li et al., 2017). Quercetin interferes in the step proteolytic cleavage of amyloid precursor protein (APP) by β- and γ-secretase enzymes involved in the production of Aβ peptide. It inhibits β-site APP cleaving enzyme-1 (BACE-1), the major β-secretase enzyme involved in the process (Naushad et al., 2019). Expression of important genes regulating the amyloidogenic pathway (such as APP, BACE1, APH1, and PSEN1 genes) is reduced by quercetin (Elfiky et al., 2021). Post-production quercetin can also inhibit Aβ aggregation by increasing macroautophagy and proteasomal degradation (Regitz et al., 2014). Furthermore, in vitro studies have demonstrated the efficacy of quercetin in inhibiting Aβ fibril formation which is exerted by increasing the bond between phenyl rings of the β-sheet structure and by decreasing the stability of preformed mature fibrils (Islam et al., 2021; Zaplatic et al., 2019). It can also competitively inhibit the activity of already formed Aβ peptides by interacting with the sites reserved for metal-binding that will ultimately result in oxidative stress, an important mechanism of action of the Aβ peptides (Tay et al., 2013).

Similar effects of quercetin on Tau peptide phosphorylation have been seen. Tau proteins play a critical role in microtubule stabilization (Barbier et al., 2019). Although tau phosphorylation by HSP 70 and glycogen synthase kinase 3β (GSK3β) is an important step in the tau physiology, hyper-phosphorylation can lead to de-polymerization and dismantling of microtubules and aggregation of fee tau peptides into insoluble filaments that accumulate in the brain as neurofibrillary tangles, a pathognomic feature of AD (Medeiros et al., 2010). Quercetin can reduce hyper-phosphorylation of the tau protein via multiple mechanisms such as AMPK downregulation and consequent suppression of endoplasmic reticular (ER) stress and NLRP3 inflammasome activation (Chen et al., 2016) or through its effect on the

PI3K/Akt/GSK3β, MAPK, and NF-κB pathways (Jiang et al., 2016) or the Ca^{2+}-calpain-p25-cyclin-dependent kinase 5 (CDK5) signaling pathway (Shen et al., 2018), all of which play important roles in tau protein regulation. Since Aβ peptides have been implicated in the onset and progression of tau pathology (Götz et al., 2001; Oddo et al., 2004), inhibitory effect of quercetin on Aβ peptide formation might also contribute to its modulating effect in tau pathology.

Quercetin can also modulate alpha-synuclein (α-Syn), a major component of Lewy bodies implicated in the pathogenesis of PD and other synucleinopathies (Meade et al., 2019). An active area of PD therapeutics research focuses on interfering with α-Syn production, aggregation/accumulation, and its cell to cell transfer (Fields et al., 2019). S129, one of the major constituents of alpha-Syn, when abnormally phosphorylated leads to agglomeration and amalgamations of alpha-Syn and subsequent neurotransmission defects (Chen and Feany, 2005). Such aggregation of alpha-Syn was commonly found in patients of PD and synucleinopathies (Meade et al., 2019). This is inhibited by quercetin administration which produces QUR-alpha-Syn complex through covalent bonds and due to high hydrophilicity of the complex stable oligomers are generated and fibrillation is prevented (Zhu et al., 2013). A molecular docking study concluded that quercetin can effectively bind and inhibit α-Syn (Baul and Rajiniraja, 2018).

8.8.3.4.2 Neurotransmitter Pathology

Neurotransmitters, whether excitatory or inhibitory, play a significant role in synaptic transmission and, therefore, functioning of the nervous system (Sheffler et al., 2021). Hence, any aberrations/pathology affecting their function are implicated in multiple neurological disorders (Sheffler et al., 2021). Neurotransmitters commonly implicated in neurodegenerative diseases include glutamate, acetylcholine, and dopamine.

Glutamate is the principal excitatory neurotransmitter and at present is more than 50% of nervous system (Mehta et al., 2013). Abnormal stimuli including oxidative stress, mitochondrial dysfunction, metabolic imbalances such as energy deficiency and excessive calcium can trigger glutamate-mediated excitotoxicity culminating in neuronal cell death that has been found to be associated with common neurodegenerative diseases such as AD, PD, ALS, Huntington's diseases and even epilepsy, ischemic brain damage (Mehta et al., 2013). Apart from this receptor-mediated excitotoxicity, glutamate can also cause non-receptor-mediated oxidative stress by decreasingcystine uptake through the cystine/glutamate antiporter leading to reduced intracellular glutathione (an anti-oxidant) (Murphy et al., 1989; Yang et al., 2013). Glutamate increases intracellular Ca^{2+} and ROS, mitochondrial dysfunction, activation of proteases like calpain, etc., and all of these changes finally culminate in the onset of apoptosis (Duchen, 2000; Yang et al., 2013). In-vitro research in HT22 cells (murine hippocampal cell line resembling neuronal precursors) showed that quercetin via suppression of Ca^{2+} influx, decreased ROS generation, calpain inhibition, etc., protects the neuronal mitochondria and reverts glutamate-mediated apoptosis (Yang et al., 2013). Quercetin blocked the release of glutamate cortical synaptosomes in adult male Sprague–Dawley rats, reduced presynaptic voltage-dependent Ca^{2+} influx and the potential role of protein kinase C (PKC) and A (PKA) in this action of quercetin (Lu et al., 2013). Similar findings were also obtained by (Silva et al., 2008).

Acetylcholine (Ach) is another important neurotransmitter of human nervous system (Picciotto et al., 2012) and its concentration at the synapse is controlled by an enzyme, acetylcholinesterase (AChE) which terminates cholinergic synaptic transmission by hydrolyzing ACh (Trang and Khandhar, 2021). Imbalance in cholinergic neurotransmission, occurring subsequent to ACh or its receptor deficiency or excessive AChE activity is the main postulate of the cholinergic hypothesis for cognitive dysfunction in AD (Hampel et al., 2019). AChE modulation has, therefore, become an attractive therapeutic target in AD research (Lazarevic-Pasti et al., 2017; McHardy et al., 2017) with studies demonstrating mild to moderate improvement in cognition in AD patients administered AChE inhibitor as has been reviewed recently (Marucci et al., 2021). Furthermore, AChE blockers have shown modulating effect on Aβ toxicity (Fancellu et al., 2020; Zhang et al., 2016). However, the use of currently available AChE blockers is limited by their safety concerns making phytochemicals with AChE modulating properties an attractive alternative (Khan et al., 2018). Quercetin is now considered a potent AChE blocker (Khan et al., 2018; Orhan et al., 2007) and a previous molecular docking study reported the higher efficacy of quercetin than conventional AChE inhibitors (Islam et al., 2013). High-dose quercetin decreased hippocampal AChE activity in in vivo PD models (Sriraksa et al., 2011). Thus, significant cognitive improvement subsequent to quercetin administration has been shown by multiple studies (Nishihira et al., 2021; Sriraksa et al., 2011; Yao et al., 2010).

8.8.3.4.3 Neurotrophic Factors

Neurotrophic factors comprising of three groups of growth factors, that is, nerve growth factor (NGF) family (alternatively referred to as neurotrophins (NTs) including the brain-derived neurotrophic factor (BDNF) and NT-3,4,5,6), glial cell line-derived neurotrophic factor (GDNF) family ligands (GFLs) (including GDNF, neurturin [NTN], persephin [PSPN], and artemin [ARTN], all of which are members of the transforming growth factor (TGF)-β superfamily] and neuropoietic cytokines (including ciliary neurotrophic factor [CNTF] and leukemia inhibitory factor [LIF]) (Kerschensteiner et al., 2003; Razavi et al., 2015). The neurotrophic factors are responsible for the regulation of blood flow and neurogenesis as well as an important contributor to the antioxidant defense system in the brain (Chiaretti et al., 2008; Razavi et al., 2015). These factors act through different Trk family of tyrosine protein kinases and the p75 receptor which is part of the tumor necrosis factor receptor (TNFR) family (Chao, 1994). Phytochemicals like quercetin are known to stimulate the generation of these neurotrophic factors resulting in synaptogenesis, neurogenesis, and regeneration (Cichon et al., 2020). Quercetin has been reported to have inductive effect on NGF and BDNF leading to neurite outgrowth via the $Na^+/K^+/2Cl^-$ cotransporter isoform (Nakajima et al., 2011) and the Rho GTPase enzyme (Palazzolo et al., 2012). Acting via the PI3/AkT pathway, quercetin can trigger neurogenesis (Tangsaengvit et al., 2013). Quercetin administration can increase neuronal survival through the generation of neurotrophic neurite-related proteins such as synaptophysin, synapsin, growth-related protein-43, microtubule-related protein, etc. (Moosavi et al., 2016). Other mechanisms of quercetin's neuroprotection are via the phosphorylation of cyclic AMP response element-binding protein (CREB) and increased expression of genes regulating the production of NGF, BDNF, CREB, etc. which in turn enhance neurogenesis (Karimipour et al., 2019; Tchantchou et al., 2009).

8.8.3.4.4 Neuroprotective Effects Against Environmental Pollutants

The anti-oxidant and anti-inflammatory effects of quercetin make it a potential candidate to counter the harmful neurological effects of exposure to environmental pollutants, most of which act primarily through these biological mechanisms. Particularly, published research provides evidence highlighting the fact that both neurodevelopmental and neurodegenerative disorders occurring secondary to exposure to particulate matter air pollution are mainly mediated through inflammation and oxidative stress (Ahadullah et al., 2021; Kang et al., 2021). Similarly, studies have shown that the administration of quercetin in experimental animals ameliorates the effects of exposure to pollutants. A recent study exposed rats to ultrafine petrol exhaust nanoparticles to induce neurotoxicity and found that pre-treatment with quercetin at a dose of 60 mg per kilogram body weight reduced the rise in the concentration of neuro-inflammatory parameters like ROS as well as that of Aβ peptide in rat brains (Durga et al., 2022).Maternal quercetin supplementation can protect against $PM_{2.5}$-induced neurodevelopmental toxicity and consequent behavioral aberration in offspring (Zhang et al., 2018). However, evidence from human studies is limited to support such therapeutic action of quercetin (Trushna et al., 2021).

8.8.3.5 QUERCETIN SAFETY

Evidence for neuroprotective action of quercetin in such neurodegenerative diseases is mostly sourced from in-vitro and in-vivo studies with fewer clinical studies (Grewal et al., 2021; Islam et al., 2021). Flavonoids including quercetin are well tolerated in animals (Ayaz et al., 2019b). For example, LD50 (dose that can kill 50% of the test population) is around 2–10 g per rat for most flavonoids (Ayaz et al., 2019b). However, it is difficult to extrapolate these values to humans (Ayaz et al., 2019b). The issue is further compounded by low bioavailability of quercetin necessitating higher doses in humans. However, since ingested quercetin undergoes high degree of first-pass metabolism, it can be argued that its toxicity potential is decreased (Batiha et al., 2020).

Very few human studies have assessed the safety profile of quercetin administration and these studies as reviewed by Dabeek and Marra 2019 have used purified aglycone supplement with a dosage range of 150–5000 mg/day for maximum of 12 weeks (Dabeek and Marra, 2019). Daily dose of 150 mg did not result in any significant adverse effect on liver and kidney function (Egert et al., 2009). Similarly, no adverse events were reported during quercetin trial conducted by McAnulty et al. where 40 athletes were administered 1000 mg quercetin per day for 6 weeks. However, the authors failed to find any significant anti-oxidant/anti-inflammatory effect of quercetin as well (McAnulty et al., 2008). Lu et al. conducted a phase I dose-escalating study to investigate safety of quercetin administration in chronic hepatitis C patients with dose ranging from 250 to 5000 mg/day for 28 days. Authors report that mild gastrointestinal side-effects were experienced by a few patients without providing further details (Lu et al., 2016). Pfeuffer et al. in their double-blind crossover study of 49 healthy male participants who were administered 150 mg per day quercetin for 8 weeks, reported significant increase in TNF-α levels (Pfeuffer et al., 2013). Older clinical trials of quercetin reporting their safety/side-effects were reviewed by Harwood et al. 2007 who concluded that no side effects were reported by published trials when participants were administered quercetin orally (3–1000 mg per day) for a maximum of 12 weeks (Harwood et al., 2007). Intravenous administration of high-dose quercetin (up to 51.3 mg kg^{-1} body weight amounting to more than 3000 mg quercetin in some cases) resulted in transient nephrotoxic side-effects (Ferry et al., 1996; Harwood et al., 2007). However, quercetin is usually ingested orally and owing to its low bioavailability, its concentration in the systemic circulation would normally not reach such high levels, making the results of this trial difficult to extrapolate to real-life conditions (Harwood et al., 2007).

Carcinogenic potential of quercetin is widely debated. According to the Ames test, quercetin has mutagenic potential. But in-vivo studies have failed to find significant carcinogenicity (Batiha et al., 2020). In 1999, the International Agency for Research on Cancer (IARC) concluded that quercetin cannot be considered as a human carcinogen (IARC Working Group on the Evaluation of Carcinogenic Risks to Humans, 1999) and in 2010, quercetin was added to the Generally Recognized as Safe (GRAS) substances list of the Food and Drug Administration (FDA) (Center for Food Safety and Applied Nutrition, n.d.; Dabeek and Marra, 2019). Similarly, teratogenicity of quercetin is ambiguous (Batiha et al., 2020).

Quercetin can affect the efficacy/safety of administration of other drugs. Co-administration of high-dose quercetin and digoxin can potentially lead to side-effects as reported by in-vivo animal model studies (Wang et al., 2004). Similarly, quercetin can modulate the human cytochrome P450 (CYP enzymes) which are responsible for metabolism of majority of commercially available pharmaceutical products (Elbarbry et al., 2017). It inhibitory effect is particularly prominent on CYP3A4 (Elbarbry et al., 2017), which is involved in metabolism of multiple common drugs like amlodipine, diazepam, simvastatin, etc. (Zhou, 2008). Hence, quercetin co-administration can potentially increase plasma concentration and elimination time of such drugs which are CYP3A4 substrates, as has been demonstrated by studies (Wang et al., 2016). Similar inhibitory action of quercetin was also noticed on CYP2C19 and CYP2D6 (Elbarbry et al., 2017; Mohos et al., 2020). Hence, caution is needed during the use of quercetin supplementation in patients using concurrent medications.

8.9 CONCLUSION: THE WAY FORWARD

Neurological diseases pose a serious threat to public health and this burden is being further worsened owing to widespread human exposure to environmental pollutants. Research has shown that environmental pollutants like particulate matter air pollution, etc., contribute to the onset and progression of neurodevelopmental and neurodegenerative diseases particularly through their potential to generate inflammation and oxidative stress in the brain. However, therapeutic strategies targeting such neurological disorders have had limited success with risk of adverse effects. Therefore, researchers are attempting to find newer therapeutic modalities. Flavonoids, because of the widespread presence in nature and human diet, have become the focus of such research. In this chapter, through the example of quercetin, a common flavonoid of the flavonol group, the mechanism of action of flavonoids in exerting beneficial health effects on the nervous system has been detailed. Although multiple experimental studies provide ample evidences regarding its benefits, human studies have been limited and thus, future research needs to focus on this gap in literature. Furthermore, the adverse effects of quercetin supplementation and its safety profile are yet ambiguous calling for further research particularly looking at its interaction with other commonly used medications.

KEYWORDS

- **environmental pollution**
- **flavonoids**
- **nervous system diseases**
- **quercetin**
- **therapeutic uses**

REFERENCES

Aghaei, M.; Janjani, H.; Yousefian, F.; Jamal, A.; Yunesian, M. Association Between Ambient Gaseous and Particulate Air Pollutants and Attention Deficit Hyperactivity Disorder (ADHD) in Children; A Systematic Review. *Environ. Res.* **2019,** *173*, 135–156. https://doi.org/10.1016/j.envres.2019.03.030

Ahadullah, Yau, S.; Lu, H.; Lee, T. M. C.; Guo, H.; Chan, C. C. H. $PM_{2.5}$ as a Potential Risk Factor for Autism Spectrum Disorder: Its Possible Link to Neuroinflammation, Oxidative Stress and Changes in Gene Expression. *Neurosci. Biobehav.Rev.* **2021,** *128*, 534–548. https://doi.org/10.1016/j.neubiorev.2021.06.043

Alexeeff, S. E.; Liao, N. S.; Liu, X.; Van DenEeden; Sidney, S. Long-Term $PM_{2.5}$ Exposure and Risks of Ischemic Heart Disease and Stroke Events: Review and Meta-Analysis. *J. Am. Heart Assoc.* **2021,** *10* (1), e016890. https://doi.org/10.1161/JAHA.120.016890

Allotey, J.; Zamora, J.; Cheong-See, F.; Kalidindi, M.; Arroyo-Manzano, D.; Asztalos, E.; van der Post, J.; Mol, B.; Moore, D.; Birtles, D.; Khan, K.; Thangaratinam, S. Cognitive, Motor, Behavioural and Academic Performances of Children Born Preterm: A Meta-Analysis and Systematic Review Involving 64 061 Children. *BJOG: Int. J. Obstetr. Gynaecol.* **2018,** *125* (1), 16–25. https://doi.org/10.1111/1471-0528.14832

Almatroodi, S. A.; Alsahli, M. A.; Almatroudi, A.; Verma, A. K.; Aloliqi, A.; Allemailem, K. S.; Khan, A. A.; Rahmani, A. H. Potential Therapeutic Targets of Quercetin, a Plant Flavonol, and Its Role in the Therapy of Various Types of Cancer through the Modulation of Various Cell Signaling Pathways. *Molecules* **2021,** *26* (5). https://doi.org/10.3390/molecules26051315

Almeida, A. F.; Borge, G. I. A.; Piskula, M.; Tudose, A.; Tudoreanu, L.; Valentová, K.; Williamson, G.; Santos, C. N. Bioavailability of Quercetin in Humans with a Focus on Interindividual Variation. *Compreh. Rev. Food Sci. Food Saf.* **2018,** *17* (3), 714–731. https://doi.org/10.1111/1541-4337.12342

Anand David, A. V.; Arulmoli, R.; Parasuraman, S. Overviews of Biological Importance of Quercetin: A Bioactive Flavonoid. *Pharmacogn. Rev.* **2016,** *10* (20), 84–89. https://doi.org/10.4103/0973-7847.194044

Ansari, M. A.; Abdul, H. M.; Joshi, G.; Opii, W. O.; Butterfield, D. A. Protective Effect of Quercetin in Primary Neurons Against A Beta (1-42): Relevance to Alzheimer's Disease. *J. Nutr. Biochem.* **2009,** *20* (4), 269–275. https://doi.org/10.1016/j.jnutbio.2008.03.002

Antonsen, S.; Mok, P. L. H.; Webb, R. T.; Mortensen, P. B.; McGrath, J. J.; Agerbo, E.; Brandt, J.; Geels, C.; Christensen, J. H.; Pedersen, C. B. Exposure to Air Pollution During Childhood and Risk of Developing Schizophrenia: A National Cohort Study. *Lancet Planet. Health* **2020,** *4* (2), e64–e73. https://doi.org/10.1016/S2542-5196 (20)30004-8

Askari, G.; Ghiasvand, R.; Feizi, A.; Ghanadian, S. M.; Karimian, J. The Effect of Quercetin Supplementation on Selected Markers of Inflammation and Oxidative Stress. *J. Res. Med. Sci.* **2012,** *17* (7), 637–641.

Attademo, L.; Bernardini, F. Air Pollution and Urbanicity: Common Risk Factors for Dementia and Schizophrenia? *Lancet Planet. Health* **2017,** *1* (3), e90–e91. https://doi.org/10.1016/S2542-5196 (17)30042-6

Ayaz, M.; Sadiq, A.; Junaid, M.; Ullah, F.; Ovais, M.; Ullah, I.; Ahmed, J.; Shahid, M. Flavonoids as Prospective Neuroprotectants and Their Therapeutic Propensity in Aging Associated Neurological Disorders. *Front. Aging Neurosci.* **2019a,** *11*, 155. https://doi.org/10.3389/fnagi.2019.00155

Ayaz, M.; Sadiq, A.; Junaid, M.; Ullah, F.; Ovais, M.; Ullah, I.; Ahmed, J.; Shahid, M. (2019b). Flavonoids as Prospective Neuroprotectants and Their Therapeutic Propensity in Aging Associated Neurological Disorders. *Front. Aging Neurosci.* **2019b,** *11*, 155. https://doi.org/10.3389/fnagi.2019.00155

Bagepally, B. S.; Balachandar, R.; Kalahasthi, R.; Tripathi, R.; Haridoss, M. Association Between Aluminium Exposure and Cognitive Functions: A Systematic Review and Meta-Analysis. *Chemosphere* **2021,** *268*, 128831. https://doi.org/10.1016/j.chemosphere.2020.128831

Bai, W.; Wang, C.; Ren, C. Intakes of Total and Individual Flavonoids by US Adults. *Int. J. Food Sci. Nutr.* **2014,** *65* (1), 9–20. https://doi.org/10.3109/09637486.2013.832170

Bakulski, K. M.; Seo, Y. A.; Hickman, R. C.; Brandt, D.; Vadari, H. S.; Hu, H.; KyunPark, S. Heavy Metals Exposure and Alzheimer's Disease and Related Dementias. *J. Alzheimer's Dis.: JAD* **2020,** *76* (4), 1215–1242. https://doi.org/10.3233/JAD-200282

Barbier, P.; Zejneli, O.; Martinho, M.; Lasorsa, A.; Belle, V.; Smet-Nocca, C.; Tsvetkov, P. O.; Devred, F.; Landrieu, I. Role of Tau as a Microtubule-Associated Protein: Structural and Functional Aspects. *Front. Aging Neurosci.* **2019,** *11*, 204. https://doi.org/10.3389/fnagi.2019.00204

Bardestani, A.; Ebrahimpour, S.; Esmaeili, A.; Esmaeili, A. Quercetin Attenuates Neurotoxicity Induced by Iron Oxide Nanoparticles. *J. Nanobiotechnol.* **2021,** *19* (1), 327–327. PubMed. https://doi.org/10.1186/s12951-021-01059-0

Basha, M. R.; Wei, W.; Bakheet, S. A.; Benitez, N.; Siddiqi, H. K.; Ge, Y.-W.; Lahiri, D. K.; Zawia, N. H. The Fetal Basis of Amyloidogenesis: Exposure to Lead and Latent Overexpression of Amyloid Precursor Protein and β-Amyloid in the Aging Brain. *J. Neurosci.* **2005,** *25* (4), 823–829. https://doi.org/10.1523/JNEUROSCI.4335-04.2005

Batiha, G. E.-S.; Beshbishy, A. M.; Ikram, M.; Mulla, Z. S.; El-Hack, M. E. A.; Taha, A. E.; Algammal, A. M.; Elewa, Y. H. A. The Pharmacological Activity, Biochemical Properties, and Pharmacokinetics of the Major Natural Polyphenolic Flavonoid: Quercetin. *Foods* **2020,** *9* (3), 374. https://doi.org/10.3390/foods9030374

Baul, H. S.; Rajiniraja, M. Favorable Binding of Quercetin to α-Synuclein as Potential Target in Parkinson Disease: An In Silico Approach. *Res. J. Pharm. Technol.* **2018,** *11* (1), 203. https://doi.org/10.5958/0974-360X.2018.00038.0

Becaria, A.; Lahiri, D. K.; Bondy, S. C.; Chen, D.; Hamadeh, A.; Li, H.; Taylor, R.; Campbell, A. Aluminum and Copper in Drinking Water Enhance Inflammatory or Oxidative Events Specifically in the Brain. *J. Neuroimmunol.* **2006,** *176* (1–2), 16–23. https://doi.org/10.1016/j.jneuroim.2006.03.025

Béjot, Y.; Reis, J.; Giroud, M.; Feigin, V. A Review of Epidemiological Research on Stroke and Dementia and Exposure to Air Pollution. *Int. J. Stroke* **2018,** *13* (7), 687–695. https://doi.org/10.1177/1747493018772800

Benameur, T.; Soleti, R.; Porro, C. The Potential Neuroprotective Role of Free and Encapsulated Quercetin Mediated by miRNA against Neurological Diseases. *Nutr.* **2021,** *13* (4), 1318. PubMed. https://doi.org/10.3390/nu13041318

Bhaskar, S.; Shalini, V.; Helen, A. Quercetin Regulates Oxidized LDL Induced Inflammatory Changes in Human PBMCs by Modulating the TLR-NF-κB Signaling Pathway. *Immunobiol.* 2**0211,** *216* (3), 367–373. https://doi.org/10.1016/j.imbio.2010.07.011

Bhat, I. U. H.; Bhat, R. Quercetin: A Bioactive Compound Imparting Cardiovascular and Neuroprotective Benefits: Scope for Exploring Fresh Produce, Their Wastes, and By-Products. *Biology* **2021,** *10* (7), 586. PubMed. https://doi.org/10.3390/biology10070586

Bhattacharjee, S.; Zhao, Y.; Hill, J. M.; Culicchia, F.; Kruck, T. P. A.; Percy, M. E.; Pogue, A. I.; Walton, J. R.; Lukiw, W. J. Selective Accumulation of Aluminum in Cerebral Arteries in Alzheimer's Disease (AD). *J. Inorg. Biochem.* **2013,** *126*, 35–37. https://doi.org/10.1016/j.jinorgbio.2013.05.007

Block, M. L.; Calderón-Garcidueñas, L. Air Pollution: Mechanisms of Neuroinflammation and CNS Disease. *Trends Neurosci.* **2009,** *32* (9), 506–516. https://doi.org/10.1016/j.tins.2009.05.009

Boelee, E.; Geerling, G.; van der Zaan, B.; Blauw, A.; Vethaak, A. D. Water and Health: From Environmental Pressures to Integrated Responses. *Acta Tropica* **2019,** *193*, 217–226. https://doi.org/10.1016/j.actatropica.2019.03.011

Bondy, S. C.; Campbell, A. Water Quality and Brain Function. *Int. J. Environ. Res. Public Health* **2018,** *15* (1), 2. https://doi.org/10.3390/ijerph15010002

Borroni, E.; Pesatori, A. C.; Bollati, V.; Buoli, M.; Carugno, M. Air Pollution Exposure and Depression: A Comprehensive Updated Systematic Review and Meta-Analysis. *Environ. Pollut.* **2022,** *292*, 118245. https://doi.org/10.1016/j.envpol.2021.118245

Bouktaib, M.; Atmani, A.; Rolando, C. Regio- and Stereo-Selective Synthesis of the Major Metabolite of Quercetin, Quercetin-3-O-β-D-Glucuronide. *Tetrahedron Lett.—* **2002,** *43*, 6263–6266. https://doi.org/10.1016/S0040-4039 (02)01264-9

Bournival, J.; Plouffe, M.; Renaud, J.; Provencher, C.; Martinoli, M.-G. Quercetin and Sesamin Protect Dopaminergic Cells from MPP+-Induced Neuroinflammation in a Microglial (N9)-Neuronal (PC12) Coculture System. *Oxidat. Med. Cell. Longevity* **2012,** *2012*, 921941. https://doi.org/10.1155/2012/921941

Braithwaite, I.; Zhang, S.; Kirkbride, J. B.; Osborn, D. P. J.; Hayes, J. F. (2019a). Air Pollution (Particulate Matter) Exposure and Associations with Depression, Anxiety, Bipolar, Psychosis and Suicide Risk: A Systematic Review and Meta-Analysis. *Environ. Health Persp.* **2019a,** *127* (12), 126002. https://doi.org/10.1289/EHP4595

Braithwaite, I.; Zhang, S.; Kirkbride, J. B.; Osborn, D. P. J.; Hayes, J. F. (2019b). Air Pollution (Particulate Matter) Exposure and Associations with Depression, Anxiety, Bipolar, Psychosis and Suicide Risk: A Systematic Review and Meta-Analysis. *Environ. Health Perspect.* **2019b,** *127* (12), 126002. PMID: 31850801 PMCID: PMC6957283. https://doi.org/10.1289/EHP4595

Brewer, G. J. Copper Toxicity in Alzheimer's Disease: Cognitive Loss from Ingestion of Inorganic Copper. *J. Trace Elements Med. Biol.* **2012,** *26* (2–3), 89–92. https://doi.org/10.1016/j.jtemb.2012.04.019

Brodowska, K. M. Natural Flavonoids: Classification, Potential Role, and Application of Flavonoid Analogues. *Eur. J. Biol. Res.* **2017,** *7* (2), 108–123.

Bush, A. I.; Masters, C. L.; Tanzi, R. E. Copper, β-amyloid, and Alzheimer's Disease: Tapping a Sensitive Connection. *Proc. Natl. Acad. Sci.* **2003,** *100* (20), 11193–11194. https://doi.org/10.1073/pnas.2135061100

Calderón-Garcidueñas, L.; Engle, R.; Mora-Tiscareño, A.; Styner, M.; Gómez-Garza, G.; Zhu, H.; Jewells, V.; Torres-Jardón, R.; Romero, L.; Monroy-Acosta, M. E.; Bryant, C.; González-González, L. O.; Medina-Cortina, H.; D'Angiulli, A. Exposure to Severe Urban Air Pollution Influences Cognitive Outcomes, Brain Volume and Systemic Inflammation in Clinically Healthy Children. *Brain Cogn.* **2011,** *77* (3), 345–355. https://doi.org/10.1016/j.bandc.2011.09.006

Cantó, C.; Auwerx, J. PGC-1alpha, SIRT1 and AMPK, an Energy Sensing Network That Controls Energy Expenditure. *Curr. Opin. Lipidol.* **2009,** *20* (2), 98–105. https://doi.org/10.1097/MOL.0b013e328328d0a4

Cassidy, A.; O'Reilly, É. J.; Kay, C.; Sampson, L.; Franz, M.; Forman, J. P.; Curhan, G.; Rimm, E. B. Habitual Intake of Flavonoid Subclasses and Incident Hypertension in Adults. *Am. J. Clin. Nutr.* **2011,** *93* (2), 338–347. https://doi.org/10.3945/ajcn.110.006783

Center for Food Safety and Applied Nutrition. *GRAS Notice Inventory—Agency Response Letter GRAS Notice No. GRN 000341* [WebContent]. Center for Food Safety and Applied Nutrition, n.d. Retrieved December 28, **2021,** from https://wayback.archive-it.

org/7993/20171031012354/https://www.fda.gov/Food/IngredientsPackagingLabeling/GRAS/NoticeInventory/ucm235935.htm

Chang, S. N.; Kim, H. J.; Kang, S. C. Quercetin Mitigates Oxidative Stress, Developmental Toxicity and Teratogenic Effects Induced by High-dose Vitamin D2 in Zebrafish Embryos. *Turk. J. Fisher. Aquat. Sci.* **2021,** *22* (3). https://doi.org/10.4194/TRJFAS20269

Chao, M. V. The p75 Neurotrophin Receptor. *J. Neurobiol.* **1994,** *25* (11), 1373–1385. https://doi.org/10.1002/neu.480251106

Chen, C. Y.; Hung, H. J.; Chang, K. H.; Hsu, C. Y.; Muo, C. H.; Tsai, C. H.; Wu, T. N. Long-Term Exposure to Air Pollution and the Incidence of Parkinson's Disease: A Nested Case-Control Study. *PLoS ONE* **2017,** *12* (8). https://doi.org/10.1371/journal.pone.0182834

Chen, J.; Deng, X.; Liu, N.; Li, M.; Liu, B.; Fu, Q.; Qu, R.; Ma, S. Quercetin Attenuates Tau Hyperphosphorylation and Improves Cognitive Disorder via Suppression of ER Stress in a Manner Dependent on AMPK Pathway. *J. Funct. Foods* **2016,** *22*, 463–476. https://doi.org/10.1016/j.jff.2016.01.036

Chen, J.-C.; Ho, F.-M.; Pei-Dawn Lee Chao, null, Chen, C.-P.; Jeng, K.-C. G.; Hsu, H.-B.; Lee, S.-T.; Wen Tung Wu, null; Lin, W.-W. Inhibition of iNOS Gene Expression by Quercetin Is Mediated by the Inhibition of IkappaB Kinase, Nuclear Factor-Kappa B and STAT1, and Depends on Heme Oxygenase-1 Induction in Mouse BV-2 Microglia. *Eur. J. Pharmacol.* **2005,** *521* (1–3), 9–20. https://doi.org/10.1016/j.ejphar.2005.08.005

Chen, L.; Feany, M. B. α-Synuclein Phosphorylation Controls Neurotoxicity and Inclusion Formation in a Drosophila Model of Parkinson Disease. *Nat. Neurosci.* **2005,** *8* (5), 657–663. https://doi.org/10.1038/nn1443

Chen, S.; Oliva, P.; Zhang, P. Air Pollution and Mental Health: Evidence from China. In *NBER Working Papers* (No. 24686; NBER Working Papers). National Bureau of Economic Research, Inc. https://ideas.repec.org/p/nbr/nberwo/24686.html

Chiaretti, A.; Antonelli, A.; Genovese, O.; Fernandez, E.; Giuda, D.; Mariotti, P.; Riccardi, R. Intraventricular Nerve Growth Factor Infusion Improves Cerebral Blood Flow and Stimulates Doublecortin Expression in Two Infants with Hypoxic-Ischemic Brain Injury. *Neurol. Res.* **2008,** *30* (3), 223–228. https://doi.org/10.1179/016164107X247948

Chin-Chan, M.; Navarro-Yepes, J.; Quintanilla-Vega, B. Environmental Pollutants as Risk Factors for Neurodegenerative Disorders: Alzheimer and Parkinson Diseases. *Front. Cell. Neurosci.* **2015,** *9*, 124–124. PubMed. https://doi.org/10.3389/fncel.2015.00124

Choi, Y.-J.; Kang, J.-S.; Park, J. H. Y.; Lee, Y.-J.; Choi, J.-S.; Kang, Y.-H. Polyphenolic Flavonoids differ in Their Antiapoptotic Efficacy in Hydrogen Peroxide-Treated Human Vascular Endothelial Cells. *J. Nutr.* **2003,** *133* (4), 985–991. https://doi.org/10.1093/jn/133.4.985

Cholanians, A. B.; Phan, A. V.; Ditzel, E. J.; Camenisch, T. D.; Lau, S. S.; Monks, T. J. From the Cover: Arsenic Induces Accumulation of α-Synuclein: Implications for Synucleinopathies and Neurodegeneration. *Toxicol. Sci.* **2016,** *153* (2), 271–281. https://doi.org/10.1093/toxsci/kfw117

Chowdhury, U. K.; Biswas, B. K.; Chowdhury, T. R.; Samanta, G.; Mandal, B. K.; Basu, G. C.; Chanda, C. R.; Lodh, D.; Saha, K. C.; Mukherjee, S. K.; Roy, S.; Kabir, S.; Quamruzzaman, Q.; Chakraborti, D. Groundwater Arsenic Contamination in Bangladesh and West Bengal, India. *Environ. Health Perspect.* **2000,** *108* (5), 393–397.

Chun, H.; Leung, C.; Wen, S. W.; McDonald, J.; Shin, H. H. Maternal Exposure to Air Pollution and Risk of Autism in Children: A Systematic Review and Meta-Analysis. *Environ. Pollut.* **2020,** *256*, 113307. https://doi.org/10.1016/j.envpol.2019.113307

Chun, O. K.; Chung, S. J.; Song, W. O. Estimated Dietary Flavonoid Intake and Major Food Sources of U.S. Adults. *J. Nutr.* **2007,** *137* (5), 1244–1252. https://doi.org/10.1093/jn/137.5.1244
Cichon, N.; Saluk-Bijak, J.; Gorniak, L.; Przyslo, L.; Bijak, M. Flavonoids as a Natural Enhancer of Neuroplasticity—An Overview of the Mechanism of Neurorestorative Action. *Antioxidants (Basel, Switzerland)* **2020,** *9* (11), E1035. https://doi.org/10.3390/antiox9111035
Ciumărnean, L.; Milaciu, M. V.; Runcan, O.; Vesa, Ştefan C.; Răchişan, A. L.; Negrean, V.; Perné, M.-G.; Donca, V. I.; Alexescu, T.-G.; Para, I.; Dogaru, G. The Effects of Flavonoids in Cardiovascular Diseases. *Molecules* **2020,** *25* (18), 4320. https://doi.org/10.3390/molecules25184320
Cook, N. C.; Samman, S. Flavonoids—Chemistry, Metabolism, Cardioprotective Effects, and Dietary Sources. *J. Nutr. Biochem.* **1996,** *7* (2), 66–76. https://doi.org/10.1016/S0955-2863(95)00168-9
Corrales, J.; Kristofco, L. A.; Steele, W. B.; Yates, B. S.; Breed, C. S.; Williams, E. S.; Brooks, B. W. Global Assessment of Bisphenol A in the Environment. *Dose-Response* **2015,** *13* (3), 1559325815598308. https://doi.org/10.1177/1559325815598308
Costa, L. G.; Cole, T. B.; Dao, K.; Chang, Y.-C.; Coburn, J.; Garrick, J. M. Effects of Air Pollution on the Nervous System and Its Possible Role in Neurodevelopmental and Neurodegenerative Disorders. *Pharmacol. Therap.* **2020,** *210*, 107523. https://doi.org/10.1016/j.pharmthera.2020.107523
Croft, K. D. The Chemistry and Biological Effects of Flavonoids and Phenolic Acids. *Ann. NY Acad. Sci.* **1998,** *854*, 435–442. https://doi.org/10.1111/j.1749-6632.1998.tb09922.x
Dabeek, W. M.; Marra, M. V. Dietary Quercetin and Kaempferol: Bioavailability and Potential Cardiovascular-Related Bioactivity in Humans. *Nutrients* **2019,** *11* (10), 2288. PubMed. https://doi.org/10.3390/nu11102288
Darband, S. G.; Sadighparvar, S.; Yousefi, B.; Kaviani, M.; Ghaderi-Pakdel, F.; Mihanfar, A.; Rahimi, Y.; Mobaraki, K.; Majidinia, M. Quercetin Attenuated Oxidative DNA Damage Through NRF2 Signaling Pathway in Rats with DMH Induced Colon Carcinogenesis. *Life Sci.* **2020,** *253*, 117584. https://doi.org/10.1016/j.lfs.2020.117584
Davoudi, M.; Barjasteh-Askari, F.; Amini, H.; Lester, D.; Mahvi, A. H.; Ghavami, V.; Rezvani Ghalhari, M. Association of Suicide with Short-Term Exposure to Air Pollution at Different Lag Times: A Systematic Review and Meta-Analysis. *Sci. Total Environ.* **2021,** *771*, 144882. https://doi.org/10.1016/j.scitotenv.2020.144882
Day, A. J.; Mellon, F.; Barron, D.; Sarrazin, G.; Morgan, M. R.; Williamson, G. Human Metabolism of Dietary Flavonoids: Identification of Plasma Metabolites of Quercetin. *Free Radic. Res.* **2001,** *35* (6), 941–952. https://doi.org/10.1080/10715760100301441
Derosa, G.; Maffioli, P.; D'Angelo, A.; Di Pierro, F. A Role for Quercetin in Coronavirus Disease 2019 (COVID-19). *Phytother. Res.* **2021,** *35* (3), 1230–1236. https://doi.org/10.1002/ptr.6887
Dias, M. C.; Pinto, D. C. G. A.; Silva, A. M. S. Plant Flavonoids: Chemical Characteristics and Biological Activity. *Molecules* **2021,** *26* (17), 5377. https://doi.org/10.3390/molecules26175377
Dixon, R. A.; Xie, D.-Y.; Sharma, S. B. Proanthocyanidins—A Final Frontier in Flavonoid Research? *New Phytol.* **2005,** *165* (1), 9–28. https://doi.org/10.1111/j.1469-8137.2004.01217.x
Dong, N.-Q.; Lin, H.-X. Contribution of Phenylpropanoid Metabolism to Plant Development and Plant–Environment Interactions. *J. Integr. Plant Biol.* **2021,** *63* (1), 180–209. https://doi.org/10.1111/jipb.13054

Duchen, M. R. Mitochondria and Calcium: From Cell Signalling to Cell Death. *J. Physiol.* **2000,** *529* (Pt 1), 57–68. https://doi.org/10.1111/j.1469-7793.2000.00057.x

Durga, M.; Vijayakumar, M.; Priya, K.; Vidhya Kanagarajan, S.; Banu, B. B.; Abraham, V. S. M.; Devasena, T.; Abdelaziz, M. A.; Mohideen, A. P.; Bahakim, N. O.; Shahid, M.; Ibrahim, S. R.; Ramesh, T. Effects of Quercetin on Ultrafine Petrol Exhaust Nanoparticles Induced DNA Damage, Oxidative Stress and Inflammation in Different Sections of Rat Brain. *J. King Saud Univ.—Sci.* **2022,** 101813. https://doi.org/10.1016/j.jksus.2021.101813

Egert, S.; Bosy-Westphal, A.; Seiberl, J.; Kürbitz, C.; Settler, U.; Plachta-Danielzik, S.; Wagner, A. E.; Frank, J.; Schrezenmeir, J.; Rimbach, G.; Wolffram, S.; Müller, M. J. Quercetin Reduces Systolic Blood Pressure and Plasma Oxidised Low-Density Lipoprotein Concentrations in Overweight Subjects with a High-Cardiovascular Disease Risk Phenotype: A Double-Blinded, Placebo-Controlled Cross-Over Study. *Br. J. Nutr.* **2009,** *102* (7), 1065–1074. https://doi.org/10.1017/S0007114509359127

Ehsanifar, M.; Jafari, A. J.; Montazeri, Z.; Kalantari, R. R.; Gholami, M.; Ashtarinezhad, A. Learning and Memory Disorders Related to Hippocampal Inflammation Following Exposure to Air Pollution. *J. Environ. Health Sci. Eng.* **2021,** *19* (1), 261–272. https://doi.org/10.1007/s40201-020-00600-x

Ejaz, H. W.; Wang, W.; Lang, M. Copper Toxicity Links to Pathogenesis of Alzheimer's Disease and Therapeutics Approaches. *International Journal of Molecular Sciences, 21* (20), 7660. https://doi.org/10.3390/ijms21207660

Elbarbry, F.; Ung, A.; Abdelkawy, K. Studying the Inhibitory Effect of Quercetin and Thymoquinone on Human Cytochrome P450 Enzyme Activities. *Pharmacogn. Mag.* **2017,** *13* (Suppl 4), S895–S899. https://doi.org/10.4103/0973-1296.224342

Elfiky, A. M.; Mahmoud, A. A.; Elreedy, H. A.; Ibrahim, K. S.; Ghazy, M. A. Quercetin Stimulates the Non-Amyloidogenic Pathway via Activation of ADAM10 and ADAM17 Gene Expression in Aluminum Chloride-Induced Alzheimer's Disease Rat Model. *Life Sci.* **2021,** *285*, 119964. https://doi.org/10.1016/j.lfs.2021.119964

Exley, C.; Mold, M. J. Aluminium in Human Brain Tissue: How Much Is Too Much? *J. Biol. Inorg. Chem.* **2019,** *24* (8), 1279–1282. https://doi.org/10.1007/s00775-019-01710-0

Falcone Ferreyra, M. L.; Rius, S.; Casati, P. Flavonoids: Biosynthesis, Biological Functions, and Biotechnological Applications. *Front. Plant Sci.* **2012,** *3*, 222. https://doi.org/10.3389/fpls.2012.00222

Fan, S.-J.; Heinrich, J.; Bloom, M. S.; Zhao, T.-Y.; Shi, T.-X.; Feng, W.-R.; Sun, Y.; Shen, J.-C.; Yang, Z.-C.; Yang, B.-Y.; Dong, G.-H. Ambient Air Pollution and Depression: A Systematic Review with Meta-Analysis Up to 2019. *Sci. Total Environ.* **2020,** *701*, 134721. PMID: 31715478. https://doi.org/10.1016/j.scitotenv.2019.134721

Fancellu, G.; Chand, K.; Tomás, D.; Orlandini, E.; Piemontese, L.; Silva, D. F.; Cardoso, S. M.; Chaves, S.; Santos, M. A. Novel Tacrine–Benzofuran Hybrids as Potential Multi-Target Drug Candidates for the Treatment of Alzheimer's Disease. *J. Enzyme Inhib. Med. Chem.* **2020,** *35* (1), 211–226. https://doi.org/10.1080/14756366.2019.1689237

Feigin, V. L.; Nichols, E.; Alam, T.; Bannick, M. S.; Beghi, E.; Blake, N.; Culpepper, W. J.; Dorsey, E. R.; Elbaz, A.; Ellenbogen, R. G.; Fisher, J. L.; Fitzmaurice, C.; Giussani, G.; Glennie, L.; James, S. L.; Johnson, C. O.; Kassebaum, N. J.; Logroscino, G.; Marin, B. et al. Global, Regional, and National Burden of Neurological Disorders, 1990–2016: A Systematic Analysis for the Global Burden of Disease Study 2016. *Lancet Neurol.* **2019,** *18* (5), 459–480. https://doi.org/10.1016/S1474-4422 (18)30499-X

Feigin, V. L.; Vos, T.; Nichols, E.; Owolabi, M. O.; Carroll, W. M.; Dichgans, M.; Deuschl, G.; Parmar, P.; Brainin, M.; Murray, C. The Global Burden of Neurological Disorders: Translating Evidence Into Policy. *Lancet Neurol.* **2020,** *19* (3), 255–265. https://doi.org/10.1016/S1474-4422 (19)30411-9

Fernyhough, P. Mitochondrial Dysfunction in Diabetic Neuropathy: A Series of Unfortunate Metabolic Events. *Curr. Diab. Rep.* **2015,** *15* (11), 89. https://doi.org/10.1007/s11892-015-0671-9

Ferry, D. R.; Smith, A.; Malkhandi, J.; Fyfe, D. W.; deTakats, P. G.; Anderson, D.; Baker, J.; Kerr, D. J. Phase I Clinical Trial of the Flavonoid Quercetin: Pharmacokinetics and Evidence for In Vivo Tyrosine Kinase Inhibition. *Clin. Cancer Res.* **1996,** *2* (4), 659–668.

Fields, C. R.; Bengoa-Vergniory, N.; Wade-Martins, R. Targeting Alpha-Synuclein as a Therapy for Parkinson's Disease. *Front. Mol. Neurosci.* **2019,** *12*, 299. https://doi.org/10.3389/fnmol.2019.00299

Foglio, E.; Buffoli, B.; Exley, C.; Rezzani, R.; Rodella, L. F. Regular Consumption of a Silicic Acid-Rich Water Prevents Aluminium-Induced Alterations of Nitrergic Neurons in Mouse Brain: Histochemical and Immunohistochemical Studies. *Histol. Histopathol.* **2012,** *27* (8), 1055–1066. https://doi.org/10.14670/HH-27.1055

Food and Agriculture Organization of the United Nations. *12.3.1 Global Food Losses—Sustainable Development Goals*, n.d. https://www.fao.org/sustainable-development-goals/indicators/1231/en/ (accessed 17 Dec 2021).

Formica, J. V.; Regelson, W. Review of the Biology of Quercetin and Related Bioflavonoids. *Food Chem. Toxicol.* **1995,** *33* (12), 1061–1080. https://doi.org/10.1016/0278-6915 (95) 00077-1

Frandsen, J. R.; Narayanasamy, P. Neuroprotection Through Flavonoid: Enhancement of the Glyoxalase Pathway. *Redox Biol.* **2018,** *14*, 465–473. https://doi.org/10.1016/j.redox.2017.10.015

Ganeshpurkar, A.; Saluja, A. K. The Pharmacological Potential of Rutin. *Saudi Pharma. J.* **2017,** *25* (2), 149–164. https://doi.org/10.1016/j.jsps.2016.04.025

Gao, X.; Coull, B.; Lin, X.; Vokonas, P.; Spiro, A.; Hou, L.; Schwartz, J.; Baccarelli, A. A. Short-Term Air Pollution, Cognitive Performance and Nonsteroidal Anti-Inflammatory Drug Use in the Veterans Affairs Normative Aging Study. *Nat. Aging* **2021,** *1* (5), 430–437. https://doi.org/10.1038/s43587-021-00060-4

García-Barrado, M. J.; Iglesias-Osma, M. C.; Pérez-García, E.; Carrero, S.; Blanco, E. J.; Carretero-Hernández, M.; Carretero, J. Role of Flavonoids in the Interactions among Obesity, Inflammation, and Autophagy. *Pharmaceuticals* **2020,** *13* (11), 342. https://doi.org/10.3390/ph13110342

García-Mediavilla, M.; Crespo, I.; Collado, P.; Esteller, A.; Sanchez-Campos, S.; Tuñón, M.; González-Gallego, J. The Anti-Inflammatory Flavones Quercetin and Kaempferol Cause Inhibition of Inducible Nitric Oxide Synthase, Cyclooxygenase-2 and Reactive C-Protein, and Down-Regulation of the Nuclear Factor Kappa B Pathway in Chang Liver Cells. *Eur. J. Pharmacol.* **2007,** *557*, 221–229. https://doi.org/10.1016/j.ejphar.2006.11.014

Gardener, S. L.; Rainey-Smith, S. R.; Weinborn, M.; Bondonno, C. P.; Martins, R. N. Intake of Products Containing Anthocyanins, Flavanols, and Flavanones, and Cognitive Function: A Narrative Review. *Front. Aging Neurosci.* **2021,** *13*, 640381–640381. PubMed. https://doi.org/10.3389/fnagi.2021.640381

Garza-Lombó, C.; Pappa, A.; Panayiotidis, M. I.; Gonsebatt, M. E.; Franco, R. Arsenic-Induced Neurotoxicity: A Mechanistic Appraisal. *J. Biol. Inorg. Chem.* **2019,** *24* (8), 1305. https://doi.org/10.1007/s00775-019-01740-8

Geraets, L.; Moonen, H.; Brauers, K.; Wouters, E. F. M.; Bast, A.; Hageman, G. Dietary Flavones and Flavonoles Are Inhibitors of Poly (ADP-ribose) polymerase-1 in Pulmonary Epithelial Cells. *J. Nutr.* **2007,** *137*, 2190–2195. https://doi.org/10.1093/jn/137.10.2190

Gillette-Guyonnet, S.; Andrieu, S.; Nourhashemi, F.; de La Guéronnière, V.; Grandjean, H.; Vellas, B. Cognitive Impairment and Composition of Drinking Water in Women: Findings of the EPIDOS Study. *Am. J. Clin. Nutr.* **2005,** *81* (4), 897–902. https://doi.org/10.1093/ajcn/81.4.897

Gitler, A. D.; Dhillon, P.; Shorter, J. Neurodegenerative Disease: Models, Mechanisms, and a New Hope. *Dis. Models Mec.* **2017,** *10* (5), 499–502. https://doi.org/10.1242/dmm.030205

Godoy, J. A.; Lindsay, C. B.; Quintanilla, R. A.; Carvajal, F. J.; Cerpa, W.; Inestrosa, N. C. Quercetin Exerts Differential Neuroprotective Effects Against H2O2 and Aβ Aggregates in Hippocampal Neurons: The Role of Mitochondria. *Mol. Neurobiol.* **2017,** *54* (9), 7116–7128. https://doi.org/10.1007/s12035-016-0203-x

Götz, J.; Chen, F.; van Dorpe, J.; Nitsch, R. M. Formation of Neurofibrillary Tangles in P3011 Tau Transgenic Mice Induced by Abeta 42 Fibrils. *Science* **2001,** *293* (5534), 1491–1495. https://doi.org/10.1126/science.1062097

Graber, M.; Mohr, S.; Baptiste, L.; Duloquin, G.; Blanc-Labarre, C.; Mariet, A. S.; Giroud, M.; Béjot, Y. Air Pollution and Stroke: A New Modifiable Risk Factor is in the Air. *Revue Neurologique* **2019,** *175* (10), 619–624. https://doi.org/10.1016/j.neurol.2019.03.003

Grewal, A. K.; Singh, T. G.; Sharma, D.; Sharma, V.; Singh, M.; Rahman, Md. H.; Najda, A.; Walasek-Janusz, M.; Kamel, M.; Albadrani, G. M.; Akhtar, M. F.; Saleem, A.; Abdel-Daim, M. M. Mechanistic Insights and Perspectives Involved in Neuroprotective Action of Quercetin. *Biomed. Pharmacother.* **2021,** *140*, 111729. https://doi.org/10.1016/j.biopha.2021.111729

Grizzetti, B.; Pretato, U.; Lassaletta, L.; Billen, G.; Garnier, J. The Contribution of Food Waste to Global and European Nitrogen Pollution. *Environ. Sci. Policy* **2013,** *33*, 186–195. https://doi.org/10.1016/j.envsci.2013.05.013

Guardia, T.; Rotelli, A. E.; Juarez, A. O.; Pelzer, L. E. Anti-Inflammatory Properties of Plant Flavonoids. Effects of Rutin, Quercetin and Hesperidin on Adjuvant Arthritis in Rat. *Farmaco (Societa Chimica Italiana: 1989)* **2001,** *56* (9), 683–687. https://doi.org/10.1016/s0014-827x (01)01111-9

Guerriero, F.; Sgarlata, C.; Francis, M.; Maurizi, N.; Faragli, A.; Perna, S.; Rondanelli, M.; Rollone, M.; Ricevuti, G. Neuroinflammation, Immune System and Alzheimer Disease: Searching for the Missing Link. *Aging Clin. Exp. Res.* **2017,** *29* (5), 821–831. https://doi.org/10.1007/s40520-016-0637-z

Gundacker, C.; Forsthuber, M.; Szigeti, T.; Kakucs, R.; Mustieles, V.; Fernandez, M. F.; Bengtsen, E.; Vogel, U.; Hougaard, K. S.; Saber, A. T. Lead (Pb) and Neurodevelopment: A Review on Exposure and Biomarkers of Effect (BDNF, HDL) and Susceptibility. *Int. J. Hygiene Environ. Health* **2021,** *238*, 113855. https://doi.org/10.1016/j.ijheh.2021.113855

Guo, L.; Huang, Z.; Huang, L.; Liang, J.; Wang, P.; Zhao, L.; Shi, Y. Surface-Modified Engineered Exosomes Attenuated Cerebral Ischemia/Reperfusion Injury by Targeting the Delivery of Quercetin Towards Impaired Neurons. *J. Nanobiotechnol.* **2021,** *19* (1), 141. https://doi.org/10.1186/s12951-021-00879-4

Guo, X.; Jiang, Q.; Tuccitto, A.; Chan, D.; Alqawlaq, S.; Won, G.-J.; Sivak, J. M. The AMPK-PGC-1α Signaling Axis Regulates the Astrocyte Glutathione System to Protect Against Oxidative and Metabolic Injury. *Neurobiol. Dis.* **2018,** *113*, 59–69. https://doi.org/10.1016/j.nbd.2018.02.004

Gustavsson, J.; Cederberg, C.; Sonesson, U.; Van Otterdijk, R.; Meybeck, A. *Global Food Losses and Food Waste*, 2011.

Hampel, H.; Mesulam, M.-M.; Cuello, A. C.; Khachaturian, A. S.; Vergallo, A.; Farlow, M. R.; Snyder, P. J.; Giacobini, E.; Khachaturian, Z. S. Revisiting the Cholinergic Hypothesis in Alzheimer's Disease: Emerging Evidence from Translational and Clinical Research. *J. Prevent. Alzheimer's Dis.* **2019,** *6* (1), 2–15. https://doi.org/10.14283/jpad.2018.43

Han, X.; Xu, T.; Fang, Q.; Zhang, H.; Yue, L.; Hu, G.; Sun, L. Quercetin Hinders Microglial Activation to Alleviate Neurotoxicity via the Interplay Between NLRP3 Inflammasome and Mitophagy. *Redox Biol.* **2021,** *44*, 102010. https://doi.org/10.1016/j.redox.2021.102010

Hansen, J. B.; Bilenberg, N.; Timmermann, C. A. G.; Jensen, R. C.; Frederiksen, H.; Andersson, A.-M.; Kyhl, H. B.; Jensen, T. K. Prenatal Exposure to Bisphenol A and Autistic- and ADHD-Related Symptoms in Children Aged 2 and 5 Years from the Odense Child Cohort. *Environ. Health* **2021,** *20* (1), 24. https://doi.org/10.1186/s12940-021-00709-y

Harborne, J. Flavonoids in the Environment: Structure-Activity Relationships. In *Progress in Clinical and Biological Research*; Cody, V., Middleton Jr E., Harborne, J., Eds., Vol. 213; Alan R. Liss, 1986;, pp 17–27.

Harborne, J. B.; Williams, C. A. Phytochemistry of the Genus Lavandula. In *Lavender*; CRC Press, 2002; pp 100–113.

Harwood, M.; Danielewska-Nikiel, B.; Borzelleca, J. F.; Flamm, G. W.; Williams, G. M.; Lines, T. C. A Critical Review of the Data Related to the Safety of Quercetin and Lack of Evidence of In Vivo Toxicity, Including Lack of Genotoxic/Carcinogenic Properties. *Food Chem. Toxicol.* **2007,** *45* (11), 2179–2205. https://doi.org/10.1016/j.fct.2007.05.015

Hasanzadeh, Z.; Nourazarian, A.; Nikanfar, M.; Laghousi, D.; Vatankhah, A. M.; Sadrirad, S. Evaluation of the Serum Dkk-1, Tenascin-C, Oxidative Stress Markers Levels and Wnt Signaling Pathway Genes Expression in Patients with Alzheimer's Disease. *J. Mol. Neurosci.* **2021,** *71* (4), 879–887. https://doi.org/10.1007/s12031-020-01710-9

Hendriks, J. J. A.; de Vries, H. E.; van der Pol, S. M. A.; van den Berg, T. K.; van Tol, E. A. F.; Dijkstra, C. D. Flavonoids Inhibit Myelin Phagocytosis by Macrophages; a Structure–Activity Relationship Study. *Biochem. Pharmacol.* **2003,** *65* (5), 877–885. https://doi.org/10.1016/S0006-2952 (02)01609-X

Hernandez-Baltazar, D.; Zavala-Flores, L. M.; Villanueva-Olivo, A. The 6-hydroxydopamine Model and Parkinsonian Pathophysiology: Novel Findings in an Older Model. *Neurologia (Barcelona, Spain)* **2017,** *32* (8), 533–539. https://doi.org/10.1016/j.nrl.2015.06.011

Hossain, M.; Abdal Dayem, A.; Han, J.; Yingfu, Y.; Kim, K.; Saha, S.; Yang, G.-M.; Choi, H.; Cho, S.-G. Molecular Mechanisms of the Anti-Obesity and Anti-Diabetic Properties of Flavonoids. *Int. J. Mol. Sci.* **2016,** *17*, 569. https://doi.org/10.3390/ijms17040569

IARC Working Group on the Evaluation of Carcinogenic Risks to Humans. Quercetin. In *Some Chemicals That Cause Tumours of the Kidney or Urinary Bladder in Rodents and Some Other Substances*; International Agency for Research on Cancer, 1999. https://www.ncbi.nlm.nih.gov/books/NBK402166/

Iside, C.; Scafuro, M.; Nebbioso, A.; Altucci, L. SIRT1 Activation by Natural Phytochemicals: An Overview. *Front. Pharmacol.* **2020,** *11*, 1225. https://doi.org/10.3389/fphar.2020.01225

Islam, M. S.; Quispe, C.; Hossain, R.; Islam, M. T.; Al-Harrasi, A.; Al-Rawahi, A.; Martorell, M.; Mamurova, A.; Seilkhan, A.; Altybaeva, N.; Abdullayeva, B.; Docea, A. O.; Calina, D.; Sharifi-Rad, J. Neuropharmacological Effects of Quercetin: A Literature-Based Review. *Front. Pharmacol.* **2021,** *12*, 665031–665031. PubMed. https://doi.org/10.3389/fphar.2021.665031

Islam, Md. R.; Zaman, A.; Jahan, I.; Chakravorty, R.; Chakraborty, S. In Silico QSAR Analysis of Quercetin Reveals Its Potential as Therapeutic Drug for Alzheimer's Disease. *J. Young Pharmacists* **2013,** *5* (4), 173–179. https://doi.org/10.1016/j.jyp.2013.11.005

Javadi, F.; Eghtesadi, S.; Ahmadzadeh, A.; Aryaeian, N.; Zabihiyeganeh, M.; Foroushani, A. R.; Jazayeri, S. The Effect of Quercetin on Plasma Oxidative Status, C-Reactive Protein and Blood Pressure in Women with Rheumatoid Arthritis. *Int. J. Prevent. Med.* **2014,** *5* (3), 293–301.

Ji, Y.; Liu, B.; Song, J.; Pan, R.; Cheng, J.; Su, H.; Wang, H. Particulate Matter Pollution Associated with Schizophrenia Hospital Re-Admissions: A Time-Series Study in a Coastal Chinese City. *Environ. Sci. Pollut. Res.* **2021,** *28* (41), 58355–58363. https://doi.org/10.1007/s11356-021-14816-3

Jiang, Q.; Flanagan, J. G.; Chan, D.; Sivak, J. M. The Protective Role of AMPK/SIRT1/PGC-1α Signalling in Retinal Astrocytes. *Investig. Ophthalmol. Visual Sci.* **2014,** *55* (13), 2270.

Jiang, W.; Luo, T.; Li, S.; Zhou, Y.; Shen, X.-Y.; He, F.; Xu, J.; Wang, H.-Q. Quercetin Protects Against Okadaic Acid-Induced Injury via MAPK and PI3K/Akt/GSK3β Signaling Pathways in HT22 Hippocampal Neurons. *PloS One* **2016,** *11* (4), e0152371. https://doi.org/10.1371/journal.pone.0152371

Johnson, C. O.; Nguyen, M.; Roth, G. A.; Nichols, E.; Alam, T.; Abate, D.; Abd-Allah, F.; Abdelalim, A.; Abraha, H. N.; Abu-Rmeileh, N. M.; Adebayo, O. M.; Adeoye, A. M.; Agarwal, G.; Agrawal, S.; Aichour, A. N.; Aichour, I.; Aichour, M. T. E.; Alahdab, F.; Ali, R. et al. Global, Regional, and National Burden of Stroke, 1990–2016: A Systematic Analysis for the Global Burden of Disease Study 2016. *Lancet Neurol.* **2019,** *18* (5), 439–458. https://doi.org/10.1016/S1474-4422 (19)30034-1

Joshi, A. U.; Minhas, P. S.; Liddelow, S. A.; Haileselassie, B.; Andreasson, K. I.; Dorn, G. W.; Mochly-Rosen, D. Fragmented Mitochondria Released from Microglia Trigger A1 Astrocytic Response and Propagate Inflammatory Neurodegeneration. *Nat. Neurosci.* **2019,** *22* (10), 1635–1648. https://doi.org/10.1038/s41593-019-0486-0

Kang, Y. J.; Tan, H.-Y.; Lee, C. Y.; Cho, H. An Air Particulate Pollutant Induces Neuroinflammation and Neurodegeneration in Human Brain Models. *Adv. Sci.* **2021,** *8* (21), 2101251. https://doi.org/10.1002/advs.202101251

Karimipour, M.; Rahbarghazi, R.; Tayefi, H.; Shimia, M.; Ghanadian, M.; Mahmoudi, J.; Bagheri, H. S. Quercetin Promotes Learning and Memory Performance Concomitantly with Neural Stem/Progenitor Cell Proliferation and Neurogenesis in the Adult Rat Dentate Gyrus. *Int. J. Dev. Neurosci.* **2019,** *74*, 18–26. https://doi.org/10.1016/j.ijdevneu.2019.02.005

Kasdagli, M.-I.; Katsouyanni, K.; Dimakopoulou, K.; Samoli, E. Air Pollution and Parkinson's Disease: A Systematic Review and Meta-Analysis Up to 2018. *Int. J.of Hygiene Environ. Health* **2019,** *222* (3), 402–409. https://doi.org/10.1016/j.ijheh.2018.12.006

Kaur, I.; Behl, T.; Aleya, L.; Rahman, Md. H.; Kumar, A.; Arora, S.; Akter, R. Role of Metallic Pollutants in Neurodegeneration: Effects of Aluminum, Lead, Mercury, and Arsenic in Mediating Brain Impairment Events and Autism Spectrum Disorder. *Environ. Sci. Pollut. Res.* **2021,** *28* (8), 8989–9001. https://doi.org/10.1007/s11356-020-12255-0

Kaur, P.; Sharma, M. A Smart and Promising Neurological Disorder Diagnostic System. In *Intelligent Data Analysis*; John Wiley & Sons, Ltd., 2020; pp. 241–264. https://doi.org/10.1002/9781119544487.ch12

Kawabata, K.; Mukai, R.; Ishisaka, A. Quercetin and Related Polyphenols: New Insights and Implications for Their Bioactivity and Bioavailability. *Food Funct.* **2015,** *6* (5), 1399–1417. https://doi.org/10.1039/C4FO01178C

Kempuraj, D.; Madhappan, B.; Christodoulou, S.; Boucher, W.; Cao, J.; Papadopoulou, N.; Cetrulo, C. L.; Theoharides, T. C. Flavonols Inhibit Proinflammatory Mediator Release, Intracellular Calcium Ion Levels and Protein Kinase C Theta Phosphorylation in Human Mast Cells. *Br. J. Pharmacol.* **2005,** *145* (7), 934–944. https://doi.org/10.1038/sj.bjp.0706246

Kempuraj, D.; Thangavel, R.; Natteru, P. A.; Selvakumar, G. P.; Saeed, D.; Zahoor, H.; Zaheer, S.; Iyer, S. S.; Zaheer, A. Neuroinflammation Induces Neurodegeneration. *J. Neurol. Neurosurg. Spine* **2016,** *1* (1), 1003. PubMed.

Kerschensteiner, M.; Stadelmann, C.; Dechant, G.; Wekerle, H.; Hohlfeld, R. Neurotrophic Cross-Talk Between the Nervous and Immune Systems: Implications for Neurological Diseases. *Ann. Neurol.* **2003,** *53* (3), 292–304. https://doi.org/10.1002/ana.10446.

Keservani, R. K.; Sharma, A. K. Flavonoids: Emerging Trends and Potential Health Benefits. *J. Chinese Pharma. Sci.* **2014,** *23* (12), 815.

Keservani, R. K.; Sharma, A. K.; Kesharwani, R. K. Nutraceutical and Functional Foods for Cardiovascular Health. In *Food Process Engineering*; Apple Academic Press, CRC Press, 2016a; pp 257–278; 291–312. ISBN: 9781771884020.

Keservani, R. K.; Sharma, A. K.; Kesharwani, R. K. Medicinal Effect of Nutraceutical Fruits for the Cognition and Brain Health. *Scientifica* **2016b,** Article ID 3109254, 10 pp. DOI: 10.1155/2016/3109254.

Keservani, R. K.; Sharma, A. K.; Kesharwani, R. K. (Eds.). *Nutraceuticals and Dietary Supplements: Applications in Health Improvement and Disease Management*; CRC Press, 2020. ISBN: 9781771888738.

Khan, A.; Plana-Ripoll, O.; Antonsen, S.; Brandt, J.; Geels, C.; Landecker, H.; Sullivan, P. F.; Pedersen, C. B.; Rzhetsky, A. Environmental Pollution Is Associated with Increased Risk of Psychiatric Disorders in the US and Denmark. *PLOS Biol.* **2019,** *17* (8), e3000353. https://doi.org/10.1371/journal.pbio.3000353

Khan, H.; Marya, Amin, S.; Kamal, M. A.; Patel, S. Flavonoids as Acetylcholinesterase Inhibitors: Current Therapeutic Standing and Future Prospects. *Biomed. Pharmacother.* **2018,** *101*, 860–870. https://doi.org/10.1016/j.biopha.2018.03.007

Khan, H.; Ullah, H.; Aschner, M.; Cheang, W. S.; Akkol, E. K. Neuroprotective Effects of Quercetin in Alzheimer's Disease. *Biomolecules* **2019,** *10* (1), 59. PubMed. https://doi.org/10.3390/biom10010059

Khedkar, R.; Singh, K. Food Industry Waste: A Panacea or Pollution Hazard? In *Paradigms in Pollution Prevention*, Jindal, T., Ed.; Springer International Publishing, 2018; pp 35–47. https://doi.org/10.1007/978-3-319-58415-7_3

Khoo, H. E.; Azlan, A.; Tang, S. T.; Lim, S. M. Anthocyanidins and Anthocyanins: Colored Pigments as Food, Pharmaceutical Ingredients, and the Potential Health Benefits. *Food Nutr. Res.* **2017**. https://foodandnutritionresearch.net/index.php/fnr/article/view/1257

Kim, H. P.; Son, K. H.; Chang, H. W.; Kang, S. S. Anti-Inflammatory Plant Flavonoids and Cellular Action Mechanisms. *J. Pharmacol. Sci.* **2004,** *96* (3), 229–245. https://doi.org/10.1254/jphs.crj04003x

Kitazawa, M.; Cheng, D.; LaFerla, F. M. Chronic Copper Exposure Exacerbates Both Amyloid and Tau Pathology and Selectively Dysregulates cdk5 in a Mouse Model of AD. *J. Neurochem.* **2009,** *108* (6), 1550–1560. https://doi.org/10.1111/j.1471-4159.2009.05901.x

Kothari, D.; Lee, W.-D.; Kim, S.-K. Allium Flavonols: Health Benefits, Molecular Targets, and Bioavailability. *Antioxid. (Basel, Switzerland)* **2020,** *9* (9), 888. PubMed. https://doi.org/10.3390/antiox9090888

Krishnamoorthy, Y.; Sarveswaran, G.; Sivaranjini, K.; Sakthivel, M.; Majella, M. G.; Kumar, S. G. Association between Indoor Air Pollution and Cognitive Impairment among Adults in

Rural Puducherry, South India. *Journal of Neurosciences in Rural Practice*, *9* (4), 529–534. PubMed. https://doi.org/10.4103/jnrp.jnrp_123_18

Kumar, S.; Pandey, A. K. Chemistry and Biological Activities of Flavonoids: An Overview. *The Scientific World Journal*, **2013,** 162750. https://doi.org/10.1155/2013/162750

Kwon, H. S.; Koh, S.-H. Neuroinflammation in Neurodegenerative Disorders: The Roles of Microglia and Astrocytes. *Transl. Neurodegen.* **2020,** *9* (1), 42. https://doi.org/10.1186/s40035-020-00221-2

Lai, W.-F.; Wong, W.-T. Design and Optimization of Quercetin-Based Functional Foods. *Crit. Rev. Food Sci. Nutr.* **2021,** 1–17. https://doi.org/10.1080/10408398.2021.1913569

Lakhanpal, P.; Rai, D. K. Quercetin: A Versatile Flavonoid. *Internet J. Med. Update——EJOURNAL* **2007,** *2* (2), Article 2. https://doi.org/10.4314/ijmu.v2i2.39851

Landrigan, P. J.; Fuller, R.; Acosta, N. J. R.; Adeyi, O.; Arnold, R.; Basu, N. (Nil), Baldé, A. B.; Bertollini, R.; Bose-O'Reilly, S.; Boufford, J. I.; Breysse, P. N.; Chiles, T.; Mahidol, C.; Coll-Seck, A. M.; Cropper, M. L.; Fobil, J.; Fuster, V.; Greenstone, M.; Haines, A. et al. The Lancet Commission on Pollution and Health. *Lancet* **2018,** *391* (10119), 462–512. https://doi.org/10.1016/S0140-6736 (17)32345-0

Lavigne, E.; Yasseen, A. S.; Stieb, D. M.; Hystad, P.; van Donkelaar, A.; Martin, R. V.; Brook, J. R.; Crouse, D. L.; Burnett, R. T.; Chen, H.; Weichenthal, S.; Johnson, M.; Villeneuve, P. J.; Walker, M. Ambient Air Pollution and Adverse Birth Outcomes: Differences by Maternal Comorbidities. *Environ. Res.* **2016,** *148*, 457–466. https://doi.org/10.1016/j.envres.2016.04.026

Lazarevic-Pasti, T.; Leskovac, A.; Momic, T.; Petrovic, S.; Vasic, V. Modulators of Acetylcholinesterase Activity: From Alzheimer's Disease to Anti-Cancer Drugs. *Curr. Med. Chem.* **2017,** *24* (30), 3283–3309. https://doi.org/10.2174/0929867324666170705123509

Lee, H.; Myung, W.; Kim, D. K.; Kim, S. E.; Kim, C. T.; Kim, H. Short-Term Air Pollution Exposure Aggravates Parkinson's Disease in a Population-Based Cohort. *Sci. Rep.* **2017,** *7*. https://doi.org/10.1038/srep44741

Lee, J.; Mitchell, A. E. Pharmacokinetics of Quercetin Absorption from Apples and Onions in Healthy Humans. *J. Agric. Food Chem.* **2012,** *60* (15), 3874–3881. https://doi.org/10.1021/jf3001857

Lee, K. K.; Miller, M. R.; Shah, A. S. V. Air Pollution and Stroke. *J. Stroke* **2018,** *20* (1), 2–11. PubMed. https://doi.org/10.5853/jos.2017.02894

Lee, M.; McGeer, E. G.; McGeer, P. L. Quercetin, Not Caffeine, Is a Major Neuroprotective Component in Coffee. *Neurobiol. Aging* **2016,** *46*, 113–123. https://doi.org/10.1016/j.neurobiolaging.2016.06.015

Lee, Y. M.; Seong, M. J.; Lee, J. W.; Lee, Y. K.; Kim, T. M.; Nam, S.-Y.; Kim, D. J.; Yun, Y. W.; Kim, T. S.; Han, S. Y.; Hong, J. T. Estrogen Receptor Independent Neurotoxic Mechanism of Bisphenol A, an Environmental Estrogen. *J. Vet. Sci.* **2007,** *8* (1), 27–38. https://doi.org/10.4142/jvs.2007.8.1.27

Lesjak, M.; Beara, I.; Simin, N.; Pintać, D.; Majkić, T.; Bekvalac, K.; Orčić, D.; Mimica-Dukić, N. Antioxidant and Anti-Inflammatory Activities of Quercetin and Its Derivatives. *J. Funct. Foods* **2018,** *40*, 68–75. https://doi.org/10.1016/j.jff.2017.10.047

Levesque, S.; Taetzsch, T.; Lull, M. E.; Kodavanti, U.; Stadler, K.; Wagner, A.; Johnson, J. A.; Duke, L.; Kodavanti, P.; Surace, M. J.; Block, M. L. Diesel Exhaust Activates and Primes Microglia: Air Pollution, Neuroinflammation, and Regulation of Dopaminergic Neurotoxicity. *Environ. Health Perspect.* **2011,** *119* (8), 1149–1155. https://doi.org/10.1289/ehp.1002986

Leyton, L.; Hott, M.; Acuña, F.; Caroca, J.; Nuñez, M.; Martin, C.; Zambrano, A.; Concha, M. I.; Otth, C. Nutraceutical Activators of AMPK/Sirt1 Axis Inhibit Viral Production and Protect Neurons from Neurodegenerative Events Triggered During HSV-1 Infection. *Virus Res.* **2015,** *205*, 63–72. https://doi.org/10.1016/j.virusres.2015.05.015

Li, C.-Y.; Li, C.-H.; Martini, S.; Hou, W.-H. Association Between Air Pollution and Risk of Vascular Dementia: A Multipollutant Analysis in Taiwan. *Environ. Int.* **2019,** *133*, 105233. https://doi.org/10.1016/j.envint.2019.105233

Li, H.; Chen, F.-J.; Yang, W.-L.; Qiao, H.-Z.; Zhang, S.-J. Quercetin Improves Cognitive Disorder in Aging Mice by Inhibiting NLRP3 Inflammasome Activation. *Food Funct.* **2021,** *12* (2), 717–725. https://doi.org/10.1039/D0FO01900C

Li, X.; Huang, S.; Jiao, A.; Yang, X.; Yun, J.; Wang, Y.; Xue, X.; Chu, Y.; Liu, F.; Liu, Y.; Ren, M.; Chen, X.; Li, N.; Lu, Y.; Mao, Z.; Tian, L.; Xiang, H. Association Between Ambient Fine Particulate Matter and Preterm Birth or Term Low Birth Weight: An Updated Systematic Review and Meta-Analysis. *Environ. Pollut.* **2017,** *227*, 596–605. https://doi.org/10.1016/j.envpol.2017.03.055

Li, Y.; Yao, J.; Han, C.; Yang, J.; Chaudhry, M. T.; Wang, S.; Liu, H.; Yin, Y. Quercetin, Inflammation and Immunity. *Nutrients* **2016,** *8* (3), 167–167. PubMed. https://doi.org/10.3390/nu8030167

Li, Y.-L.; Guo, H.; Zhao, Y.-Q.; Li, A.-F.; Ren, Y.-Q.; Zhang, J.-W. Quercetin Protects Neuronal Cells from Oxidative Stress and Cognitive Degradation Induced by Amyloid β-Peptide Treatment. *Mol. Med. Rep.* **2017,** *16* (2), 1573–1577. https://doi.org/10.3892/mmr.2017.6704

Liddelow, S. A.; Guttenplan, K. A.; Clarke, L. E.; Bennett, F. C.; Bohlen, C. J.; Schirmer, L.; Bennett, M. L.; Münch, A. E.; Chung, W.-S.; Peterson, T. C.; Wilton, D. K.; Frouin, A.; Napier, B. A.; Panicker, N.; Kumar, M.; Buckwalter, M. S.; Rowitch, D. H.; Dawson, V. L.; Dawson, T. M. et al. Neurotoxic Reactive Astrocytes Are Induced by Activated Microglia. *Nature* **2017,** *541* (7638), 481–487. https://doi.org/10.1038/nature21029

Liu, R.; Young, M. T.; Chen, J. C.; Kaufman, J. D.; Chen, H. Ambient Air Pollution Exposures and Risk of Parkinson Disease. *Environ. Health Perspect.* **2016,** *124* (11), 1759–1765. https://doi.org/10.1289/EHP135

Liu, X.; Yu, Z.; Huang, X.; Gao, Y.; Wang, X.; Gu, J.; Xue, S. Peroxisome Proliferator-Activated Receptor γ (PPARγ) Mediates the Protective Effect of Quercetin Against Myocardial Ischemia-Reperfusion Injury via Suppressing the NF-κB Pathway. *Am. J. Transl. Res.* **2016,** *8* (12), 5169–5186.

Liu, Z.; Zhou, T.; Ziegler, A. C.; Dimitrion, P.; Zuo, L. Oxidative Stress in Neurodegenerative Diseases: From Molecular Mechanisms to Clinical Applications. *Oxid. Med. Cell. Longevity* **2017,** *2017*, 2525967. https://doi.org/10.1155/2017/2525967

Livingston, G.; Sommerlad, A.; Orgeta, V.; Costafreda, S. G.; Huntley, J.; Ames, D.; Ballard, C.; Banerjee, S.; Burns, A.; Cohen-Mansfield, J.; Cooper, C.; Fox, N.; Gitlin, L. N.; Howard, R.; Kales, H. C.; Larson, E. B.; Ritchie, K.; Rockwood, K.; Sampson, E. L. et al. Dementia Prevention, Intervention, and Care. *Lancet* **2017,** *390* (10113), 2673–2734. https://doi.org/10.1016/S0140-6736 (17)31363-6

Lu, C.-W.; Lin, T.-Y.; Wang, S.-J. Quercetin Inhibits Depolarization-Evoked Glutamate Release in Nerve Terminals from Rat Cerebral Cortex. *NeuroToxicology* **2013,** *39*, 1–9. https://doi.org/10.1016/j.neuro.2013.07.009

Lu, D.; Yu, L.; Li, M.; Zhai, Q.; Tian, F.; Chen, W. Behavioral Disorders Caused by Nonylphenol and Strategies for Protection. *Chemosphere* **2021,** *275*, 129973. https://doi.org/10.1016/j.chemosphere.2021.129973

Lu, N. T.; Crespi, C. M.; Liu, N. M.; Vu, J. Q.; Ahmadieh, Y.; Wu, S.; Lin, S.; McClune, A.; Durazo, F.; Saab, S.; Han, S.; Neiman, D. C.; Beaven, S.; French, S. W. A Phase I Dose Escalation Study Demonstrates Quercetin Safety and Explores Potential for Bioflavonoid Antivirals in Patients with Chronic Hepatitis C. *Phytother. Res.* **2016,** *30* (1), 160–168. https://doi.org/10.1002/ptr.5518

Lu, X.; Xu, S.; Zhang, Y.; Gao, D.; Yang, H.; Song, J.; Wang, L.; Zhang, Q.; Shang, N.; Niu, Q. Longitudinal Study of the Effects of Occupational Aluminium Exposure on Workers' Cognition. *Chemosphere* **2021,** *271*, 129569. https://doi.org/10.1016/j.chemosphere.2021.129569

Magalingam, K. B.; Radhakrishnan, A.; Haleagrahara, N. Protective Effects of Quercetin Glycosides, Rutin, and Isoquercetrin Against 6-Hydroxydopamine (6-OHDA)-Induced Neurotoxicity in Rat Pheochromocytoma (PC-12) Cells. *Int. J. Immunopathol. Pharmacol.* **2016,** *29* (1), 30–39. https://doi.org/10.1177/0394632015613039

Makino, R.; Takano, K.; Kita, K.; Nishimukai, M. Influence of Long-Term Feeding of High-Fat Diet on Quercetin and Fat Absorption from the Small Intestine in Lymph Duct-Cannulated Rats. *Biosci. Biotechnol. Biochem.* **2018,** *82* (11), 2007–2011. https://doi.org/10.1080/09168451.2018.1498726

Mamani-Matsuda, M.; Kauss, T.; Al-Kharrat, A.; Rambert, J.; Fawaz, F.; Thiolat, D.; Moynet, D.; Coves, S.; Malvy, D.; Mossalayi, M. D. Therapeutic and Preventive Properties of Quercetin in Experimental Arthritis Correlate with Decreased Macrophage Inflammatory Mediators. *Biochem. Pharmacol.* **2006,** *72* (10), 1304–1310. https://doi.org/10.1016/j.bcp.2006.08.001

Manach, C.; Regerat, F.; Texier, O.; Agullo, G.; Demigne, C.; Remesy, C. Bioavailability, Metabolism and Physiological Impact of 4-Oxo-Flavonoids. *Nutr. Res.* **1996,** *16* (3), 517–544. https://doi.org/10.1016/0271-5317 (96)00032-2

Manasfi, T.; Coulomb, B.; Boudenne, J.-L. Occurrence, Origin, and Toxicity of Disinfection Byproducts in Chlorinated Swimming Pools: An Overview. *Int. J. Hygiene Environ. Health* **2017,** *220* (3), 591–603. https://doi.org/10.1016/j.ijheh.2017.01.005

Manzoor, M. F.; Hussain, A.; Sameen, A.; Sahar, A.; Khan, S.; Siddique, R.; Aadil, R. M.; Xu, B. Novel Extraction, Rapid Assessment and Bioavailability Improvement of Quercetin: A Review. *Ultrason. Sonochem.* **2021,** *78*, 105686. https://doi.org/10.1016/j.ultsonch.2021.105686

Marucci, G.; Buccioni, M.; Ben, D. D.; Lambertucci, C.; Volpini, R.; Amenta, F. Efficacy of Acetylcholinesterase Inhibitors in Alzheimer's Disease. *Neuropharmacology* **2021,** *190*, 108352. https://doi.org/10.1016/j.neuropharm.2020.108352

Mazumdar, M.; Xia, W.; Hofmann, O.; Gregas, M.; Sui, S. H.; Hide, W.; Yang, T.; Needleman, H. L.; Bellinger, D. C. Prenatal Lead Levels, Plasma Amyloid β Levels, and Gene Expression in Young Adulthood. *Environ. Health Perspect.* **2012,** *120* (5), 702–707. https://doi.org/10.1289/ehp.1104474

McAnulty, S. R.; McAnulty, L. S.; Nieman, D. C.; Quindry, J. C.; Hosick, P. A.; Hudson, M. H.; Still, L.; Henson, D. A.; Milne, G. L.; Morrow, J. D.; Dumke, C. L.; Utter, A. C.; Triplett, N. T.; Dibarnardi, A. Chronic Quercetin Ingestion and EXERCISE-INDUCED Oxidative Damage and Inflammation. *Appl. Physiol. Nutr. Metabol.* **2008,** *33* (2), 254–262. https://doi.org/10.1139/H07-177

McHardy, S. F.; Wang, H.-Y. L.; McCowen, S. V.; Valdez, M. C. Recent Advances in Acetylcholinesterase Inhibitors and Reactivators: An Update on the Patent Literature (2012–2015). *Expert Opin. Therap. Patents* **2017,** *27* (4), 455–476. https://doi.org/10.1080/13543776.2017.1272571

Meade, R. M.; Fairlie, D. P.; Mason, J. M. Alpha-Synuclein Structure and Parkinson's Disease—Lessons and Emerging Principles. *Mol. Neurodegen.* **2019,** *14* (1), 29. https://doi.org/10.1186/s13024-019-0329-1

Medeiros, R.; Baglietto-Vargas, D.; LaFerla, F. M. The Role of Tau in Alzheimer's Disease and Related Disorders. *CNS Neurosci. Therap.* **2010,** *17* (5), 514–524. https://doi.org/10.1111/j.1755-5949.2010.00177.x

Mehta, A.; Prabhakar, M.; Kumar, P.; Deshmukh, R.; Sharma, P. L. Excitotoxicity: Bridge to Various Triggers in Neurodegenerative Disorders. *Eur. J. Pharmacol.* **2013,** *698* (1), 6–18. https://doi.org/10.1016/j.ejphar.2012.10.032

Mekonnen, A.; Tessema, A.; Ganewo, Z.; Haile, A. Climate Change Impacts on Household Food Security and Farmers Adaptation Strategies. *J. Agric. Food Res.* **2021,** *6*, 100197. https://doi.org/10.1016/j.jafr.2021.100197

Merelli, A.; Repetto, M.; Lazarowski, A.; Auzmendi, J. Hypoxia, Oxidative Stress, and Inflammation: Three Faces of Neurodegenerative Diseases. *J. Alzheimer's Dis.* **2021,** *82* (s1), S109–S126. https://doi.org/10.3233/JAD-201074

Meyer, P.-F.; Tremblay-Mercier, J.; Leoutsakos, J.; Madjar, C.; Lafaille-Maignan, M.-É.; Savard, M.; Rosa-Neto, P.; Poirier, J.; Etienne, P.; Breitner, J.; PREVENT-AD Research Group. INTREPAD: A Randomized Trial of Naproxen to Slow Progress of Presymptomatic Alzheimer Disease. *Neurology* **2019,** *92* (18), e2070–e2080. https://doi.org/10.1212/WNL.0000000000007232

Mezzaroba, L.; Alfieri, D. F.; Colado Simão, A. N.; Vissoci Reiche, E. M. The Role of Zinc, Copper, Manganese and Iron in Neurodegenerative Diseases. *NeuroToxicology* **2019,** *74*, 230–241. https://doi.org/10.1016/j.neuro.2019.07.007

Mierziak, J.; Kostyn, K.; Kulma, A. Flavonoids as Important Molecules of Plant Interactions with the Environment. *Molecules (Basel, Switzerland)* **2014,** *19* (10), 16240–16265. PubMed. https://doi.org/10.3390/molecules191016240

Mirza, A.; King, A.; Troakes, C.; Exley, C. Aluminium in Brain Tissue in Familial Alzheimer's Disease. *J. Trace Elements Med. Biol.* **2017,** *40*, 30–36. https://doi.org/10.1016/j.jtemb.2016.12.001

Mohos, V.; Fliszár-Nyúl, E.; Ungvári, O.; Kuffa, K.; Needs, P. W.; Kroon, P. A.; Telbisz, Á.; Özvegy-Laczka, C.; Poór, M. Inhibitory Effects of Quercetin and Its Main Methyl, Sulfate, and Glucuronic Acid Conjugates on Cytochrome P450 Enzymes, and on OATP, BCRP and MRP2 Transporters. *Nutrients* **2020,** *12* (8), 2306. https://doi.org/10.3390/nu12082306

Moosavi, F.; Hosseini, R.; Saso, L.; Firuzi, O. Modulation of Neurotrophic Signaling Pathways by Polyphenols. *Drug Design Dev. Ther.* **2016,** *10*, 23–42. https://doi.org/10.2147/DDDT.S96936

Moser, V. C.; Phillips, P. M.; Levine, A. B.; McDaniel, K. L.; Sills, R. C.; Jortner, B. S.; Butt, M. T. Neurotoxicity Produced by Dibromoacetic Acid in Drinking Water of Rats. *Toxicol. Sci.* **2004,** *79* (1), 112–122. https://doi.org/10.1093/toxsci/kfh081

Murphy, T. H.; Miyamoto, M.; Sastre, A.; Schnaar, R. L.; Coyle, J. T. Glutamate Toxicity in a Neuronal Cell Line Involves Inhibition of Cystine Transport Leading to Oxidative Stress. *Neuron* **1989,** *2* (6), 1547–1558. https://doi.org/10.1016/0896-6273 (89)90043-3

Nabavi, S. F.; Russo, G. L.; Daglia, M.; Nabavi, S. M. Role of Quercetin as an Alternative for Obesity Treatment: You Are What You Eat! *Food Chem.* **2015,** *179*, 305–310. https://doi.org/10.1016/j.foodchem.2015.02.006

Nair, M. P.; Mahajan, S.; Reynolds, J. L.; Aalinkeel, R.; Nair, H.; Schwartz, S. A.; Kandaswami, C. The Flavonoid Quercetin Inhibits Proinflammatory Cytokine (Tumor Necrosis Factor

Alpha) Gene Expression in Normal Peripheral Blood Mononuclear Cells via Modulation of the NF-κβ System. *Clin. Vaccine Immunol.* **2006,** *13* (3), 319–328. https://doi.org/10.1128/CVI.13.3.319-328.2006

Nakajima, K.; Niisato, N.; Marunaka, Y. Quercetin Stimulates NGF-Induced Neurite Outgrowth in PC12 Cells via Activation of Na (+)/K (+)/2Cl (−) Co-transporter. *Cell. Physiol. Biochem.* **2011,** *28* (1), 147–156. https://doi.org/10.1159/000331723

Narayana, K. R.; Reddy, M. S.; Chaluvadi, M.; Krishna, D. Bioflavonoids Classification, Pharmacological, Biochemical Effects and Therapeutic Potential. *Indian J. Pharmacol.* **2001,** *33* (1), 2–16.

Naushad, M.; Durairajan, S. S. K.; Bera, A. K.; Senapati, S.; Li, M. Natural Compounds with Anti-BACE1 Activity as Promising Therapeutic Drugs for Treating Alzheimer's Disease. *Planta Medica* **2019,** *85* (17), 1316–1325. https://doi.org/10.1055/a-1019-9819

Needham, L. L.; Barr, D. B.; Caudill, S. P.; Pirkle, J. L.; Turner, W. E.; Osterloh, J.; Jones, R. L.; Sampson, E. J. Concentrations of Environmental Chemicals Associated with Neurodevelopmental Effects in U.S. Population. *Neurotoxicology* **2005,** *26* (4), 531–545. https://doi.org/10.1016/j.neuro.2004.09.005

Nishihira, J.; Nishimura, M.; Kurimoto, M.; Kagami-Katsuyama, H.; Hattori, H.; Nakagawa, T.; Muro, T.; Kobori, M. The Effect of 24-Week Continuous Intake of Quercetin-Rich Onion on age-Related Cognitive Decline in Healthy Elderly People: A Randomized, Double-Blind, Placebo-Controlled, Parallel-Group Comparative Clinical Trial. *J. Clin. Biochem. Nutr.* **2021,** *69* (2), 203–215. https://doi.org/10.3164/jcbn.21-17

Obrador, E.; Salvador, R.; López-Blanch, R.; Jihad-Jebbar, A.; Vallés, S. L.; Estrela, J. M. Oxidative Stress, Neuroinflammation and Mitochondria in the Pathophysiology of Amyotrophic Lateral Sclerosis. *Antioxidants* **2020,** *9* (9), 901. https://doi.org/10.3390/antiox9090901

Oddo, S.; Billings, L.; Kesslak, J. P.; Cribbs, D. H.; LaFerla, F. M. Abeta Immunotherapy Leads to Clearance of Early, But Not Late, Hyperphosphorylated Tau Aggregates via the Proteasome. *Neuron* **2004,** *43* (3), 321–332. https://doi.org/10.1016/j.neuron.2004.07.003

Orhan, I.; Kartal, M.; Tosun, F.; Şener, B. Screening of Various Phenolic Acids and Flavonoid Derivatives for their Anticholinesterase Potential. *Zeitschrift Für Naturforschung C* **2007,** *62* (11–12), 829–832. https://doi.org/10.1515/znc-2007-11-1210

Ouellette, J.; Lacoste, B. From Neurodevelopmental to Neurodegenerative Disorders: The Vascular Continuum. *Front. Aging Neurosci.* **2021**. https://doi.org/10.3389/fnagi.2021.749026

Padureanu, R.; Albu, C. V.; Mititelu, R. R.; Bacanoiu, M. V.; Docea, A. O.; Calina, D.; Padureanu, V.; Olaru, G.; Sandu, R. E.; Malin, R. D.; Buga, A.-M. Oxidative Stress and Inflammation Interdependence in Multiple Sclerosis. *J. Clin. Med.* **2019,** *8* (11), 1815. PubMed. https://doi.org/10.3390/jcm8111815

Palacios, N. Air Pollution and Parkinson's Disease—Evidence and Future Directions. *Rev. Environ. Health* **2017,** *32* (4), 303–313. https://doi.org/10.1515/reveh-2017-0009

Palacios, N.; Fitzgerald, K. C.; Hart, J. E.; Weisskopf, M. G.; Schwarzschild, M. A.; Ascherio, A.; Laden, F. Particulate Matter and Risk of Parkinson Disease in a Large Prospective Study of Women. *Environ. Health: Glob. Access Sci. Source* **2014,** *13* (1). https://doi.org/10.1186/1476-069X-13-80

Palacios, N.; Fitzgerald, K. C.; Hart, J. E.; Weisskopf, M.; Schwarzschild, M. A.; Ascherio, A.; Laden, F. Air Pollution and Risk of Parkinson's Disease in a Large Prospective Study of Men. *Environ. Health Perspect.* **2017,** *125* (8). https://doi.org/10.1289/EHP259

Palazzolo, G.; Horvath, P.; Zenobi-Wong, M. The Flavonoid Isoquercitrin Promotes Neurite Elongation by Reducing RhoA Activity. *PLoS ONE*, *7* (11), e49979. https://doi.org/10.1371/journal.pone.0049979

Panche, A. N.; Diwan, A. D.; Chandra, S. R. Flavonoids: An Overview. *J. Nutr. Sci.* **2016,** *5*, e47. PubMed Central. https://doi.org/10.1017/jns.2016.41

Park, M. W.; Cha, H. W.; Kim, J.; Kim, J. H.; Yang, H.; Yoon, S.; Boonpraman, N.; Yi, S. S.; Yoo, I. D.; Moon, J.-S. NOX4 Promotes Ferroptosis of Astrocytes by Oxidative Stress-Induced Lipid Peroxidation via the Impairment of Mitochondrial Metabolism in Alzheimer's Diseases. *Redox Biol.* **2021,** *41*, 101947. https://doi.org/10.1016/j.redox.2021.101947

Park, S.-H.; Lee, C. W.; Cho, S. M.; Lee, H.; Park, H.; Lee, J.; Lee, J. H. Crystal Structure and Enzymatic Properties of Chalcone Isomerase from the Antarctic Vascular Plant Deschampsia Antarctica Desv. *PLoS ONE* **2018,** *13* (2), e0192415. https://doi.org/10.1371/journal.pone.0192415

Pateiro, M.; Gómez, B.; Munekata, P. E. S.; Barba, F. J.; Putnik, P.; Kovačević, D. B.; Lorenzo, J. M. Nanoencapsulation of Promising Bioactive Compounds to Improve Their Absorption, Stability, Functionality and the Appearance of the Final Food Products. *Molecules (Basel, Switzerland)* **2021,** *26* (6), 1547. PubMed. https://doi.org/10.3390/molecules26061547

Paul, K. C.; Haan, M.; Mayeda, E. R.; Ritz, B. R. Ambient Air Pollution, Noise, and Late-Life Cognitive Decline and Dementia Risk. *Annu. Rev. Public Health* **2019,** *40*, 203–220. PubMed. https://doi.org/10.1146/annurev-publhealth-040218-044058

Pfeuffer, M.; Auinger, A.; Bley, U.; Kraus-Stojanowic, I.; Laue, C.; Winkler, P.; Rüfer, C. E.; Frank, J.; Bösch-Saadatmandi, C.; Rimbach, G.; Schrezenmeir, J. Effect of Quercetin on Traits of the Metabolic Syndrome, Endothelial Function and Inflammation in Men with Different APOE Isoforms. *Nutr. Metabol. Cardiovasc. Dis.* **2013,** *23* (5), 403–409. https://doi.org/10.1016/j.numecd.2011.08.010

Picciotto, M. R.; Higley, M. J.; Mineur, Y. S. Acetylcholine as a Neuromodulator: Cholinergic Signaling Shapes Nervous System Function and Behavior. *Neuron* **2012,** *76* (1), 116–129. https://doi.org/10.1016/j.neuron.2012.08.036

Pontifex, M. G.; Malik, M. M. A. H.; Connell, E.; Müller, M.; Vauzour, D. Citrus Polyphenols in Brain Health and Disease: Current Perspectives. *Front. Neurosci.* **2021,** *15*. https://doi.org/10.3389/fnins.2021.640648

Power, M. C.; Weisskopf, M. G.; Alexeeff, S. E.; Coull, B. A.; Spiro, A., 3rd; Schwartz, J. Traffic-Related Air Pollution and Cognitive Function in a Cohort of Older Men. *Environ. Health Perspect.* **2011,** *119* (5), 682–687. PubMed. https://doi.org/10.1289/ehp.1002767

PubChem. *PubChem Compound Summary for CID* **72281,** *Hesperetin*. Bethesda (MD): National Library of Medicine (US), National Center for Biotechnology Information; **2004,** n.d.-a. https://pubchem.ncbi.nlm.nih.gov/compound/72281 (accessed 12 Jan 2022).

PubChem. *PubChem Compound Summary for CID* **128861,** *Cyanidin*. Bethesda (MD): National Library of Medicine (US), National Center for Biotechnology Information; **2004,** n.d.-b. https://pubchem.ncbi.nlm.nih.gov/compound/128861 (accessed 12 Jan 2022).

PubChem. *PubChem Compound Summary for CID* **182232,** *(+)-Epicatechin*. Bethesda (MD): National Library of Medicine (US), National Center for Biotechnology Information; **2004,** n.d.-c. https://pubchem.ncbi.nlm.nih.gov/compound/182232 (accessed 12 Jan 2022).

PubChem. *PubChem Compound Summary for CID* **439533,** *Taxifolin*. Bethesda (MD): National Library of Medicine (US), National Center for Biotechnology Information; **2004,** n.d.-d. https://pubchem.ncbi.nlm.nih.gov/compound/439533 (accessed 12 Jan 2022).

PubChem. *PubChem Compound Summary for CID* **5280343,** *Quercetin*. Bethesda (MD): National Library of Medicine (US), National Center for Biotechnology Information; **2004,** n.d.-e. https://pubchem.ncbi.nlm.nih.gov/compound/5280343 (accessed 12 Jan 2022).

PubChem. *PubChem Compound Summary for CID 5280443, Apigenin.* Bethesda (MD): National Library of Medicine (US), National Center for Biotechnology Information; **2004,** n.d.-f. https://pubchem.ncbi.nlm.nih.gov/compound/5280443 (accessed 12 Jan 2022).

Rahdar, A.; Hasanein, P.; Bilal, M.; Beyzaei, H.; Kyzas, G. Z. Quercetin-Loaded F127 Nanomicelles: Antioxidant Activity and Protection Against Renal Injury Induced by Gentamicin in Rats. *Life Sci.* **2021,** *276*, 119420. https://doi.org/10.1016/j.lfs.2021.119420

Rajasärkkä, J.; Pernica, M.; Kuta, J.; Lašňák, J.; Šimek, Z.; Bláha, L. Drinking Water Contaminants from Epoxy Resin-Coated Pipes: A Field Study. *Water Res.* **2016,** *103*, 133–140. https://doi.org/10.1016/j.watres.2016.07.027

Ransohoff, R. M. How Neuroinflammation Contributes to Neurodegeneration. *Science* **2016,** *353* (6301), 777–783. https://doi.org/10.1126/science.aag2590

Razavi, S.; Nazem, G.; Mardani, M.; Esfandiari, E.; Salehi, H.; Esfahani, S. H. Z. Neurotrophic Factors and Their Effects in the Treatment of Multiple Sclerosis. *Adv. Biomed. Res.* **2015,** *4*, 53. https://doi.org/10.4103/2277-9175.151570

Regitz, C.; Dußling, L. M.; Wenzel, U. Amyloid-Beta ($A\beta_{1-42}$)-Induced Paralysis in Caenorhabditis Elegans Is Inhibited by the Polyphenol Quercetin Through Activation of Protein Degradation Pathways. *Mol. Nutr. Food Res.* **2014,** *58* (10), 1931–1940. https://doi.org/10.1002/mnfr.201400014

Rice-Evans, C. A.; Miller, N. J.; Bolwell, P. G.; Bramley, P. M.; Pridham, J. B. The Relative Antioxidant Activities of Plant-Derived Polyphenolic Flavonoids. *Free Radic. Res.* **1995,** *22* (4), 375–383. https://doi.org/10.3109/10715769509145649

Riskiyanti, D.; Agustiningsih, D. Effects of Quercetin on the Nicotine-Induced Oxidative Status in Male Wistar Rats: Study on C-Reactive Protein (CRP) and Malondialdehyde (MDA) Concentrations. *J. Med. Sci.* **2016,** *48*, 81–88. https://doi.org/10.19106/JMedSci004802201602

Rius-Pérez, S.; Torres-Cuevas, I.; Millán, I.; Ortega, Á. L.; Pérez, S. PGC-1α, Inflammation, and Oxidative Stress: An Integrative View in Metabolism. *Oxid. Med. Cell. Longevity* **2020,** *2020*, 1452696–1452696. PubMed. https://doi.org/10.1155/2020/1452696

Rodríguez-Arce, E.; Saldías, M. Antioxidant Properties of Flavonoid Metal Complexes and Their Potential Inclusion in the Development of Novel Strategies for the Treatment Against Neurodegenerative Diseases. *Biomed. Pharmacother.* **2021,** *143*, 112236. https://doi.org/10.1016/j.biopha.2021.112236

Rondeau, V.; Commenges, D.; Jacqmin-Gadda, H.; Dartigues, J.-F. Relation Between Aluminum Concentrations in Drinking Water and Alzheimer's Disease: An 8-Year Follow-Up Study. *Am. J. Epidemiol.* **2000,** *152* (1), 59–66.

Russo, M. V.; McGavern, D. B. Inflammatory Neuroprotection Following Traumatic Brain Injury. *Science* **2016,** *353* (6301), 783–785. PubMed. https://doi.org/10.1126/science.aaf6260

Saigo, T.; Wang, T.; Watanabe, M.; Tohge, T. Diversity of Anthocyanin and Proanthocyanin Biosynthesis in Land Plants. *Curr. Opin. Plant Biol.* **2020,** *55*, 93–99. https://doi.org/10.1016/j.pbi.2020.04.001

Saini, N.; Gahlawat, S. K.; Lather, V. Flavonoids: A Nutraceutical and Its Role as Anti-inflammatory and Anticancer Agent. In *Plant Biotechnology: Recent Advancements and Developments*; Gahlawat, S., Salar, R., Siwach, P., Duhan, J., Kumar, S., Kaur, P., Eds.; Springer, 2017; pp. 255–270. https://doi.org/10.1007/978-981-10-4732-9_13

Salustri, C.; Barbati, G.; Ghidoni, R.; Quintiliani, L.; Ciappina, S.; Binetti, G.; Squitti, R. Is Cognitive Function Linked to Serum Free Copper Levels? A Cohort Study in a Normal Population. *Clin. Neurophysiol.* **2010,** *121* (4), 502–507. https://doi.org/10.1016/j.clinph.2009.11.090

Santos, E. L.; Maia, B. H. L. N. S.; Teixeira, A. P. F. and S. D. Flavonoids: Classification, Biosynthesis and Chemical Ecology. In *Flavonoids—From Biosynthesis to Human Health*; IntechOpen, 2017. https://doi.org/10.5772/67861

Schikowski, T.; Altuğ, H. The Role of Air Pollution in Cognitive Impairment and Decline. *Neurochem. Int.* **2020,** *136*, 104708. https://doi.org/10.1016/j.neuint.2020.104708

Shah, A. S. V.; Lee, K. K.; McAllister, D. A.; Hunter, A.; Nair, H.; Whiteley, W.; Langrish, J. P.; Newby, D. E.; Mills, N. L. Short Term Exposure to Air Pollution and Stroke: Systematic Review and Meta-Analysis. *BMJ* **2015,** *350*, h1295. https://doi.org/10.1136/bmj.h1295

Shaji, E.; Santosh, M.; Sarath, K. V.; Prakash, P.; Deepchand, V.; Divya, B. V. Arsenic Contamination of Groundwater: A Global Synopsis with Focus on the Indian Peninsula. *Geosci. Front.* **2021,** *12* (3), 101079. https://doi.org/10.1016/j.gsf.2020.08.015

Sharma, P.; Gaur, V. K.; Sirohi, R.; Varjani, S.; Hyoun Kim, S.; Wong, J. W. C. Sustainable Processing of Food Waste for Production of Bio-Based Products for Circular Bioeconomy. *Bioresour. Technol.* **2021,** *325*, 124684. https://doi.org/10.1016/j.biortech.2021.124684

Sharma, V.; Mishra, M.; Ghosh, S.; Tewari, R.; Basu, A.; Seth, P.; Sen, E. Modulation of Interleukin-1beta Mediated Inflammatory Response in Human Astrocytes by Flavonoids: Implications in Neuroprotection. *Brain Res. Bull.* **2007,** *73* (1–3), 55–63. https://doi.org/10.1016/j.brainresbull.2007.01.016

Sheffler, Z. M.; Reddy, V.; Pillarisetty, L. S. Physiology, Neurotransmitters. In *StatPearls*. StatPearls Publishing. http://www.ncbi.nlm.nih.gov/books/NBK539894/

Shen, X.-Y.; Luo, T.; Li, S.; Ting, O.-Y.; He, F.; Xu, J.; Wang, H.-Q. Quercetin Inhibits Okadaic Acid-Induced Tau Protein Hyperphosphorylation Through the Ca2+-Calpain-p25-CDK5 Pathway in HT22 Cells. *Int. J. Mol. Med.* **2018,** *41* (2), 1138–1146. https://doi.org/10.3892/ijmm.2017.3281

Sheykhansari, S.; Kozielski, K.; Bill, J.; Sitti, M.; Gemmati, D.; Zamboni, P.; Singh, A. V. Redox Metals Homeostasis in Multiple Sclerosis and Amyotrophic Lateral Sclerosis: A Review. *Cell Death Dis.* **2018,** *9* (3), 348. https://doi.org/10.1038/s41419-018-0379-2

Shi, L.; Steenland, K.; Li, H.; Liu, P.; Zhang, Y.; Lyles, R. H.; Requia, W. J.; Ilango, S. D.; Chang, H. H.; Wingo, T.; Weber, R. J.; Schwartz, J. A national Cohort Study (2000–2018) of Long-Term Air Pollution Exposure and Incident Dementia in Older Adults in the United States. *Nat. Commun.* **2021,** *12* (1), 6754. https://doi.org/10.1038/s41467-021-27049-2

Shi, Z.-H.; Li, N.-G.; Tang, Y.-P.; Shi, Q.-P.; Tang, H.; Li, W.; Zhang, X.; Fu, H.-A.; Duan, J.-A. Biological Evaluation and SAR Analysis of O-Methylated Analogs of Quercetin as Inhibitors of Cancer Cell Proliferation. *Drug Dev. Res.* **2014,** *75* (7), 455–462. https://doi.org/10.1002/ddr.21181

Shi, Z.-H.; Li, N.-G.; Tang, Y.-P.; Wei-Li, null, Lian-Yin, null, Yang, J.-P.; Hao-Tang, null; Duan, J.-A. Metabolism-Based Synthesis, Biologic Evaluation and SARs Analysis of O-Methylated Analogs of Quercetin as Thrombin Inhibitors. *Eur. J. Med. Chem.* **2012,** *54*, 210–222. https://doi.org/10.1016/j.ejmech.2012.04.044

Shokoohinia, Y.; Rashidi, M.; Hosseinzadeh, L.; Jelodarian, Z. Quercetin-3-O-β-D-Glucopyranoside, a Dietary Flavonoid, Protects PC12 Cells from H_2O_2-Induced Cytotoxicity Through Inhibition of Reactive Oxygen Species. *Food Chem.* **2015,** *167*, 162–167. https://doi.org/10.1016/j.foodchem.2014.06.079

Silva, B.; Oliveira, P. J.; Dias, A.; Malva, J. O. Quercetin, Kaempferol and Biapigenin from *Hypericum perforatum* Are Neuroprotective Against Excitotoxic Insults. *Neurotoxicity Res.* **2008,** *13* (3–4), 265–279. https://doi.org/10.1007/BF03033510

Silva, R. F. M.; Pogačnik, L. Polyphenols from Food and Natural Products: Neuroprotection and Safety. *Antioxidants* **2020,** *9* (1), 61. PubMed. https://doi.org/10.3390/antiox9010061

Singh, G.; Sharma, M.; Kumar, G. A.; Rao, N. G.; Prasad, K.; Mathur, P.; Pandian, J. D.; Steinmetz, J. D.; Biswas, A.; Pal, P. K.; Prakash, S.; Sylaja, P. N.; Nichols, E.; Dua, T.; Kaur, H.; Alladi, S.; Agarwal, V.; Aggarwal, S.; Ambekar, A. et al. The Burden of Neurological Disorders Across the States of India: The Global Burden of Disease Study 1990–2019. *Lancet Global Health* **2021,** *9* (8), e1129–e1144. https://doi.org/10.1016/S2214-109X (21)00164-9

Smith, J. A.; Das, A.; Ray, S. K.; Banik, N. L. Role of Pro-Inflammatory Cytokines Released from Microglia in Neurodegenerative Diseases. *Brain Res. Bull.* **2012,** *87* (1), 10–20. https://doi.org/10.1016/j.brainresbull.2011.10.004

Solanki, I.; Parihar, P.; Mansuri, M. L.; Parihar, M. S. Flavonoid-Based Therapies in the Early Management of Neurodegenerative Diseases. *Adv. Nutrition* **2015,** *6* (1), 64–72. PubMed. https://doi.org/10.3945/an.114.007500

Song, J.; Du, G.; Wu, H.; Gao, X.; Yang, Z.; Liu, B.; Cui, S. Protective Effects of Quercetin on Traumatic Brain Injury Induced Inflammation and Oxidative Stress in Cortex Through Activating Nrf2/HO-1 Pathway. *Restorative Neurol. Neurosci.* **2021,** *39* (1), 73–84. https://doi.org/10.3233/RNN-201119

Soto, C. Unfolding the Role of Protein Misfolding in Neurodegenerative Diseases. *Nat. Rev. Neurosci.* **2003,** *4* (1), 49–60. https://doi.org/10.1038/nrn1007

Soto, C.; Pritzkow, S. Protein Misfolding, Aggregation, and Conformational Strains in Neurodegenerative Diseases. *Nat. Neurosci.* **2018,** *21* (10), 1332–1340. https://doi.org/10.1038/s41593-018-0235-9

Sriraksa, N.; Wattanathorn, J.; Muchimapura, S.; Tiamkao, S.; Brown, K.; Chaisiwamongkol, K. Cognitive-Enhancing Effect of Quercetin in a Rat Model of Parkinson's Disease Induced by 6-Hydroxydopamine. *Evid.-Based Complement. Altern. Med.* **2011,** *2012*, e823206. https://doi.org/10.1155/2012/823206

Stein, T. P.; Schluter, M. D.; Steer, R. A.; Guo, L.; Ming, X. Bisphenol A Exposure in Children with Autism Spectrum Disorders. *Autism Res.* **2015,** *8* (3), 272–283. https://doi.org/10.1002/aur.1444

Suganthy, N.; Devi, K. P.; Nabavi, S. F.; Braidy, N.; Nabavi, S. M. Bioactive Effects of Quercetin in the Central Nervous System: Focusing on the Mechanisms of Actions. *Biomed. Pharmacother.* **2016a,** *84*, 892–908. https://doi.org/10.1016/j.biopha.2016.10.011

Suganthy, N.; Devi, K. P.; Nabavi, S. F.; Braidy, N.; Nabavi, S. M. Bioactive Effects of Quercetin in the Central Nervous System: Focusing on the Mechanisms of Actions. *Biomed. Pharmacother.* **2016b,** *84*, 892–908. https://doi.org/10.1016/j.biopha.2016.10.011

Sul, O.-J.; Ra, S. W. Quercetin Prevents LPS-Induced Oxidative Stress and Inflammation by Modulating NOX2/ROS/NF-kB in Lung Epithelial Cells. *Molecules* **2021,** *26* (22), 6949. https://doi.org/10.3390/molecules26226949

Tangsaengvit, N.; Kitphati, W.; Tadtong, S.; Bunyapraphatsara, N.; Nukoolkarn, V. Neurite Outgrowth and Neuroprotective Effects of Quercetin from *Caesalpinia mimosoides* Lamk. on Cultured P19-Derived Neurons. *Evid.-Based Complement. Altern. Med.* 2013, *2013*, 838051. https://doi.org/10.1155/2013/838051

Taoufik, E.; Kouroupi, G.; Zygogianni, O.; Matsas, R. Synaptic Dysfunction in Neurodegenerative and Neurodevelopmental Diseases: An Overview of Induced Pluripotent Stem-Cell-Based Disease Models. *Open Biol.* **2018,** *8* (9), 180138. https://doi.org/10.1098/rsob.180138

Tapiero, H.; Townsend, D. M.; Tew, K. D. Trace Elements in Human Physiology and Pathology. Copper. *Biomed. Pharmacother.* **2003,** *57* (9), 386–398.

Tauheed, J.; Sanchez-Guerra, M.; Lee, J. J.; Paul, L.; Ibne Hasan, M. O. S.; Quamruzzaman, Q.; Selhub, J.; Wright, R. O.; Christiani, D. C.; Coull, B. A.; Baccarelli, A. A.; Mazumdar,

M. Associations Between Post Translational Histone Modifications, Myelomeningocele Risk, Environmental Arsenic Exposure, and Folate Deficiency Among Participants in a Case Control Study in Bangladesh. *Epigenetics* **2017,** *12* (6), 484–491. https://doi.org/10.1080/15592294.2017.1312238

Tay, W. M.; da Silva, G. F. Z.; Ming, L.-J. Metal Binding of Flavonoids and Their Distinct Inhibition Mechanisms Toward the Oxidation Activity of Cu2+-β-Amyloid: Not Just Serving as Suicide Antioxidants! *Inorg. Chem.* **2013,** *52* (2), 679–690. https://doi.org/10.1021/ic301832p

Tchantchou, F.; Lacor, P. N.; Cao, Z.; Lao, L.; Hou, Y.; Cui, C.; Klein, W. L.; Luo, Y. Stimulation of Neurogenesis and Synaptogenesis by Bilobalide and Quercetin via Common Final Pathway in Hippocampal Neurons. *J. Alzheimer's Dis.* **2009,** *18* (4), 787–798. https://doi.org/10.3233/JAD-2009-1189

Thomson, E. M. Air Pollution, Stress, and Allostatic Load: Linking Systemic and Central Nervous System Impacts. *J. Alzheimer's Dis.* **2019,** *69* (3), 597–614. https://doi.org/10.3233/JAD-190015

Thurston, G. D.; Kipen, H.; Annesi-Maesano, I.; Balmes, J.; Brook, R. D.; Cromar, K.; De Matteis, S.; Forastiere, F.; Forsberg, B.; Frampton, M. W.; Grigg, J.; Heederik, D.; Kelly, F. J.; Kuenzli, N.; Laumbach, R.; Peters, A.; Rajagopalan, S. T.; Rich, D.; Ritz, B. et al. A Joint ERS/ATS Policy Statement: What Constitutes an Adverse Health Effect of Air Pollution? An Analytical Framework. *Eur. Respirat. J.* **2017,** *49* (1), 1600419. PubMed. https://doi.org/10.1183/13993003.00419-2016

Thygesen, M.; Holst, G. J.; Hansen, B.; Geels, C.; Kalkbrenner, A.; Schendel, D.; Brandt, J.; Pedersen, C. B.; Dalsgaard, S. Exposure to Air Pollution in Early Childhood and the Association with Attention-Deficit Hyperactivity Disorder. *Environ. Res.* **2020,** *183*, 108930–108930. PubMed. https://doi.org/10.1016/j.envres.2019.108930

Tohge, T.; de Souza, L. P.; Fernie, A. R. Current Understanding of the Pathways of Flavonoid Biosynthesis in Model and Crop Plants. *J. Exp. Bot.* **2017,** *68* (15), 4013–4028. https://doi.org/10.1093/jxb/erx177

Trang, A.; Khandhar, P. B. Physiology, Acetylcholinesterase. In *StatPearls*; StatPearls Publishing, 2021. http://www.ncbi.nlm.nih.gov/books/NBK539735/

Trushna, T.; Dhiman, V.; Raj, D.; Tiwari, R. R. Effects of Ambient Air Pollution on Psychological Stress and Anxiety Disorder: A Systematic Review and Meta-Analysis of Epidemiological Evidence. *Rev. Environ. Health* **2020**. https://doi.org/10.1515/reveh-2020-0125

Trushna, T.; Tripathi, A. K.; Tiwari, R. R.; Sindhuprava, R. Nutraceuticals with Anti-Inflammatory and Anti-Oxidant Properties as Intervention for Reducing the Health Effects of Fine Particulate Matter: Potential and Prospects. *Accelerated Technol. Biotechnol. Bioassays Med. Chem. Nat. Products Res.* **2021,** *24*, 1–1. https://doi.org/10.2174/1386207324666210412121226

Tyler, C. R.; Allan, A. M. The Effects of Arsenic Exposure on Neurological and Cognitive Dysfunction in Human and Rodent Studies: A Review. *Curr. Environ. Health Rep.* **2014,** *1* (2), 132–147. https://doi.org/10.1007/s40572-014-0012-1

Van Dyke, N.; Yenugadhati, N.; Birkett, N. J.; Lindsay, J.; Turner, M. C.; Willhite, C. C.; Krewski, D. Association Between Aluminum in Drinking Water and Incident Alzheimer's Disease in the Canadian Study of Health and Aging Cohort. *NeuroToxicology* **2021,** *83*, 157–165. https://doi.org/10.1016/j.neuro.2020.04.002

Vanani, A. R.; Mahdavinia, M.; Shirani, M.; Alizadeh, S.; Dehghani, M. A. Protective Effects of Quercetin Against Oxidative Stress Induced by Bisphenol-A in Rat Cardiac Mitochondria. *Environ. Sci. Pollut. Res.* **2020,** *27* (13).

Volk, H. E.; Perera, F.; Braun, J. M.; Kingsley, S. L.; Gray, K.; Buckley, J.; Clougherty, J. E.; Croen, L. A.; Eskenazi, B.; Herting, M.; Just, A. C.; Kloog, I.; Margolis, A.; McClure, L. A.; Miller, R.; Levine, S.; Wright, R. Prenatal Air Pollution Exposure and Neurodevelopment: A Review and Blueprint for a Harmonized Approach Within ECHO. *Environ. Res.* **2020,** 110320. https://doi.org/10.1016/j.envres.2020.110320

von Ehrenstein, O.; Aralis, H.; Cui, X.; Wu, J.; Li, L.; Cockburn, M.; Ritz, B. Exposure to Traffic Related Air Pollution in Pregnancy and Infancy in Relation to Autism Spectrum Disorder in California. *Environ. Epidemiol.* **2019,** *3*, 417. https://doi.org/10.1097/01.EE9.0000610672.40487.98

Wang, B.; Du, Y. Cadmium and Its Neurotoxic Effects. *Oxid. Med. Cell. Longevity* **2013,** *2013*, e898034. https://doi.org/10.1155/2013/898034

Wang, R.; Zhang, H.; Sun, S.; Wang, Y.; Chai, Y.; Yuan, Y. Effect of Ginkgo Leaf Tablets on the Pharmacokinetics of Amlodipine in Rats. *Eur. J. Drug Metabol. Pharmacokinetics* **2016,** *41* (6), 825–833. https://doi.org/10.1007/s13318-015-0312-3

Wang, Y.-H.; Chao, P.-D. L.; Hsiu, S. L.; Wen, K.-C.; Hou, Y.-C. Lethal Quercetin-Digoxin Interaction in Pigs. *Life Sci.* **2004,** *74* (10), 1191–1197. https://doi.org/10.1016/j.lfs.2003.06.044

WHO. *9 Out of 10 People Worldwide Breathe Polluted Air, But More Countries Are Taking Action*; World Health Organisation, n.d. https://www.who.int/news-room/detail/02-05-2018-9-out-of-10-people-worldwide-breathe-polluted-air-but-more-countries-are-taking-action (accessed 1 July 2020).

Wilker, E. H.; Preis, S. R.; Beiser, A. S.; Wolf, P. A.; Au, R.; Kloog, I.; Li, W.; Schwartz, J.; Koutrakis, P.; DeCarli, C.; Seshadri, S.; Mittleman, M. A. Long-Term Exposure to Fine Particulate Matter, Residential Proximity to Major Roads and Measures of Brain Structure. *Stroke* **2015,** *46* (5), 1161–1166. PubMed. https://doi.org/10.1161/STROKEAHA.114.008348

Wu, J.; Basha, Md. R.; Brock, B.; Cox, D. P.; Cardozo-Pelaez, F.; McPherson, C. A.; Harry, J.; Rice, D. C.; Maloney, B.; Chen, D.; Lahiri, D. K.; Zawia, N. H. Alzheimer's Disease (AD)-Like Pathology in Aged Monkeys after Infantile Exposure to Environmental Metal Lead (Pb): Evidence for a Developmental Origin and Environmental Link for AD. *J, Neurosci.* **2008,** *28* (1), 3–9. https://doi.org/10.1523/JNEUROSCI.4405-07.2008

Xie, A.; Gao, J.; Xu, L.; Meng, D. Shared Mechanisms of Neurodegeneration in Alzheimer's Disease and Parkinson's Disease. *BioMed Res. Int.* **2014,** *2014*, 648740. https://doi.org/10.1155/2014/648740

Xue, F.; Nie, X.; Shi, J.; Liu, Q.; Wang, Z.; Li, X.; Zhou, J.; Su, J.; Xue, M.; Chen, W.-D.; Wang, Y.-D. Quercetin Inhibits LPS-Induced Inflammation and ox-LDL-Induced Lipid Deposition. *Front. Pharmacol.* **2017,** *8*, 40. https://doi.org/10.3389/fphar.2017.00040

Yang, D.; Liu, X.; Liu, M.; Chi, H.; Liu, J.; Han, H. Protective Effects of Quercetin and Taraxasterol Against H_2O_2-Induced Human Umbilical Vein Endothelial Cell Injury In Vitro. *Exp. Therap. Med.* **2015,** *10* (4), 1253–1260. https://doi.org/10.3892/etm.2015.2713

Yang, E.-J.; Kim, G.-S.; Kim, J. A.; Song, K.-S. Protective Effects of Onion-Derived Quercetin on Glutamate-Mediated Hippocampal Neuronal Cell Death. *Pharmacogn. Mag.* **2013,** *9* (36), 302–308. https://doi.org/10.4103/0973-1296.117824

Yang, R.; Shen, Y.-J.; Chen, M.; Zhao, J.-Y.; Chen, S.-H.; Zhang, W.; Song, J.-K.; Li, L.; Du, G.-H. Quercetin Attenuates Ischemia Reperfusion Injury by Protecting the Blood-Brain Barrier Through Sirt1 in MCAO Rats. *J. Asian Nat. Products Res.* **2021,** 1–12. https://doi.org/10.1080/10286020.2021.1949302

Yang, X.-Y.; Li, Q.-J.; Zhang, W.-C.; Zheng, S.-Q.; Qu, Z.-J.; Xi, Y.; Wang, G. AMPK-SIRT1-PGC1α Signal Pathway Influences the Cognitive Function of Aged Rats in

Sevoflurane-Induced Anesthesia. *J. Mol. Neurosci.* **2020,** *70* (12), 2058–2067. https://doi.org/10.1007/s12031-020-01612-w

Yao, Y.; Han, D. D.; Zhang, T.; Yang, Z. Quercetin Improves Cognitive Deficits in Rats with Chronic Cerebral Ischemia and Inhibits Voltage-Dependent Sodium Channels in Hippocampal CA1 Pyramidal Neurons. *Phytother. Res.* **2010,** *24* (1), 136–140. https://doi.org/10.1002/ptr.2902

Yu, J.; Wu, Y.; Yang, P. High Glucose-Induced Oxidative Stress Represses Sirtuin Deacetylase Expression and Increases Histone Acetylation Leading to Neural Tube Defects. *J. Neurochem.* **2016,** *137* (3), 371–383. https://doi.org/10.1111/jnc.13587

Zaheer, F.; Slevin, J. T. Trichloroethylene and Parkinson Disease. *Neurol. Clin.* **2011,** *29* (3), 657–665. https://doi.org/10.1016/j.ncl.2011.05.001

Zamora-Ros, R.; Knaze, V.; Luján-Barroso, L.; Slimani, N.; Romieu, I.; Fedirko, V.; de Magistris, M. S.; Ericson, U.; Amiano, P.; Trichopoulou, A.; Dilis, V.; Naska, A.; Engeset, D.; Skeie, G.; Cassidy, A.; Overvad, K.; Peeters, P. H. M.; Huerta, J. M.; Sánchez, M.-J. et al. Estimated Dietary Intakes of Flavonols, Flavanones and Flavones in the European Prospective Investigation Into Cancer and Nutrition (EPIC) 24 Hour Dietary Recall Cohort. *Br. J. Nutr.* **2011,** *106* (12), 1915–1925. https://doi.org/10.1017/S000711451100239X

Zaplatic, E.; Bule, M.; Shah, S. Z. A.; Uddin, Md. S.; Niaz, K. Molecular Mechanisms Underlying Protective Role of Quercetin in Attenuating Alzheimer's Disease. *Life Sci.* **2019,** *224*, 109–119. https://doi.org/10.1016/j.lfs.2019.03.055

Zhang, C.; Du, Q.-Y.; Chen, L.-D.; Wu, W.-H.; Liao, S.-Y.; Yu, L.-H.; Liang, X.-T. Design, Synthesis and Evaluation of Novel Tacrine-Multialkoxybenzene Hybrids as Multi-Targeted Compounds Against Alzheimer's Disease. *Eur. J. Med. Chem.* **2016,** *116*, 200–209. https://doi.org/10.1016/j.ejmech.2016.03.077

Zhang, C.; Wang, Y.; Wang, D.; Zhang, J.; Zhang, F. NSAID Exposure and Risk of Alzheimer's Disease: An Updated Meta-Analysis From Cohort Studies. *Front. Aging Neurosci.* **2018,** *10*, 83. https://doi.org/10.3389/fnagi.2018.00083

Zhang, J.; Ning, L.; Wang, J. Dietary Quercetin Attenuates Depressive-Like Behaviors by Inhibiting Astrocyte Reactivation in Response to Stress. *Biochem. Biophys. Res. Commun.* **2020,** *533* (4), 1338–1346. https://doi.org/10.1016/j.bbrc.2020.10.016

Zhang, M.; Liu, W.; Zhou, Y.; Li, Y.; Qin, Y.; Xu, Y. Neurodevelopmental Toxicity Induced by Maternal $PM_{2.5}$ Exposure and Protective Effects of Quercetin and Vitamin C. *Chemosphere* **2018,** *213*, 182–196. https://doi.org/10.1016/j.chemosphere.2018.09.009

Zhang, M.; Swarts, S.; Yin, L.; Liu, C.; Tian, Y.; Cao, Y.; Swarts, M.; Yang, S.; Zhang, S.; Zhang, K. Oxygen Transport to Tissue XXXII. In La Manna, J. C., Puchowicz, M. A., Xu, K., Harrison, D. K., Bruley, D. F., Eds, 2011; pp 283–289.

Zhang, Q.; Song, W.; Zhao, B.; Xie, J.; Sun, Q.; Shi, X.; Yan, B.; Tian, G.; Liang, X. Quercetin Attenuates Diabetic Peripheral Neuropathy by Correcting Mitochondrial Abnormality via Activation of AMPK/PGC-1α Pathway In Vivo and In Vitro. *Front. Neurosci.* **2021,** *15*, 84. https://doi.org/10.3389/fnins.2021.636172

Zhao, L.; Wang, H.; Du, X. The Therapeutic Use of Quercetin in Ophthalmology: Recent Applications. *Biomed. Pharmacother.* **2021,** *137*, 111371. https://doi.org/10.1016/j.biopha.2021.111371

Zhao, P.; Hu, Z.; Ma, W.; Zang, L.; Tian, Z.; Hou, Q. Quercetin Alleviates Hyperthyroidism-Induced Liver Damage via Nrf2 Signaling Pathway. *BioFactors* **2020,** *46* (4), 608–619. https://doi.org/10.1002/biof.1626

Zhao, Y.; Pan, Y.; Wang, W.; Guo, L. A Brain-Inspired Dynamic Environmental Emergency Response Framework for Sudden Water Pollution Accidents. *Water* **2021,** *13* (21), 3097. https://doi.org/10.3390/w13213097

Zheng, Z.; White, C.; Lee, J.; Peterson, T. S.; Bush, A. I.; Sun, G.; Weisman, G. A.; Petris, M. J. Altered Microglial Copper Homeostasis in a Mouse Model Of Alzheimer's Disease. *J. Neurochem.* **2010,** *114* (6), 1630–1638. https://doi.org/10.1111/j.1471-4159.2010.06888.x

Zhou, S.-F. Drugs Behave as Substrates, Inhibitors and Inducers of Human Cytochrome P450 3A4. *Curr. Drug Metabol.* **2008,** *9* (4), 310–322. https://doi.org/10.2174/138920008784220664

Zhu, M.; Han, S.; Fink, A. L. Oxidized Quercetin Inhibits α-Synuclein Fibrillization. *Biochim. Biophys. Acta* **2013,** *1830* (4), 2872–2881. https://doi.org/10.1016/j.bbagen.2012.12.027

CHAPTER 9

ROLE OF HERBS IN METABOLIC SYNDROME

SANDEEP O. WAGHULDE, NILESH GORDE, AJAY KHARCHE, and MOHAN KALE

Konkan Gyanpeeth Rahul Dharkar College of Pharmacy & Research Institute, Dahivali, Maharashtra, India

ABSTRACT

When many metabolic abnormalities are present, an individual's chance of developing type 2 diabetes and cardiovascular disease increases. Traditional medicinal herbs, many of which have been used for thousands of years, include a wide array of phytochemicals that may help treat metabolic syndrome. Botanical preparations, on the other hand, must be described and standardized via the discovery of their active components and their distinct mechanisms of action, followed by controlled clinical studies with well-defined objectives in order to be successful in combating metabolic syndrome. Medicinal herbs having various therapeutic targets are widely available and may be used in alternative treatment techniques for risk factors of metabolic syndrome. Various physiologically active chemicals are found in medicinal plants, all of which have some kind of therapeutic effect. Long-term health and wellness effects of phytochemicals found in indigenous medicinal plants have been studied and shown to include a wide range of complicated biochemical, metabolic, and physiologic systems. It is important to Identify new medicinal plants that may be utilized to treat metabolic syndrome components and describe their putative mechanisms of action. This chapter examines the efficacy of an ethnomedicinal plant to protect against

The Flavonoids: Extraction and Applications. Deepika Saini, Rajesh K. Kesharwani, & Raj K. Keservani (Eds.)

the emergence of health issues associated with metabolic syndrome. Using current developments in the usage of medicinal plants and their therapeutically important phytochemicals, we aim to provide up-to-date information on medicinal plants and their therapeutically vital phytochemicals.

9.1 INTRODUCTION

Metabolic syndrome is a group of disorders caused by risk factors that typically co-exist with three of the four basic metabolic control medical conditions: diabetes, obesity, dyslipidemia (high triglyceride levels or low HDL cholesterol), and hypertension. One in four people in high-income countries suffers from metabolic syndrome, according to the World Health Organization (Miller et al., 2014). Metabolic syndrome has also become more common in low-income countries (Li et al., 2018). All these medical conditions, as well as cancer (Esposito et al., 2012), stroke (Rodriguez-Colon et al., 2009), polycystic ovary syndrome (Mendrick et al., 2018), non-alcoholic steatohepatitis (Mendrick et al., 2018), type 2 diabetes (Ramrez-Vélez et al., 2019), and cardiovascular disease (CVD) (Tune et al., 2017; Isomaa et al., 2001) are associated with metabolic syndrome. Metabolic syndrome may lead to heavy economic and social problems. As a result, developing and implementing an effective plan is critical to fight against metabolic syndrome. Improvement in insulin resistance, managing weight gain, regulation of blood glucose level and blood pressure, and treatment of dyslipidemia can be helpful to treat metabolic syndrome. It has been shown in several recent studies (Dommermuth et al., 2018; Wei et al., 2018; Ng et al., 2018); In addition, as the condition progresses, the efficacy of each medication may be evaluated, as standard therapies lose their effectiveness. Sodium-glucose transporter-2 inhibitors that have been clinically tested and shown to be effective should be evaluated as possible new therapies for metabolic syndrome (Lim and Eckel, 2014).

The incidence of metabolic syndrome-related metabolic illnesses has grown worldwide. The International Diabetes Federation (IDF) predicts that in the next several decades, the incidence and prevention of diabetes will grow by a quarter of the worldwide adult population (O'Neill and O'Driscoll, 2015). According to Nolan et al. (2017), as a public health problem, metabolic syndrome is on the rise in resource-constrained nations like sub-Saharan Africa.). Governments in developing countries are spending huge amount of money to tackle the rise in metabolic syndrome and its associated

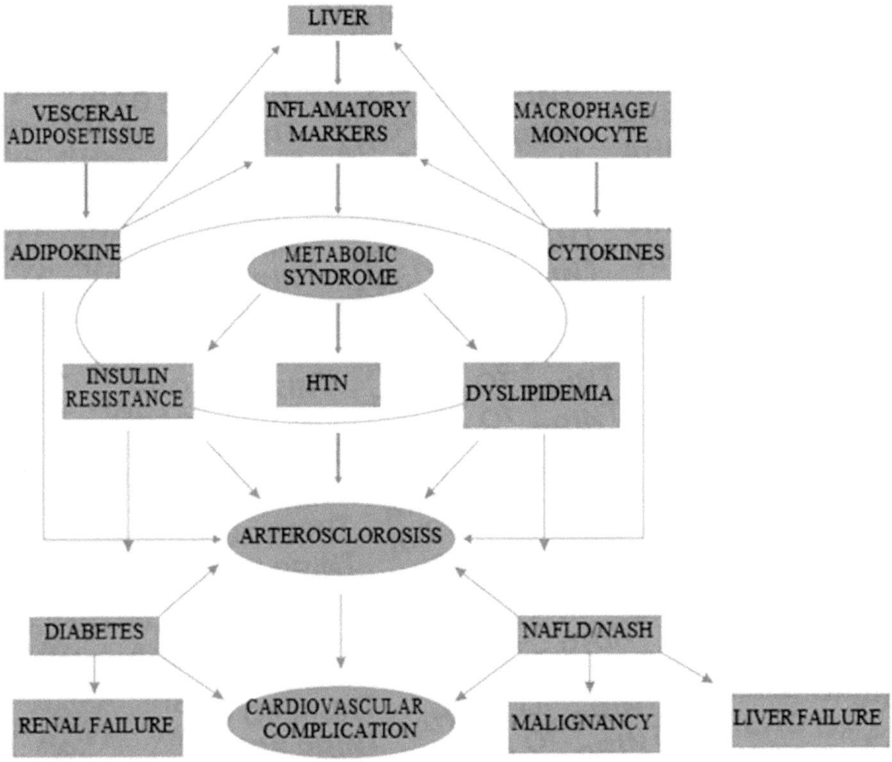

FIGURE 9.1 Adipokines, cytokines, and inflammatory markers are all implicated in the evolution of the metabolic disease and the difficulties that might arise as a result. These acronyms for hypertension, NAFLD/NASH, and non-alcoholic steatohepatitis all refer to the same condition.

risk factors (Boudreau et al., 2009). Epigenetic, environmental, and genetic factors contribute to the development of metabolic syndrome and related risk factors (Gluckman et al., 2009). It is believed that this epidemic is being fueled by inactivity, a diet heavy in fat and sugar, and a sedentary way of life. An important aspect of current treatment methods is to combine pharmacological drugs and lifestyle adjustments to target specific biochemical pathways linked with risk factors for metabolic syndrome (Moller, 2001). Long-term usage of pharmaceutical medications by low-income patients may lead to a number of unwanted side effects, which can be pricey. Only a few adverse health effects are linked to metabolic dysfunction when it is targeted as a single treatment. Metabolic problems need investigation and development of alternative and supplement treatments. The usage of drug stores should be a part of the treatment methods for metabolic development hazard concerns. A medical shop (s) is defined as any store or drug shop (s) that provides health-enhancing medicines and short-term relief, or that has medicines that possess good preventative skills (Sofowora, 1996; Keservani et al., 2016). Medical practitioners may now recognize that drug stores have a role to play in the therapy and prevention of metabolic disorders. According to McGaw et al. (2005), The use of natural drugs is not always constrained to growing nations, however, has turned out to be a multibillion-greenback enterprise spanning throughout all demographic and socioeconomic strata.

Metabolic diseases benefit from the therapeutic effects of bioactive chemicals found in medicinal plants, both additive and synergistic (Dillard and German, 2000; Rodriguez-Casado et al., 2016). People extracted and identified the majority of pharmaceutical medications derived from medicinal plants using local knowledge, and they then isolated the plants' primary active constituents (Altemimi et al., 2017). The plant components utilized in medical treatments might serve as prospective targets for pharmaceutical drug synthesis. To reduce the need for pricey synthetic pharmaceutical drugs, medicinal plants have been found to possess useful phytochemical components. The WHO reports that even in countries where pharmaceuticals are readily available, the majority of the world's rural and urban people still rely on traditional medicine therapy for their primary healthcare (Tomlinson and Akerele, 2015). When it comes to treating metabolic disorders, medicinal plants are increasingly used due to the belief that medicinal plants have no side effects or acute toxicity when compared with the existing pharmaceutical therapies (Tabasum and Khare, 2016).

Even though people choose to utilize medicinal plants and believe that their preparations are safe, still scientific confirmation is the most significant factor in determining their safety. Herbal remedies formerly relied on dubious evidence culled from folklore (Naghdi, 2018). Herbal products and therapeutic plants are frequently discussed in the media and on public forums. According to a lack of scientifically accurate assessments and studies, therapeutic herbs are believed to be ineffective and safe. The identification of physiologically active phytochemical components in medicinal plants and the systematic study of these plants has greatly improved the treatment of metabolic disorders. Research works to study the mechanisms of action of phytochemical components of medicinal plants have recently been aided by new advancements in molecular methodologies. For the treatment of health outcomes associated with metabolic illnesses, we believe that this chapter will provide medical experts and healthcare practitioners with a plethora of information about medicinal plants as safe, accessible, and aesthetic alternatives. This chapter discusses the role of medicinal plants and phytochemicals in the treatment and prevention of obesity, cardiovascular, and metabolic diseases as well as non-alcoholic fatty liver disease.

9.2 MEDICINAL PLANTS IN THE MANAGEMENT OF OBESITY

Obesity and other forms of metabolic syndrome are on the rise all over the world. Overweight is defined as having a body mass index (BMI) of 25 or more; obese is defined as having a BMI of 30 or more. An increasing number of people worldwide are becoming obese in large part due to changes in eating habits and lack of physical exercise (Caballero, 2007). Between 1989 and 2011, obesity cases are believed to have nearly doubled. According to the WHO, obesity-related causes of death claim the lives of 2.8 million individuals each year. The life expectancy of those who are overweight or obese is reduced by 7 years at the age of 40 (Peeters et al., 2003). According to the latest statistics, there were about 1.9 billion adults and 650 million obese people in 2016. According to a 1998 study, almost 21% of the world's population was determined to be obese (Nishida and Mucavele, 2005). There were 57% of women and 29% of males in the whole population who weighed more than their ideal weight (Van Der Merwe and Pepper, 2006). Obesity and overweight are on the rise in the United States, as evidenced by the fact that 68% of women and 31% of men worldwide are either overweight or obese. At times of increased intake of high-energy meals (carbohydrates), coupled with a reduction in physical activity, Obesity is a risk factor for developing diabetes (Kopelman, 2000; Goedecke et al., 1995–2005). Many health problems have been linked to being overweight, including many types of cancer (kidney, breast, colon, prostate), type 2 diabetes, and heart disease (Field et al., 2001; Keservani et al., 2017). Maintaining a healthy lifestyle, getting regular exercise, reducing added sugar and sodium intake, and increasing your consumption of whole grains, veggies, and legumes are some of the best ways to lose or maintain healthy body weight (Keservani et al., 2010a, 2010b). Pharmacological treatments and surgical interventions are also recommended (Rane et al., 2020a, 2020b, 2020c; Bray, 2004). Obesity therapy is complicated by the fact that only 5–10% of people who lose weight do so for the long term (McTigue et al., 2003). When medication is stopped or a healthy lifestyle is maintained, the weight loss is reversed (Howard, 1981; Lowe, 2003). In addition, several synthetic pharmaceuticals have negative side effects (Curioni and Lourenço, 2005; Abdollahi and Afshar-Imani, 2003; Yanovski and Yanovski, 2014). Herbal supplements, which are more effective and less expensive than medicines, can help people lose or maintain their weight. Some of the findings of this study are summarized in the following lines. Some plants that are frequently used for weight reduction throughout the globe, their active ingredients, and their method of action will be included in this section of the review (Table 9.1).

There were only two trials that failed to demonstrate weight loss (Garrison and Chambliss, 2006; Hoeger et al., 1998) and one that showed a substantial reduction in body fat (both studies indicate weight loss) (Table 9.2). Slimax (a plant extract from several plants, including *Zingiber officinale*) and other weight loss products, such as those containing caffeine and ephedra (Boozer et al., 2001; Greenway et al., 2004; Hackman et al., 2006), have all been shown to be effective in the treatment of obesity and associated conditions (Ignjatovic et al., 2000)."HCA-SX with or without *Gymnema sylvestre* (Preuss et al., 2004) successfully lowered intake levels of food and energy. However, a dietary supplement, including caffeine, ephedra, and *Caralluma fimbriata* had no effect. Capsicum and other naturally occurring lipotropic substances did not have a substantial effect on energy intake.

9.3 MEDICINAL PLANTS FOR THE MANAGEMENT OF DIABETES MELLITUS

Chronic diabetes is a long-term metabolic condition that is defined by high fasting and postprandial blood glucose levels, as well as elevated insulin levels. Diabetes will affect around 5.4% of the global population by 2025, up from 4% in 1995. According to the WHO, poorer nations will bear the bulk of the cost. Researchers in India have found that not only is diabetes prevalent, but it is also rising at an alarming rate among urban residents in the previous decade (Grover et al., 2002; Scartezzini and Speroni, 2000; Seth and Sharma, 2004). In India, it has been estimated that there are 33 million people with diabetes. This number is likely to rise to 57.2 million by 2025. Even today, diabetes mellitus remains a complex illness that can occur due to insulin shortage or insulin dysfunction. Absence of beta-cells results in type I diabetes (also known as "‘insulin-dependent diabetes"). Because of this, the afflicted patients must rely on an external supply of insulin. Dietary changes, physical activity, and medication are the sole options for those with type II diabetes (insulin-dependent). About 90% of diabetics have type II diabetes, which is the most common form of the condition. Diabetic symptoms include (i) high blood sugar levels; (ii) frequent urination; (iii) unusual thirst; (iv)

TABLE 9.1 Human Studies Considering the Anti-Obesity Effects of Herbal Medicines.

Target	Herbs (scientific name)	Study	Dose/duration	Groups	Main outcome	Other relevant effects and complications	References
Healthy volunteers	Slimax: *Polygonatum multiflorum* extract, *Dimocarpus longan* extract, *Ligusticum sinense* extract, *Lilium brownii* extract, and *Zingiber officinale* extract	RCT	6 wk	C: Placebo I: Compound	Significant weight loss, as well as a reduction in waist and hip circumferences Circumstances and BMI	Lipid metabolism is altered with a significant impact on adipose tissue buildup and the release of lipids	Ignjatovic et al. (2000)
Over wt. (n = 35)	The use of a natural supplement: (Ma Huang and Guarana)	RCT (double-blind)	72 mg (ephedra) 240 mg (caffeine)/ 8 wk	C: Placebo (n = 24) I: Compound (n = 24)	Body weight and total body fat decreased significantly, as well as hip and waist circumferences.	The active group had a 23% decline in serum TG, whereas the placebo group saw no change in TG levels. They also said that they had insomnia and a headache	Boozer et al., (2001)
Healthy	Inulin, L-phenylalanine, and other lipotropic elements make up this natural nutritional supplement	RCT (double-blind)	4 wk	C: wt. loss program (n = 67) I: wt. loss program + compound (n = 56)	Body fat percentage, mass, and FFM all decreased significantly, but there was no discernible change in weight, BMI, or caloric intake		Hoeger et al. (1998)
Hhyper-lipidemic (n = 30)	*Terminalia arjuna* Roxb	Before-after CT			Obesity reduction is very evident. Some people experience the loss of body weight	The amount of total lipids in the blood has decreased. Significant alleviation from palpitation, dyspnea, chest and joint discomfort, and nausea. Some instances of lowered blood pressure	Ziauddin et al. (2004)

TABLE 9.1 *(Continued)*

Target	Herbs (scientific name)	Study	Dose/duration	Groups	Main outcome	Other relevant effects and complications	References
Obese nondiabetic women (n = 32)	ARALOX is a mixture of *Aralia mandshurica* (A) and *Engelhardtia chrysolepis* (E) extracts	RCT	450 mg (A) & 450 mg (E)/d	C: Diet + placebo I: Diet + compound	Decreased total body weight and fat mass	Perilipin content in fatty tissue and TG levels in the blood decreased. Increased the activity of hormone-sensitive lipase	Abidov et al. (2006)
Human (obese and over wt.) healthy	Herbal supplement containing caffeine and ephedra	RCT (double-blind)	210 mg (e) & 72 mg (c)/12 wk	C: Placebo I: Compound	Significant loss of body weight and fat mass	In terms of lipids or blood pressure, no variations were seen. There were no major side effects	Greenway et al. (2004)
Obese women with IGT (n = 80)	Bofu-tsusho-san includes *Ephedrae herba*, *Glycyrrhizae radix*, *Forsythiae fructus*, *Schizonepetae spica*).	RCT (double-blind)	Equivalent of (24 mg/ephedrine and 280 mg caffeine/24 wk)	C: wt. loss program I: wt. loss program + compound	Body weight and visceral fat were dramatically reduced in the group I compared with baseline, whereas the placebo group was losing weight but no change in visceral fat	RMR has stayed unchanged. In comparison to week 0, a significant decrease in insulin resistance was seen. Having loose bowel motions led in the withdrawal of three times	Hioki et al., (2004)
Human (obese and over wt.)	An *Irvingia gabonensis* and *Cissus quadrangularis* combination (IG)	RCT (double-blind)	300 mg (CQ) and 500 mg (IG) per day/10 wk	C: Placebo I: CQ CQ + IG	Both the CQ and IG groups had substantial reductions in body weight, body fat percentage, and waist circumference, but the combined group (CQ + IG) saw the most dramatic reductions	There was a significant reduction in plasma Chol and LDL and fasting blood glucose levels	Oben et al., (2008)
Healthy (n = 80)	*Sambucus nigra* (S) and *Asparagus officinalis* are used in conjunction (A)	Before-after CT	(S): 1 mg anthocyanin, 370 mg flavonol, 150 mg hydroxycinnamate (A): 19 mg saponin per day	-	Sig. decrease in mean of the wt.	Significance of an increase in blood pressure, general health, and psychological and emotional well-being	Chrubasik et al. (2008)

TABLE 9.1 *(Continued)*

Target	Herbs (scientific name)	Study	Dose/duration	Groups	Main outcome	Other relevant effects and complications	References
Healthy (n = 25)	White bean fractionated extract, a proprietary product.	RCT (double-blind)	2000 mg/14 wk	C: Placebo + wt. loss program I: Extract + wt. loss program	Significant decreases in body weight and waist size were seen in both groups at baseline. However, there seems to be no correlation between groups	There were no adverse effect	Udani and Singh (2007)
Obese women	*Garcinia atroviridis* is rich in calcium hydroxycitrate.	RCT	2 mo	C: Diet I: Diet + extract	Significant weight loss and better BMI reduction. a significant reduction in the thickness of the triceps skin folds		Roong-pisuthipong et al. (2007)
Over wt. (n = 50)	*Caralluma fimbriata*	RCT	1 g/60 d	C: wt. loss program I: wt. loss program + extract	Significantly reduced hunger and waist circumference. Improvements in weight, BMI and hip circumference but not a significant drop in body fat or calorie consumption		Kuriyan et al. (2007)
Obese and over wt. women (n = 41)	Ephedra (e) and caffeine (c) are included in this multi-nutrient supplement (c)	RCT (double-blind)	40 mg (e) and 100 mg (c)/9 mo	C: Control supplement I: Multinutrient supplement	A significant drop in weight and a decrease in hunger	Serum cholesterol, triglyceride, glucose, insulin, and leptin levels were all reduced, as were mild side effects such dry mouth, sleeplessness, anxiety, and palpitation	Hackman et al. (2006)
Over wt. women	*Magnolia officinalis* and *Phellodendron amurense* proprietary extracts	RCT	750 mg/6 wk	C: Placebo I: Extract	No sig. wt. gain for the I group but sig. wt. gain	In the evening, the I groups had lower levels of cortisol	Garrison and Chambliss (2006)
Human (over wt. and	Ephedrine, caffeine, and other stimulants are present in this product	RCT (double-blind)	12 wk	C: Placebo I: Compound	In C. groups added weight loss of 1/5 kg and a higher decrease in BMI, and waist circumference	It's an unique white bean fractionation method	Coffey et al., (2004)

TABLE 9.1 *(Continued)*

Target	Herbs (scientific name)	Study	Dose/duration	Groups	Main outcome	Other relevant effects and complications	References
obese) (n = 102)					No difference in fat mass or body fat percentage		
Obese (n = 60)	An HCA-SX product, niacin, bound chromium (NBC), or *Gymnema sylvestre* extract are all examples of hydroxycitric acid (HCA) (GSE)	RCT (double-blind)	HCA-SX: 4667 mg GSE: 400 mg NBC: 4 mg/8 wk	C: Placebo I1 = HCA-SX I2 = GSE + NBC + HCA-SX All groups had wt. loss program	Percent was shown 5–6% decrease in body wt. and BMI & sig. decrease in food intake in both I groups	In both I groups, serum lipids and leptin dropped, but HDL levels climbed and fat metabolites were eliminated in the urine. Although there were a few minor side effects, they were not life-altering	Preuss et al., (2004)
Obese (n = 24)	It is an unique white bean fractionation method. (*Phaseolus vulgaris*)	RCT (double-blind)	3000 mg/8 wk	C: Placebo I: Extract	129% decrease in body weight	Between groups TG was reduced by a factor three times that of the C. group. No adverse	Udani et al., (2004)
Healthy (n = 58)	Guggulu (Medohar)	RCT	1/5, 3 g/30 d	C: wt. loss program I: wt. loss program + extract	Mean losing weight was greater in the I group. In the I group, all patients who weighed more than 90 kg lost weight, whereas just three patients in the C group did it	effect was shown	Bhatt et al., (1995)
Over wt. and obese	*Cissus quadrangularis*	RCT (double-blind)	300, 1028 mg	C: Placebo I: Two extract formulation: CQR-300, CORE	Did not lose wt. Body weight and body fat have decreased significantly	Serum lipids and glucose levels were reduced to a significant extent. HDL-C plasma 5-HT and creatinine levels increased a substantial portion in response to the therapy	Oben et al., (2007)

TABLE 9.2 Animal Studies on the Anti-Obesity of Herbal Medicines.

Target	Herbs (scientific name)	Dose/duration	Groups	Main outcome	Other potential side effects and problems are included	References
Rat (obese)	*Laminaria japonica* Aresch and *Benincasa hispida* (Thunb.) Cogn.	(2.5 g/kg)	–	Significant drop in Lee's index and cell size	Obesity can be prevented and weight loss achieved by reducing subcutaneous fat mass significantly	Wang et al. (2000)
Mouse	*Rhus vemiciflua* Stokes	8 wk	C: HFD I: HFD + extract	Obesity can be prevented and weight loss achieved by reducing subcutaneous fat mass substantially	Lowered plasma TG	Jeon et al. (2003)
Mouse	*Hibiscus sabdariffa*	120 mg/kg; 60 d	C: Healthy and obese (by MSG) + placebo I: Same groups + extract	Obese mice's weight growth was drastically decreased, and both groups drank more liquids	There is no signature. TG and Chol levels have changed. Although ALT levels rose, the rise was insignificant	Alarcon-Aguilar et al. (2007)
Mouse	*Agave tequilana* (TEQ) and *Dasylirion* spp. *fructans* (DAS)	10% supplement	C: STD I: STD + Raftilose/DAS/TEQ	Body weight increase and food consumption seem to be slowing down The group (TEQ) had the lowest score	There was a significant drop in TG levels in the Raftilose group, but a decrease in glucose and cholesterol levels. GLP-1, its precursor, and proglucagon mRNA are more abundant in I groups	Urías-Silvas et al. (2008)
Rat	*Platycodon grandiflorum*	150 mg/kg 7 wk	C: NLD/HFD I: Same groups + extract	Weight and size of subcutaneous adipose tissue and adipocytes significantly decreased in the I group	HFD greatly reduces plasma TG and chol levels and significantly increases FABP mRNA expression	Park et al. (2007)

MSG, monosodium glutamate; FABO, fatty acid β oxidation; STD, standard diet; LFD, low-fat diet; NLD, normal diet; HFD, high-fat diet; FABP, fatty acid-binding protein; FFM, fat-free mass; sig.: significant; *AST*, aspartate transaminase; *ALT*, alanine transaminase; *C*, control; *I*, intervention; *FAS*, fatty acid synthetase; *UCP*, uncoupling protein; *GDP*, guanosine 5' diphosphate; *FAS*, fatty acid synthetase; *TG*, triglyceride; *HDL*, high-density lipoprotein; *LDL*, low-density lipoprotein; *FBG*, fasting blood glucose, *ip*, intraperitoneal; *iv*, intravenous. Caps HT2 is a herbal formulation containing methanolic extract of selected parts of plants: *Commiphora mukul*; *Allium sativum*; *Plumbago indica/Semicarpus anacardium/Hemidesmus indicus/Terminalia arjuna/Tinospora cordifolia/Withania somnifera/Ocimum sanctum*.

nausea and vomiting; (v) blurred vision; (vi) extreme tiredness and weakness; and (vii) irritability, mood changes, and other symptoms.

While diabetes pathophysiology remains uncertain, experimental data show that free radicals may be involved in its development (Ramachandran et al., 2002; Matteucci and Giampietro, 2000; Oberley et al., 1988; Baynes and Thorpe, 1997). Damage to biological components (DNA, proteins, and lipids) is caused by free radicals. Antioxidants that neutralize free radicals have been found to lessen the severity of diabetic complications in animal models (Lipinski, 2001; Kubisch et al., 1997; Singh et al., 2017). Naziroglu et al. (2001) for the most part, diabetic problems are the result of lipid and protein abnormalities. Elastin, laminin, and collagen are frequently attacked by free radicals in diabetics because these proteins have a long shelf-life outside of the cell. These proteins are transformed into glycoproteins by hyperglycemia. Patients with diabetes are more likely to suffer from problems, such as cataracts or microangiopathy because of alterations in proteins in tissues, such as the lens, the vascular wall, and basement membranes (Giugliano et al., 1996). In diabetes, free radicals oxidize the lipoproteins. As a result, lipoprotein metabolism exhibits a number of abnormalities (VLDL, low-density lipoprotein, and HDL). Patients with diabetes are more vulnerable to lipid peroxidation because they have higher levels of oxidative stress in their bodies. It is because nonenzymatic glycosylation of proteins produces AGEs and hence they are so dangerous. In tissues, long-lived molecules are more likely to accumulate AGEs, which may lead to problems in cell and tissue function (Brownlee, 1996; Elgawish et al., 1999). In addition, AGEs have a direct effect on both micro and macrovascular permeability because they bind specifically to the macrophage receptors. Endothelial dysfunction occurs as a result of free radical production. AGEs can also disrupt nucleic acids and histones, which can lead to mutations and altered organic sensitivity in these molecules. Diabetes may be a complicated condition with numerous side effects and needs a wide range of treatment options. Patients with diabetes are either insulin-deficient or insulin-resistant. In the event of a complete deficiency of insulin, the diabetic patients are treated by giving insulin injections. For those cases when insulin does not reach the cells, numerous medications have been established that take into account the problems of carbohydrate metabolism. For example, glucosidase inhibitors, such as acarbose, miglitol, and voglibose are used to treat gut-based postprandial hyperglycemia. By preventing carbs from degradation, they reduce the amount of glucose absorbed by cells. Metformin (a biguanide) is used to improve glucose absorption by peripheral cells.

Insulinotropic and pancreatic cell secretory stimulant, glibenclamide is a sulphonylurea used in treating diabetes.

Existing therapies have limitations because of their high price and side effects, such as hypoglycemia (low blood sugar), weight gain, digestive problems, and liver damage (Dey et al., 2002). Antioxidant and antidiabetic treatments are being sought after in light of new discoveries and the significance of oxidative stress in diabetes complications should be understood.

There is renewed interest in finding a cure for diabetes mellitus via the use of medicinal herbs. Conventional medications produced from medicinal plant prototypic chemicals are many. Oral glucose-lowering drug metformin has been shown to be effective. *Galega officinalis*, a plant used to treat diabetes, helped spur its creation. Guanidine, a hypoglycemic agent, is found in *G. officinalis*. Synalin A and Synalin B are the alkyl biguanides developed as oral antidiabetic medications in Europe in the 1920s, and due to the toxicity of guanidine, were eventually withdrawn off the market as insulin became more widely accessible. Metformin was developed as a result of the guanidine and biguanide research. Traditional remedies for diabetes have been recorded in the past, but only a handful have been evaluated scientifically and medically to determine their efficacy. Human and animal studies demonstrate that several plant extracts can lower blood sugar levels. The World Health Organization's Expert Committee on Diabetes has called for additional research into traditional medicinal herbs.

Researchers have researched only a small number of people with diabetes mellitus in order to understand the mechanism of action of herbal supplements. As listed in Table 9.3, many plants have been demonstrated to increase glucose absorption in cultured muscle cells, liver cells, or preadipocytes, indicating that they may have hypoglycemic effects. *Euclea undulatus*, *Strychnos henningsii*, *Cissampelos capensis*, *Elaeodendron transvaalense*, and *Hypoxis argentea* are some of the other plants used in the treatment of diabetes. *Senna alexandri*, *Cymbopogon citratus*, *Nuxia floribunda*, and *Curcubita pepo* include inhibitors of -amylase and glucosidase, which may lower postprandial hyperglycemia. To increase the insulin release, the regeneration of pancreatic beta-cells may be maintained by antidiabetic agents found in *Carica papaya* and *H. argentea* (Sasidharan et al., 2011). Several plants, such as *Vernonia amygdalina*, *Hypoxis hemerocallidea*, *Mimusops zeyheri*, *Catharanthus roseus*, and *Sutherlandia frutescens*, are well-known around the globe for their ability to cure diabetes. The absence of pharmacological proof has led traditional practitioners

TABLE 9.3 Indian Medicinal Plants with Antidiabetic and Related Beneficial Properties.

Plant name	Ayurvedic/common name/ Herbal formulation	Antidiabetic and other beneficial effects in traditional medicine	References
Annona squamosa	Sugar apple	Ethanolic leaf extract contains hypoglycemic and antihyperglycemic effects, as well as a higher plasma insulin level	Kaleem et al. (2006), Gupta, et al. (2005), Gupta, M et al. (2005)
Artemisia pallens	Davana	Glucose utilization is increased or inhibited in hypoglycemic patients	Subramoniam et al. (1996)
Areca catechu	Supari	Hypoglycemic	Chempakam (1993)
Beta vulgaris	Chukkander	Increases glucose tolerance in OGTT	Yoshikawa et al. (1996)
Boerhaavia diffusa	Punarnava	Glucose-6-phosphatases are reduced, oxidative stress and hexokinase activity are raised, and insulin levels are enlarged	Pari and Venkateswaran (2004), Pari and Amarnath Satheesh and Pari (2004a, 2004b)
Bombax ceiba	Semul	Hypoglycemic	Saleem et al. (1999)
Butea monosperma	Palasa	Antihyperglycemic	Somani et al. (2006)
Camellia sinensis	Tea	Antioxidant and hypoglycemic activity	Gomes et al. (1995), Devasagayam et al. (1996)
Capparis decidua	Karir or Pinju	Hypolipidemic and low-glycemic properties	Agarwal and Chauhan (1988)
Caesalpinia bonducella	Sagarghota, Fevernut	Insulin secretory, hypolipidemic hypoglycemic	Chakrabarti et al. (2005), Chakrabarti et al. (2003), Sharma et al. (1997)
Coccinia indica	Bimb or Kanturi	Hypoglycemic	Kamble et al. (1998)
Emblica officinalis	Amla, Dhatriphala, a component of the "Triphala" herbal formulation	Antioxidant, hypoglycemic, and reduces lipid peroxidation	Bhattacharya et al. (1999), Kumar and Müller (1999)
Eugenia uniflora	Pitanga	Lipase activity is inhibited when hypoglycemic	Devasagayam et al. (1995)

TABLE 9.3 *(Continued)*

Plant name	Ayurvedic/common name/ Herbal formulation	Antidiabetic and other beneficial effects in traditional medicine	References
Enicostema littorale	Krimihrita	Reduce the activity of glucose 6-phosphatase and fructose 1,6 bisphosphatase by increasing the hexokinase enzyme's activity. The hypoglycemic effect is dose-dependent	Arai et al. (1999), Maroo et al. (2003)
Ficus bengalensis	Bur	Antioxidant and hypoglycemic	Vijayvargia et al. (2000)
Gymnema sylvestre	Gudmar or Merasingi	Antihyperglycemic and hypolipidemic effects	Augusti et al. (1994), Chattopadhyay (1999)
Hemidesmus indicus	Anantamul	Antivenom and anti-inflammatory properties	Preuss et al. (1998)
Hibiscus rosasinesis	Gudhal or Jasson	Beta-cells in the pancreas begin to secrete insulin as a result of this substance..	Alam and Gomes (1998)
Ipomoea batatas	Sakkargand	Insulin resistance is lessened	Sachdewa et al. (1999)
Momordica cymbalaria	Kadavanchi	Hypolipidemic and hypoglycemic	Kusano and Abe (2000), Nagaraju (1992)
Murraya koenigii	Curry patta	Glycogenolysis and gluconeogenesis are decreased, but glycogenesis is increased	Rao and Rao (1999)
Musa sapientum	Banana	Antioxidant and antihyperglycemic	Khan et al. (1995), Dhanabal et al. (2005)
Phaseolus vulgaris Hulga	white kidney bean	The alpha-amylase activity is inhibited by an antioxidative, hypoglycemic, hypolipidemic drug. There has been evidence of abnormal mRNA expression in the skeletal muscle of insulin receptors and GLUT-4	Pari and Umamaheswari (2000), Pari and Maheswari et al. (1999), Tormo et al. (2004)
Punica granatum	Anar	Antioxidant and hypoglycemic properties	Venkateswaran et al. (2004), Knott et al. (1992)

TABLE 9.3 *(Continued)*

Plant name	Ayurvedic/common name/ Herbal formulation	Antidiabetic and other beneficial effects in traditional medicine	References
Salacia reticulata	Vairi	Inhibitory action against sucrase and α-glucosidase	Jafri et al. (2000)
Scoparia dulcis	Sweet broomweed	Hypoglycemic, antihyperlipidic, and antioxidant properties make it a good insulin-secretagogue	Yoshikawa et al. (1998), Pari and Latha (2005), Pari and Latha (2006)
Swertia chirayita	Chirata	Stimulates insulin release from islets	Latha et al. (2004), Saxena et al. (1993)
Syzygium alternifolium	Shahajire	Hypoglycemic and antihyperglycemic	Rao et al. (2001)
Terminalia belerica	Behada, a constituent of "Triphala"	Hypoglycemia and an antibacterial	Sabu and Kuttan (2002)
Terminalia chebula	Hirda	Hypoglycemic and antibacterial	Sabu and Kuttan (2002)
Tinospora crispa		A hypoglycemic medication that increases insulin release from the pancreatic	Noor and Ashcroft (1998)
Vinca rosea	Sadabahar	Antihyperglycemic	Chattopadhyay et al. (1991)
Withania somnifera	Ashvagandha, winter cherry	Dietary diuretic and cholesterol-lowering medication	Andallu et al. (2000)

to overlook certain plants traditionally used to treat diabetes, such as *M. zeyheri*. Many of these botanicals' effective dosages, toxicity, and possible combinations with contemporary antidiabetic pharmaceutical medicines are yet undetermined.

9.4 THE USE OF MEDICINAL PLANTS IN THE MANAGEMENT OF CARDIOVASCULAR DISORDERS

The greatest cause of death in the general population is CVD (Hu et al., 2004; Mayosi et al., 2009). CVDs and their risk factors associated include hypertension, myocardial infarction, angina pectoris, stroke, atherosclerosis, peripheral artery disease, and transient ischemic attack (TIA) (Michalsky et al., 2018). Low- and middle-income nations bear a significant share of the burden of CVD. Middle-aged persons are most at risk for CVD (Lawes et al., 2008). The incidence of CVD has increased significantly during the last several decades (Alberts et al., 2005).

CVD problems may be prevented in part by lowering blood pressure using antihypertensive medications. Allopurinol, calcium channel blockers, beta-blockers, and thiazide diuretics are among the most often prescribed antihypertensive medications in the world (Tasos et al., 2019). CVDs, such as hypertension and atherosclerosis can be treated using plant extracts similar to ischemic heart disease and congestive heart failure (Rastogi et al., 2016; Rouhi-Boroujeni et al., 2017). Notable medicinal plants used across the globe to treat hypertension are discussed in the next section. As a treatment for hypertension, some traditional healers have suggested the oral decoctions of *Helichrysum ceres* (Musabayane et al., 2008). Bioactive natriuretic and diuretic phytochemicals are considered to be responsible for the plant extract's hypotensive effects (Musabayane, 2012). Ethanolic leaf extracts of *H. ceres* have also been demonstrated in vivo to show a vasodilatory effect on the vascular smooth muscles, resulting in a reduction in TPR (total peripheral resistance) and so decreasing blood pressure (Kamadyaapa et al., 2009). To reduce hypertension in mouse models, ethanolic extracts of *Ekebergia capensis* leaves were successfully used (Bwititi et al., 2000). The hypotensive effects of *E. capensis* is believed to be due to its effect on the vascular smooth muscle TPR. An innovative diabetic rat model has shown that the inability of the kidney to eliminate salt from the blood may be repaired by using the leaves of *Opuntia megacantha* (Al-Qattan et al., 2008). This shows that the kidney's ability to regulate blood volume may be a therapeutic advantage of plant extracts in the treatment of hypertension. In addition to the phenols and flavonoids found in *A. sativum*, *Sclerocarya birrea*, *Ficus thonningii*, and *Olea europea*, more medicinal plants that are often used in the treatment of hypertension are also known for their vasorrelaxant, bradycardic, and cardioprotective properties (Somova et al., 2003). The leaves of Wild African olive trees (*O. europea*) have diuretic, anti-atherosclerotic, and antihypertensive properties (Cooper et al., 1997). This medicinal plant has been shown to be effective against hypertension and atherosclerosis in animal trials utilizing an insulin-resistant rat model after 6 weeks of therapy with extracts of *O. europea*. Two important blood pressure regulation signaling pathways that are targeted by numerous antihypertensive herbal treatments include RAAS and renin–angiotensin–aldosterone. When angiotensin I is converted to angiotensin II, this enzyme plays a critical role in the development of hypertension (Morgan et al., 2001). Screening for antihypertensive drugs is based on the ACE inhibition. *Asystasia gangetica* and *Asystasia dubius* have been shown to possess ACE inhibitor action similar to *Galinsoga parviflora* and *Tulbagia violacea*, as well as the extracts of *Stangeria eriopus* and *Oxygonum* and *Tulbagia dubius* from both water and ethanol, accordingly (Table 9.4). For hypertension, the ACE inhibition activity of *T. violacea* was the highest (Ramesar et al., 2008), suggesting its potential. This ACE suppression is due to tannins interfering with the ACE activity found in many of these medicinal plants (Duncan et al., 1999).

9.5 MEDICINAL PLANTS FOR NON-ALCOHOLIC FATTY LIVER DISEASE

NAFLD-related deaths and diseases may be on the rise over the world. Research demonstrates that NAFLD can develop independently of the obesity-related metabolic syndrome, which has long been associated with obesity (Lonardo et al., 2015). Advanced technology or diagnostic biopsies are necessary for the diagnosis of NAFLD. Traditional medical practitioners, on the other hand, have a lot of work to do before they can accurately identify NAFLD. This study will focus on medicinal herbs from South Africa that have been shown to reduce NAFLD and hepatic steatosis. The existing material is mostly derived from animal research in the lab. High-fructose diets have been associated with a higher risk of NAFLD. For 12 weeks, Wistar rats fed a high-fructose diet were protected from NAFLD by aqueous extracts of *Terminalia sericea* leaves.

TABLE 9.4 Indian Medicinal Plants with Cardiovascular and Related Beneficial Properties.

Plant Name	Ayurvedic/common name/Herbal formulation	Cardio vascular and other beneficial effects in traditional medicine	References
Helichrysum ceres	Blombos straw flower	Hypotensive	Musabayane (2012)
Ekebergia capensis	Cape ash, dogplum (English)	Hypotensive	Musabayane (2012)
Opuntia megacantha	Sweet prickly pear (English)	Hypotensive	Kamadyaapa et al. (2009)
Allium sativum	Garlic (English)	Attenuation of structural nephropathy progression	Bwititi et al. (2000)
Sclerocarya birrea	Marula (English)	Phenolic compounds hypotensive	Al-Qattan et al. (2008)
Ficus thonningii	Stem bark	Decrease in mean arterial pressure	(Braca et al., 2003)
Olea europea	Swati	Diuretic, anti-atherosclerotic, and antihypertensive effects	Musabayane et al. (2007), Bennani-Kabchi et al. (2000)
Tulbaghia violacea	Wild garlic	ACE inhibitors	Somova et al., (2003), Ramesar et al., (2008)

Rooibos (*Aspalathus linearis*) herbal tea, prepared from the plant's aerial parts, has been proved to offer several health benefits. The C-glucosyl dihydrochalcone aspalathin present in fermented tea, which is more concentrated than the normal tea, is often attributed to these health benefits. Adipocyte-judged 3T3-L cells exhibited improved lipid metabolism in the presence of aspalathin and aspalathin-enriched tea, which was confirmed in vitro tests (Mazibuko et al., 2015)". Green Rooibos extract rich in aspalathin was found to lower insulin resistance in the liver and to reduce obesity and insulin resistance in obese, insulin-resistant rats in vitro by processes that involved modulation of AMPK pathways (Mazibuko-Mbeje et al., 2019).

Obesity and NAFLD treatment may benefit from dietary modifications, including a limited food diet. As an appetite suppressor, it has been utilized by the Khoi San people for many years. *Hoodia gordonii* is a succulent shrub. *H. gordonii* may be a viable therapy option for NAFLD because of its appetite-suppressing properties. According to a recent study conducted on obese rats (Van Heerden, 2008), treatment of obesity with *H. gordonii* extract reduced body weight, muscle mass, and adipocyte size, despite the fact that studies on human individuals have shown mixed outcomes in this aspect (Smith and Krygsman, 2014). Using *H. gordonii* to treat NAFLD will require more research. *S. frutescens*, a legume, is well-known for its medicinal benefits. Recently, it has been demonstrated to have antidiabetic characteristics, and it has also been shown to modify the lipid metabolism in 3T3 adipocytes and insulin-resistant rats (MacKenzie et al., 2012). The ability of hot aqueous extract to reverse fructose-induced liver steatosis has also been shown in vivo (Clarke, 2010). It has been shown that *Aloe vera* gel worsened hepatic steatosis in rats when administered in the form of a hot aqueous extract. Kaempferol, one of *A. vera*'s primary bioactive agents, is thought to be responsible for the plant's hepatoprotective properties (Bhalla and Chauhan, 2015).

When given to Zucker diabetic fatty rats, the phytosterols, such as lophenol and cycloartanol in *A. vera* dramatically decreased the lipogenic gene expression and reduced the hepatic lipid accumulation (Misawa et al., 2012). Experimenting with a high-fructose diet on rats, we discovered that *Moringa oleifera* leaves might prevent the development of NAFLD (Muhammad et al., 2020). *M. oleifera* fermented extract may have decreased the obesity-inducing effects of a high-fat diet in rats. Inflammatory cytokine mRNA levels may have decreased as a result of the activation of genes associated with lipid metabolism in the liver and the subsequent drop in hepatic mRNA levels of these genes (Joung et al., 2017). DWJ504 was shown to disrupt the signaling pathways for Toll-like 4 receptors, tumor necrosis factor alpha, and interleukin 6 in the final 4 weeks of a high-fat diet in mice, according to a study in mice (Kang et al., 2016). Oil from *Syzigium aromaticum* was previously found to be hepatoprotective against chemicals that were intended to harm the liver, and oleanolic acid from this plant was found to protect neonatal rats from developing the disease-causing condition known as NAFLD when it was given to them while they were nursing. *S. aromaticum*'s cold-pressed oil (Nyakudya et al., 2018). Many plants have been found to be useful in experimental models to prevent NAFLD and/or therapy as a consequence (Table 9.5). For human consumption, however, plants and their physiologically active components must be examined.

9.6 CONCLUSIONS

South African health care costs are mostly driven by non-communicable illnesses and metabolic disease risk factors. South Africa has a unique variety of medicinal plants with well-documented and scientifically proven therapeutic properties that may be used to reduce the costs associated with health care. The significance of gut microorganisms in metabolic diseases has gained increased attention. Bacteroidetes-to-Firmicutes ratios may be influenced by phytochemicals found in several plants reviewed in this study (Kumar Singh et al., 2019). Plant polyphenols and nondigestible fiber may potentially have a favorable prebiotic effect on the microbiome (Carrera-Quintanar et al., 2018). It is possible to treat certain gastrointestinal issues with conventional medicine, but it does not have the ability to evaluate how it affects our gut bacteria. There is a need for further research, and most definitely for a review specifically dedicated to the current crucial topic. There is a need to preserve plant diversity and ensure that plants are not over-exploited, as has been seen in China (Li et al., 2015). According to a study, new plant-derived medications for treating metabolic syndrome hold enormous promise and should not be ignored in the battle to prevent their adverse influence on healthcare.

TABLE 9.5 Lists of Plants with Laboratory-Tested Potential for Use in the Treatment of Non-Alcoholic Fatty Liver Disease AMPK, AMP-Activated Protein Kinase

Plant Name	Ayurvedic/common name/Herbal formulation	Beneficial effects in traditional medicine	References
Moringa oleifera	Moringa, Drumstick tree	Pro-inflammatory pathways are suppressed by upregulation of lipid metabolism genes in the brain	Joung et al. (2017)
Aspalathus linearis (Burm. F.) R. Dahlgren	Rooibos, red bush (English)	Improved adipocyte lipid metabolism prevents AMPK from being triggered by insulin resistance in the liver	Mazibuko et al. (2015)
Hoodia gordonii	Bitter ghaap (English); khobab (Khoi)	Appetite suppression, decrease adipocytes	Smith and Krygsman (2014)
Sutherlandia frutescens	Cancer bush (English)	Regulation of adipocytes and lipid metabolism	MacKenzie et al. (2012)
Aloe vera	Aloe (English)	Phytosterols, decrease hepatic lipid accumulation	Bhalla and Chauhan (2015)
Opuntia ficus-indica (L) Mill	Prickly pear (English)	Decrease inflammation, prevent steatosis	Kang et al. (2016)
Syzigium aromaticum	Clove (English)	Regulating pathways in the liver responsible for the metabolism of lipids	Nyakudya et al. (2018)

KEYWORDS

- **medicinal plants**
- **herbs**
- **metabolic syndrome**
- **safety**
- **diabetes**
- **risk factors**
- **cardiovascular risk**
- **bioactive substances**

REFERENCES

Abdollahi, M.; Afshar-Imani, B. A Review on Obesity and Weight Loss Measures. *Middle East Pharm.* **2003,** *11*, 6–10.

Abidov, M. T.; del Rio, M. J.; Ramazanov, T. Z.; Klimenov, A. L.; Dzhamirze, Sh; Kalyuzhin, O. V. Effects of *Aralia mandshurica* and Engelhardtia Chrysolepis Extracts on Some Parameters of Lipid Metabolism in Women with Nondiabetic Obesity. *Bull. Exp. Biol. Med.* **2006,** *141* (3), 343–346. https://doi.org/10.1007/s10517-006-0167-3

Andallu, B.; Radhika, B. Hypoglycemic, Diuretic and Hypocholesterolemic Effect of Winter Cherry (Withania somnifera, Dunal) Root. *Indian J. Exp. Biol.* **2000,** *38* (6), 607–609.

Agarwal, V.; Chauhan, B. M. A Study on Composition and Hypolipidemic Effect of Dietary Fiber from Some Plant Foods. *Plant Foods Human Nutr.* **1988,** *38* (2), 189–197. https://doi.org/10.1007/BF01091723

Akinrinde, A.; Koekemoer, T.; Van De Venter, M. V.; Bradley, G. In Vitro Investigation of Potential Anti-Diabetic Activity of the Corm Extract of Hypoxis argentea Harv. ex Baker. *Acta Pharmaceutica* **2018,** *68* (4), 389–407. https://doi.org/10.2478/acph-2018-0023

Alam, M. I.; Gomes, A. Viper Venom-Induced Inflammation and Inhibition of Free Radical Formation by Pure Compound (2-Hydroxy-4-Methoxybenzoic Acid) Isolated and Purified from Anantamul (Hemidesmus indicus R. *Toxicon* **1998,** *36* (1), 207–215. https://doi.org/10.1016/s0041-0101 (97)00070-6

Alarcon-Aguilar, F. J.; Zamilpa, A.; Perez-Garcia, M. D.; Almanza-Perez, J. C.; Romero-Nuñez, E.; Campos-Sepulveda, E. A.; Vazquez-Carrillo, L. I.; Roman-Ramos, R. Effect of *Hibiscus sabdariffa* on Obesity in MSG Mice. *J. Ethnopharmacol.* **2007,** *114* (1), 66–71. https://doi.org/10.1016/j.jep.2007.07.020

Alberts, M.; Urdal, P.; Steyn, K.; Stensvold, I.; Tverdal, A.; Nel, J. H.; Steyn, N. P. Prevalence of Cardiovascular Diseases and Associated Risk Factors in a Rural Black Population of South Africa. *Eur. J. Cardiovasc. Prev. Rehabil.* **2005,** *12* (4), 347–354. https://doi.org/10.1097/01.hjr.0000174792.24188.8e

Al-Qattan, K.; Thomson, M.; Ali, M. Garlic (*Allium sativum*) and Ginger (*Zingiber officinale*) Attenuate Structural Nephropathy Progression in Streptozotocin-Induced Diabetic Rats. *e-SPEN, Eur. J. Clin. Nutr. Metabol.* **2008,** *3* (2), e62–e71. https://doi.org/10.1016/j.eclnm.2007.12.001

Altemimi, A.; Lakhssassi, N.; Baharlouei, A.; Watson, D. G.; Lightfoot, D. A. Phytochemicals: Extraction, Isolation and Identification of Bioactive Compounds from Plant Extracts. *Plants* **2017,** *6* (4), 42. https://doi.org/10.3390/plants6040042

Arai, I.; Amagaya, S.; Komatsu, Y.; Okada, M.; Hayashi, T.; Kasai, M.; Arisawa, M.; Momose, Y. Improving Effects of the Extracts from *Eugenia uniflora* on Hyperglycemia and Hypertriglyceridemia in Mice. *J. Ethnopharmacol.* **1999,** *68* (1–3), 307–314. https://doi.org/10.1016/s0378-8741 (99)00066-5

Augusti, K. T.; Daniel, R. S.; Cherian, S.; Sheela, C. G.; Nair, C. R. Effect of Leucoperalgonin Derivative from *Ficus bengalensis* Linn. on Diabetic Dogs. *Indian J. Med. Res.* **1994,** *99*, 82–86.

Baynes, J. W.; Thorpe, S. R. The Role of Oxidative Stress in Diabetic Complications. *Curr. Opin. Endocrinol. Diab.* **1996,** *3*(4), 277–284. https://doi.org/10.1097/00060793-199608000-00001

Bennani-Kabchi, N.; Fdhil, H.; Cherrah, Y.; El Bouayadi, F. B.; Kehel, L.; Marquie, G. Therapeutic Effect of Olea europea var. Oleaster Leaves on Carbohydrate and Lipid Metabolism in Obese and Prediabetic Sand Rats (Psammomysobesus). *Ann. Pharm. Franc* **2000,** *58* (4), 271–277.

Bhalla, A.; Chauhan, U. Identification of Anti-hyperlipidemic Components in Aloe Vera Through Reverse Phase HPlC. *J. Biol. Sci. Med.* **2015,** *1*, 21–27.

Bhatt, A. D.; Dalal, D. G.; Shah, S. J.; Joshi, B. A.; Gajjar, M. N.; Vaidya, R. A.; Vaidya, A. B.; Antarkar, D. S. Conceptual and Methodologic Challenges of Assessing the Short-Term Efficacy of Gugguluin Obesity: Data Emergent from a Naturalistic Clinical Trial. *J. Postgraduate Med.* **1995,** *41* (1), 5–7.

Bhattacharya, A.; Chatterjee, A.; Ghosal, S.; Bhattacharya, S. K. Antioxidant Activity of Active Tannoid Principles of *Emblica officinalis* (Amla). *Indian J. Experimental Biol.* **1999,** *37* (7), 676–680.

Boaduo, N. K. K.; Katerere, D.; Elo, J. N.; Naidoo, V. Evaluation of Six Plant Species Used Traditionally in the Treatment and Control of Diabetes Mellitus in South Africa Using in Vitro Methods. *Pharma. Biol.* **2014,** *52* (6), 756–761. https://doi.org/10.3109/13880209.2013.869828

Boozer, C. N.; Nasser, J. A.; Heymsfield, S. B.; Wang, V.; Chen, G.; Solomon, J. L. An Herbal Supplement Containing Ma Huang-Guarana for Weight Loss: A Randomized, Double-Blind Trial. *Int. J. Obes. Relat. Metabol. Disord.* **2001,** *25* (3), 316–324. https://doi.org/10.1038/sj.ijo.0801539

Boudreau, D. M.; Malone, D. C.; Raebel, M. A.; Fishman, P. A.; Nichols, G. A.; Feldstein, A. C.; Boscoe, A. N.; Ben-Joseph, R. H.; Magid, D. J.; Okamoto, L. J. Health Care Utilization and Costs by Metabolic Syndrome Risk Factors. *Metabol. Synd. Relat. Disord.* **2009,** *7* (4), 305–314. https://doi.org/10.1089/met.2008.0070

Braca, A.; Politi, M.; Sanogo, R.; Sanou, H.; Morelli, I.; Pizza, C.; De Tommasi, N. Chemical Composition and Antioxidant Activity of Phenolic Compounds from Wild and Cultivated Sclerocaryabirrea (Anacardiaceae) Leaves. *J. Agric. Food Chem.* **2003,** *51* (23), 6689–6695. https://doi.org/10.1021/jf030374m

Bray, G. A. Medical Consequences of Obesity. *J. Clin. Endocrinol. Metabol.* **2004,** *89* (6), 2583–2589.https://doi.org/10.1210/jc.2004-0535

Brownlee, M. Advanced Protein Glycosylation in Diabetes in Diabetes and Ageing. *Annu. Rev. Med.* **1996,** *46*, 223–234.

Bwititi, P.; Musabayane, C. T.; Nhachi, C. F. Effects of *Opuntia megacantha* on Blood Glucose and Kidney Function in Streptozotocin Diabetic Rats. *J. Ethnopharmacol.* **2000,** *69* (3), 247–252. https://doi.org/10.1016/s0378-8741 (99)00123-3

Caballero, B. The Global Epidemic of Obesity: An Overview. *Epidemiol. Rev.* **2007,** *29*, 1–5. https://doi.org/10.1093/epirev/mxm012

Carrera-Quintanar, L.; López Roa, R. I.; Quintero-Fabián, S.; Sánchez-Sánchez, M. A.; Vizmanos, B.; Ortuño-Sahagún, D. Phytochemicals That Influence Gut Microbiota as Prophylactics and for the Treatment of Obesity and Inflammatory Diseases. *Mediators Inflamm.* **2018,** *2018*, 9734845.https://doi.org/10.1155/2018/9734845

Chakrabarti, S.; Biswas, T. K.; Rokeya, B.; Ali, L.; Mosihuzzaman, M.; Nahar, N.; Khan, A. K.; Mukherjee, B. Advanced Studies on the Hypoglycemic Effect of Caesalpiniabonducella F. in Type 1 and 2 Diabetes in Long Evans Rats. *J. Ethnopharmacol.* **2003,** *84* (1), 41–46. https://doi.org/10.1016/s0378-8741 (02)00262-3

Chakrabarti, S.; Biswas, T. K.; Seal, T.; Rokeya, B.; Ali, L.; Azad Khan, A. K.; Nahar, N.; Mosihuzzaman, M.; Mukherjee, B. Antidiabetic Activity of Caesalpinia bonducella F. in Chronic Type 2 Diabetic Model in Long-Evans Rats and Evaluation of Insulin Secretagogue Property of Its Fractions on Isolated Islets. *J. Ethnopharmacol.* **2005,** *97* (1), 117–122. https://doi.org/10.1016/j.jep.2004.10.025

Chattopadhyay, R. R. A Comparative Evaluation of Some Blood Sugar Lowering Agents of Plant Origin. *J. Ethnopharmacol.* **1999,** *67* (3), 367–372. https://doi.org/10.1016/s0378-8741 (99)00095-1

Chattopadhyay, S. R.; Sarkar, S. K.; Ganguly, S.; Banerjee, R. N.; Basu, T. K. Hypoglycemic and Anti-Hyperglycemic Effect of *Vinca rosea* Linn. *Indian J. Physiol. Pharmacol.* **1991,** *35*, 145–151.

Chempakam, B. Hypoglycemic Activity of Arecoline in Betel Nut *Areca catechu* L. *Indian J. Exp. Biol.* **1993,** *31* (5), 474–475.

Chen, A. K.; Roberts, C. K.; Barnard, R. J. Effect of a Short-Term Diet and Exercise Intervention on Metabolic Syndrome in Overweight Children. *Metabol.: Clin. Exp.* **2006,** *55* (7), 871–878. https://doi.org/10.1016/j.metabol.2006.03.001

Chinese Diabetes Society. Standards of Medical Care for Type 2 Diabetes in China 2017. *Chin. J. Pract. Intern. Med.* **2018,** *38*, 292–344.

Chrubasik, C.; Maier, T.; Dawid, C.; Torda, T.; Schieber, A.; Hofmann, T.; Chrubasik, S. An Observational Study and Quantification of the Actives in a Supplement with *Sambucus nigra* and *Asparagus officinalis* Used for Weight Reduction. *Phytother. Res.* **2008,** *22* (7), 913–918. https://doi.org/10.1002/ptr.2415

Church, T. Exercise in obesity, metabolic syndrome, and diabetes. *Progress Cardiovasc. Dis.* **2011,** *53* (6), 412–418. https://doi.org/10.1016/j.pcad.2011.03.013

Clarke, D. B. Glucosinolates, Structures and Analysis in Food. *Analyt. Chem.* **2010,** *9660*, 310–325.

Coffey, C. S.; Steiner, D.; Baker, B. A.; Allison, D. B. A Randomized Double-Blind Placebo-Controlled Clinical Trial of a Product Containing Ephedrine, Caffeine, and Other Ingredients from Herbal Sources for Treatment of Overweight and Obesity in the Absence of Lifestyle Treatment. *Int. J. Obes. Relat. Metabol. Disord.* **2004,** *28* (11), 1411–1419. https://doi.org/10.1038/sj.ijo.0802784

Cooper, R.; McFarlane-Anderson, N.; Bennett, F. I.; Wilks, R.; Puras, A.; Tewksbury, D.; Ward, R.; Forrester, T. ACE Angiotensinogen and Obesity: A Potential Pathway Leading to Hypertension. *J. Human Hypertens.* **1997,** *11* (2), 107–111. https://doi.org/10.1038/sj.jhh.1000391

Curioni, C. C.; Lourenço, P. M. Long-Term Weight Loss After Diet and Exercise: A Systematic Review. *Int. J. Obes.* **2005,** *29* (10), 1168–1174. https://doi.org/10.1038/sj.ijo.0803015

Deutschländer, M. S.; Lall, N.; Van de Venter, M.; Hussein, A. A. Hypoglycemic Evaluation of a New Triterpene and Other Compounds Isolated from Euclea undulata Thunb.var. myrtina (Ebenaceae) Root Bark. *J. Ethnopharmacol.* **2011,** *133* (3), 1091–1095. https://doi.org/10.1016/j.jep.2010.11.038

Devasagayam, T. P. A.; Kamat, J. P.; Mohan, H.; Kesavan, P. C. Caffeine as an Antioxidant: Inhibition of Lipid Peroxidation Induced by Reactive Oxygen Species. *Biochim. Biophys. Acta* **1996,** *1282* (1), 63–70. https://doi.org/10.1016/0005-2736 (96)00040-5

Devasagayam, T. P. A.; Subramanian, M.; Singh, B. B.; Ramanathan, R.; Das, N. P. Protection of Plasmid pBR322 DNA by Flavonoids Against Single-Strand Breaks Induced by Singlet Molecular Oxygen. *J. Photochem. Photobiol. B, Biol.* **1995,** *30* (2–3), 97–103. https://doi.org/10.1016/1011-1344 (95)07159-y

Dey, L.; Attele, A. S.; Yuan, C. S. Alternative Therapies for Type 2diabetes. *Altern. Med. Rev.* **2002,** *7* (1), 45–58.

Dhanabal, S. P.; Sureshkumar, M.; Ramanathan, M.; Suresh, B. Hypoglycemic Effect of Ethanolic Extract of *Musa sapientum* on Alloxan Induced Diabetes Mellitus in Rats and Its Relation with Antioxidant Potential. *J. Herbal Pharmacother.* **2005,** *5* (2), 7–19. https://doi.org/10.1080/J157v05n02_02

Dillard, C. J.; German, J. B. Phytochemicals: Nutraceuticals and Human Health. *J. Sci. Food Agric.* **2000,** *80* (12), 1744–1756. https://doi.org/10.1002/1097-0010 (20000915)80:12<1744::AID-JSFA725>3.0.CO;2-W

Dommermuth, R.; Ewing, K. Metabolic Syndrome: Systems Thinking in Heart Disease. *Primary Care* **2018,** *45* (1), 109–129. https://doi.org/10.1016/j.pop.2017.10.003

Duncan, A. C.; Jäger, A. K.; van Staden, J. Screening of Zulu Medicinal Plants for Angiotensin Converting Enzyme (ACE) Inhibitors. *J. Ethnopharmacol.* **1999,** *68* (1–3), 63–70. https://doi.org/10.1016/s0378-8741 (99)00097-5

Elgawish, A.; Glomb, M.; Friendlander, M.; Monnier, V. M. Involvement of Hydrogen Peroxide in Collagen Crosslinking by High Glucose In Vitro and In Vivo. *J. Biol. Chem.* **1999,** *271*, 12964–12971.

El-Hadary, A. E.; Ramadan Hassanien, M. F. Hepatoprotective Effect of Cold-Pressed Syzygium aromaticum Oil Against Carbon Tetrachloride (CCl_4)-Induced Hepatotoxicity in Rats. *Pharma. Biol.* **2016,** *54* (8), 1364–1372. https://doi.org/10.3109/13880209.2015.1078381

Esposito, K.; Chiodini, P.; Colao, A.; Lenzi, A.; Giugliano, D. Metabolic Syndrome and Risk of Cancer: A Systematic Review and Meta-Analysis. *Diab. Care* **2012,** *35* (11), 2402–2411. https://doi.org/10.2337/dc12-0336

Feng, J. F.; Tang, Y. N.; Ji, H.; Xiao, Z. G.; Zhu, L.; Yi, T. Biotransformation of Dioscoreanipponica by Rat Intestinal Microflora and Cardio Protective Effects of Diosgenin. *Oxid. Med. Cell. Longevity* **2017,** *2017*, 4176518. https://doi.org/10.1155/2017/4176518

Field, A. E.; Coakley, E. H.; Must, A.; Spadano, J. L.; Laird, N.; Dietz, W. H.; Rimm, E.; Colditz, G. A. Impact of Overweight on the Risk of Developing Common Chronic Diseases During a 10-Year Period. *Arch. Intern. Med.* **2001,** *161* (13), 1581–1586. https://doi.org/10.1001/archinte.161.13.1581

Garrison, R.; Chambliss, W. G. Effect of a Proprietary Magnolia and Phellodendron Extract on Weight Management: A Pilot, Double-Blind, Placebo-Controlled Clinical Trial. *Altern. Ther. Health Med.* **2006,** *12* (1), 50–54.

Gluckman, P. D.; Hanson, M. A.; Buklijas, T.; Low, F. M.; Beedle, A. S. Epigenetic Mechanisms That Underpin Metabolic and Cardiovascular Diseases. *Nat. Rev. Endocrinol.* **2009,** *5* (7), 401–408. https://doi.org/10.1038/nrendo.2009.102

Giugliano, D.; Ceriello, A.; Paolisso, G. Oxidative Stress and Diabetic Vascular Complications. *Diab. Care* **1996,** *19* (3), 257–267. https://doi.org/10.2337/diacare.19.3.257

Goedecke, J. H.; Jenning, C. L.; Lambertc, E. V. *Chronic Diseases of Lifestyle in South Africa*, 1995–2005.

Gomes, A.; Vedasiromoni, J. R.; Das, M.; Sharma, R. M.; Ganguly, D. K. Antihyperglycemic Effect of Black Tea (*Camellia sinensis*) in Rats. *J. Ethnopharmacol.* **1995,** *45* (3), 223–226. https://doi.org/10.1016/0378-8741 (95)01223-z

Greenway, F. L.; De Jonge, L.; Blanchard, D.; Frisard, M.; Smith, S. R. Effect of a Dietary Herbal Supplement Containing Caffeine and Ephedra on Weight, Metabolic Rate, and Body Composition. *Obes. Res.* **2004,** *12* (7), 1152–1157. https://doi.org/10.1038/oby.2004.144

Grover, J. K.; Yadav, S.; Vats, V. Medicinal Plants of India with Antidiabetic Potential. *J. Ethnopharmacol.* **2002,** *81* (1), 81–100. https://doi.org/10.1016/s0378-8741 (02)00059-4

Guan, X.-R.; Zhu, L.; Xiao, Z.-G.; Zhang, Y.; Chen, H.; Yi, T. Bioactivity, Toxicity and Detoxification Assessment of Dioscorea bulbifera L.: A Comprehensive Review. *Phytochem. Rev.* **2017,** *16* (3), 573–601. https://doi.org/10.1007/s11101-017-9505-5

Gupta, R. K.; Kesari, A. N.; Murthy, P. S.; Chandra, R.; Tandon, V.; Watal, G. Hypoglycemic and Antidiabetic Effect of Ethanolic Extract of Leaves of *Annona squamosa* L. in Experimental Animals. *J. Ethnopharmacol.* **2005,** *99* (1), 75–81. https://doi.org/10.1016/j.jep.2005.01.048

Gupta, R. K.; Kesari, A. N.; Watal, G.; Murthy, P. S.; Chandra, R.; Tandon, V. Nutritional and Hypoglycemic Effect of Fruit Pulp of *Annona squamosa* in Normal Healthy and Alloxan-Induced Diabetic Rabbits. *Ann. Nutr. Metabol.* **2005,** *49* (6), 407–413. https://doi.org/10.1159/000088987

Hackman, R. M.; Havel, P. J.; Schwartz, H. J.; Rutledge, J. C.; Watnik, M. R.; Noceti, E. M.; Stohs, S. J.; Stern, J. S.; Keen, C. L. Multinutrient Supplement Containing Ephedra and Caffeine Causes Weight Loss and Improves Metabolic Risk Factors in Obese Women: A Randomized Controlled Trial. *Int. J. Obes.* **2006,** *30* (10), 1545–1556. https://doi.org/10.1038/sj.ijo.0803283

Hioki, C.; Yoshimoto, K.; Yoshida, T. Efficacy of Bofu-Tsushosan, an Oriental Herbal Medicine, in Obese Japanese Women with Impaired Glucose Tolerance. *Clin. Exp. Pharmacol. Physiol.* **2004,** *31* (9), 614–619. https://doi.org/10.1111/j.1440-1681.2004.04056.x

Hoeger, W. W.; Harris, C.; Long, E. M.; Hopkins, D. R. Four Week Supplementation with a Natural Dietary Compound Produces Favorable Changes in Body Composition. *Adv. Ther.* **1998,** *15* (5), 305–314.

Howard, A. N. The Historical Development, Efficacy and Safety of Very-Low -Calorie Diets. *Int. J. Obes.* **1981,** *5* (3), 195–208.

Hu, G.; Qiao, Q.; Tuomilehto, J.; Balkau, B.; Borch-Johnsen, K.; Pyorala, K.; DECODE Study Group. Prevalence of the Metabolic Syndrome and Its Relation to All-Cause and Cardiovascular Mortality in Nondiabetic European Men and Women. *Arch. Intern. Med.* **2004,** *164* (10), 1066–1076. https://doi.org/10.1001/archinte.164.10.1066

Ignjatovic, V.; Ogru, E.; Heffernan, M.; Libinaki, R.; Lim, Y.; Ng, F. Studies on the Use of "slimax", a Chinese Herbal Mixture, in the Treatment of Human Obesity. *Pharma. Biol.* **2000,** *38* (1), 30–35. https://doi.org/10.1076/1388-0209 (200001)3811-BFT030

In Medical Research Council. *Technical Report*; Tygerberg: Western Cape, 2006; pp 65–79.

Isomaa, B.; Almgren, P.; Tuomi, T.; Forsén, B.; Lahti, K.; Nissén, M.; Taskinen, M. R.; Groop, L. Cardiovascular Morbidity and Mortality Associated with the Metabolic Syndrome. *Diab. Care* **2001,** *24* (4), 683–689. https://doi.org/10.2337/diacare.24.4.683

Jafri, M. A.; Aslam, M.; Javed, K.; Singh, S. Effect of *Punica granatum* Linn. (Flowers) on Blood Glucose Level in Normal and Alloxan Induced Diabetic Rats. *J. Ethnopharmacol.* **2000,** *70* (3), 309–314. https://doi.org/10.1016/s0378-8741 (99)00170-1

Jeon, W. K.; Kim, J. H.; Lee, H. W.; Ko, B. S.; Kim, H. K. Effects of Rhus Verniciflua Stokes (RVS) Extract on Diet-Induced Obesity in C57BL/6 Mouse. *Kor. J. Pharmacogn.* **2003,** *34*, 339–343.

Joung, H.; Kim, B.; Park, H.; Lee, K.; Kim, H. H.; Sim, H. C.; Do, H. J.; Hyun, C. K.; Do, M. S. Fermented *Moringa oleifera* Decreases Hepatic Adiposity and Ameliorates Glucose Intolerance in High-Fat Diet-Induced Obese Mice. *J. Med. Food* **2017,** *20* (5), 439–447. https://doi.org/10.1089/jmf.2016.3860

Kaleem, M.; Asif, M.; Ahmed, Q. U.; Bano, B. Antidiabetic and Antioxidant Activity of *Annona squamosal* Extract in Streptozotocin-Induced Diabetic Rats. *Singapore Med. J.* **2006,** *47* (8), 670–675.

Kamadyaapa, D. R.; Gondwe, M. M.; Moodley, K.; Ojewole, J. A.; Musabayane, C. T. Cardiovascular Effects of Ekebergia capensis Sparrm (Meliaceae) Ethanolic Leaf Extract in Experimental Animal Paradigms. *Cardiovasc. J. Africa* **2009,** *20* (3), 162–7, 169.

Kamble, S. M.; Kamlakar, P. L.; Vaidya, S, and Bambole. V (D).: Influence of *Coccinia indica* on Certain Enzymes in Glycolytic and Lipolytic Pathway in Human Diabetes. *Indian J. Med. Sci.* **1998,** *52*, 143–146.

Kang, J. W.; Shin, J. K.; Koh, E. J.; Ryu, H.; Kim, H. J.; Lee, S. M. Opuntia Ficus-Indica Seed Attenuates Hepatic Steatosis and Promotes M2 Macrophage Polarization in High-Fat Diet-Fed Mice. *Nutr. Res.* **2016,** *36* (4), 369–379. https://doi.org/10.1016/j.nutres.2015.12.007

Kanwar, P.; Sharma, N.; Rekha, A. Medicinal Plants Use in Traditional Healthcare Systems Prevalent in Western Himalayas. *Ind. J (Know (Trans.))* **2006,** *5*, 300–309.

Keservani, R. K.; Kesharwani, R. K.; Vyas, N.; Jain, S.; Raghuvanshi, R.; Sharma, A. K. Nutraceuticals and Functional Food as Future Food. *Der Pharm. Lett.* **2010a.,** *2* (1), 106–116.

Keservani, R. K.; Kesharwani, R. K.; Sharma, A. K.; Vyas, N.; Chadoker, A. Nutritional Supplements: An Overview. *Int. J. Curr. Pharm. Rev. Res.* **2010b,** *1* (1), 59–75.

Keservani, R. K.; Sharma A. K. Flavonoids: Emerging Trends and Potential Health Benefits. *J. Chin. Pharm. Sci.* **2014,** *23* (12), 815-822.

Keservani, R. K.; Kesharwani, R. K.; Sharma, A. K. Medicinal Effect of Nutraceutical Fruits for the Cognition and Brain Health. *Scientifica* **2016,** 1–10. Article ID 31**09254,** DOI: 10.1155/2016/3109254 citrusbb

Keservani, R. K.; Sharma, A. K.; Kesharwani, R. K. An Overview and Therapeutic Applications of Nutraceutical & Functional Foods. In: *Recent Advances in Drug Delivery Technology*; Keservani, R. K., Sharma, A. K., Kesharwani, R. K., Eds.; IGI Global Publisher, 2017; Ch 6; pp 160–201. ISBN:9781522507543

Khan, B. A.; Abraham, A.; Leelamma, S. Hypoglycemic Action of Murrayakoenigii (Curry Leaf) and *Brassica juncea* (Mustard) Mechanism of Action. *Indian J. Biochem. Biophys.* **1995,** *32* (2), 106–108.

Knott, R. M.; Grant, G.; Bardocz, S.; Pusztai, A.; de Carvalho, A. F.; Hesketh, J. E. Alterations in the Level of Insulin Receptor and GLUT-4 mRNA in Skeletal Muscle from Rats Fed a Kidney Bean (*Phaseolus vulgaris*) Diet. *Int. J. Biochem.* **1992,** *24* (6), 897–902. https://doi.org/10.1016/0020-711x (92)90094-h

Kopelman, P. G. Obesity as a Medical Problem. *Nature* **2000,** *404* (6778), 635–643. https://doi.org/10.1038/35007508

Kubisch, H. M.; Wang, J.; Bray, T. M.; Phillips, J. P. Targeted Over Expression of Cu/Zn Superoxide Dismutase Protects Pancreatic Beta Cells Against Oxidative Stress. *Diabetes* **1997,** *46* (10), 1563–1566. https://doi.org/10.2337/diabetes.46.10.1563

Kumar Singh, A.; Cabral, C.; Kumar, R.; Ganguly, R.; Kumar Rana, H.; Gupta, A.; RosariaLauro, M.; Carbone, C.; Reis, F.; Pandey, A. K. Beneficial Effects of Dietary Polyphenols on Gut Microbiota and Strategies to Improve Delivery Efficiency. *Nutrients* **2019,** *11* (9), 2216. https://doi.org/10.3390/nu11092216

Kumar, K. C. S.; Müller, K. Medicinal Plants from Nepal, II: Evaluation as Inhibitors of Lipid Peroxidation in Biological Membranes. *J. Ethnopharmacol.* **1999,** *64* (2), 135–139. https://doi.org/10.1016/s0378-8741 (98)00117-2

Kuriyan, R.; Raj, T.; Srinivas, S. K.; Vaz, M.; Rajendran, R.; Kurpad, A. V. Effect of *Caralluma fimbriata* Extract on Appetite, Food Intake and Anthropometry in Adult Indian Men and Women. *Appetite* **2007,** *48* (3), 338–344. https://doi.org/10.1016/j.appet.2006.09.013

Kusano, S.; Abe, H. Antidiabetic Activity of Whites Skinned Potato (*Ipomoea batatas*) in Obese Zucker Fatty Rats. *Biol. Pharma. Bull.* **2000,** *23* (1), 23–26. https://doi.org/10.1248/bpb.23.23

Latha, M.; Pari, L.; Sitasawad, S.; Bhonde, R. Insulin Secretagogue Activity and Cytoprotective Role of the Traditional Antidiabetic Plant *Scoparia dulcis* (Sweet Broom Weed). *Life Sci.* **2004,** *75* (16), 2003–2014. https://doi.org/10.1016/j.lfs.2004.05.012

Lawes, C. M.; Vander Hoorn, S.; Rodgers, A.; International Society of Hypertension. Global Burden of Blood-Pressure-Related Disease, 2001. *Lancet* **2008,** *371* (9623), 1513–1518. https://doi.org/10.1016/S0140-6736 (08)60655-8

Lembede, B. W. *Effect of Dietary Terminalia Sericea Aqueous Leaf Extracts on High-Fructose Diet Fed Growing Wistar Rats* [Master's Thesis]; University of the Witwatersrand, 2014.

Li, W.; Song, F.; Wang, X.; Wang, L.; Wang, D.; Yin, X.; Cao, S.; Gong, Y.; Yue, W.; Yan, F.; Zhang, H.; Sheng, Z.; Wang, Z.; Lu, Z. Prevalence of Metabolic Syndrome Among Middle-Aged and Elderly Adults in China: Current Status and Temporal Trends. *Ann. Med.* **2018,** *50* (4), 345–353. https://doi.org/10.1080/07853890.2018.1464202

Li, X.; Chen, Y.; Lai, Y.; Yang, Q.; Hu, H.; Wang, Y. Sustainable Utilization of Traditional Chinese Medicine Resources: Systematic Evaluation on Different Production Modes. *Evid.-Based Complem. Altern. Med* **2015,** *2015*, 218901. https://doi.org/10.1155/2015/218901

Lim, S.; Eckel, R. H. Pharmacological Treatment and Therapeutic Perspectives of Metabolic Syndrome. *Rev. Endocr. Metabol. Disord.* **2014,** *15* (4), 329–341. https://doi.org/10.1007/s11154-014-9298-4

Lipinski, B. Pathophysiology of Oxidative Stress in Diabetes Mellitus. *J. Diab. Complications* **2001,** *15* (4), 203–210. https://doi.org/10.1016/s1056-8727 (01)00143-x

Lonardo, A.; Ballestri, S.; Marchesini, G.; Angulo, P.; Loria, P. Nonalcoholic Fatty Liver Disease: A Precursor of the Metabolic Syndrome. *Digest. Liver Dis.* **2015,** *47* (3), 181–190. https://doi.org/10.1016/j.dld.2014.09.020

Lowe, M. R. Self-Regulation of Energy Intake in the Prevention and Treatment of Obesity: Is It Feasible? *Obes. Res.* **2003,** *11*, 44S–59S. https://doi.org/10.1038/oby.2003.223

MacKenzie, J.; Koekemoer, T. C.; Roux, S.; van de Venter, M.; Dealtry, G. B. Effect of *Sutherlandia frutescens* on the Lipid Metabolism in an Insulin Resistant Rat Model and 3T3-L1 Adipocytes. *Phytother. Res.* **2012,** *26* (12), 1830–1837. https://doi.org/10.1002/ptr.4653

Maroo, J.; Vasu, V. T.; Gupta, S. Dose Dependent Hypoglycemic Effect of Aqueous Extract of Enicostemalittoraleblume in Alloxan Induced Diabetic Rats. *Phytomedicine* **2003,** *10* (2–3), 196–199. https://doi.org/10.1078/094471103321659933

Matteucci, E.; Giampietro, O. Oxidative Stress in Families of Type 1 Diabetic Patients. *Diab. Care* **2000,** *23* (8), 1182–1186. https://doi.org/10.2337/diacare.23.8.1182

Mayosi, B. M.; Flisher, A. J.; Lalloo, U. G.; Sitas, F.; Tollman, S. M.; Bradshaw, D. The Burden of Non-Communicable Diseases in South Africa. *Lancet* **2009,** *374* (9693), 934–947. https://doi.org/10.1016/S0140-6736 (09)61087-4

Mazibuko, S. E.; Joubert, E.; Johnson, R.; Louw, J.; Opoku, A. R.; Muller, C. J. Aspalathin Improves Glucose and Lipid Metabolism in 3T3-L1 Adipocytes Exposed to Palmitate. *Mol. Nutr. Food Res.* **2015,** *59* (11), 2199–2208. https://doi.org/10.1002/mnfr.201500258

Mazibuko-Mbeje, S. E.; Dludla, P. V.; Roux, C.; Johnson, R.; Ghoor, S.; Joubert, E.; Louw, J.; Opoku, A. R.; Muller, C. J. F. Aspalathin-Enriched Green Rooibos Extract Reduces Hepatic Insulin Resistance by Modulating PI3K/AKT and AMPK Pathways. *Int. J. Mol. Sci.* **2019,** *20* (3), 633.https://doi.org/10.3390/ijms20030633

McGaw, L.; Jäger, A.; Grace, O.; Fennel, C.; van Staden, J. Medicinal Plants. In *Ethics in Agriculture—An African Perspective*; Springer, 2005; pp 67–83.

McTigue, K. M.; Harris, R.; Hemphill, B.; Lux, L.; Sutton, S.; Bunton, A. J.; Lohr, K. N. Screening and Interventions for Obesity in Adults: Summary of the Evidence for the U. S. Preventive Services Task Force. *Ann. Intern. Med.* **2003,** *139* (11), 933–949. https://doi.org/10.7326/0003-4819-139-11-200312020-00013

Mendrick, D. L.; Diehl, A. M.; Topor, L. S.; Dietert, R. R.; Will, Y.; La Merrill, M. A.; Bouret, S.; Varma, V.; Hastings, K. L.; Schug, T. T.; Emeigh Hart, S. G.; Burleson, F. G. Metabolic Syndrome and Associated Diseases: From the Bench to the Clinic. *Toxicol. Sci.* **2018,** *162* (1), 36–42. https://doi.org/10.1093/toxsci/kfx233

Michalsky, M. P.; Inge, T. H.; Jenkins, T. M.; Xie, C.; Courcoulas, A.; Helmrath, M.; Brandt, M. L.; Harmon, C. M.; Chen, M.; Dixon, J. B.; Urbina, E. M.; Teen-LABS Consortium. Cardiovascular Risk Factors After Adolescent Bariatric Surgery. *Pediatrics* **2018,** *141* (2), e20172485.https://doi.org/10.1542/peds.2017-2485

Miller, J. M.; Kaylor, M. B.; Johannsson, M.; Bay, C.; Churilla, J. R. Prevalence of Metabolic Syndrome and Individual Criterion in US Adolescents: 2001–2010 National Health and Nutrition Examination Survey. *Metabol. Synd. Relat. Disord.* **2014,** *12* (10), 527–532. https://doi.org/10.1089/met.2014.0055

Misawa, E.; Tanaka, M.; Nabeshima, K.; Nomaguchi, K.; Yamada, M.; Toida, T.; Iwatsuki, K. Administration of Dried Aloe Vera Gel Powder Reduced Body Fat Mass in Diet-Induced Obesity (DIO) Rats. *J. Nutr. Sci. Vitaminol.* **2012,** *58* (3), 195–201. https://doi.org/10.3177/jnsv.58.195

Moller, D. E. New Drug Targets for Type 2 Diabetes and the Metabolic Syndrome. *Nat.* **2001,** *414* (6865), 821–827. https://doi.org/10.1038/414821a

Morgan, T. O.; Anderson, A. I.; MacInnis, R. J. ACE Inhibitors, Beta-Blockers, Calcium Blockers, and Diuretics for the Control of Systolic Hypertension. *Am. J. Hypertens.* **2001,** *14* (3), 241–247. https://doi.org/10.1016/s0895-7061 (00)01266-8

Muhammad, N.; Ibrahim, K. G.; Ndhlala, A. R.; Erlwanger, K. H. Moringa oleifera Lam. Prevents the Development of High Fructose Diet-Induced Fatty Liver. *South African J. Bo.* **2020,** 129, 32–39. https://doi.org/10.1016/j.sajb.2018.12.003

Musabayane, C. T. The Effects of Medicinal Plants on Renal Function and Blood Pressure in Diabetes Mellitus. *Cardiovasc. J. Africa* **2012,** *23* (8), 462–468. https://doi.org/10.5830/CVJA-2012-025

Musabayane, C. T.; Gondwe, M.; Kamadyaapa, D. R.; Chuturgoon, A. A.; Ojewole, J. A. Effects of *Ficus thonningii* (Blume)[Morarceae] Stem-Bark Ethanolic Extract on Blood

Glucose, Cardiovascular and Kidney Functions of Rats, and on Kidney Cell Lines of the Proximal (LLC-PK1) and Distal Tubules (MDBK). *Renal Failure* **2007,** *29* (4), 389–397. https://doi.org/10.1080/08860220701260735

Musabayane, C. T.; Kamadyaapa, D. R.; Gondwe, M.; Moodley, K.; Ojewole, J. A. Cardiovascular Effects of Helichrysum ceres S. Moore [Asteraceae] Ethanolic Leaf Extract in Some Experimental Animal Paradigms. *Cardiovasc. J. Afr.* **2008,** *19* (5), 246–253.

Nagaraju, N. *Biochemical Studies on Some Medicinal Plants of Rayalaseema Region* [PhDThesis]; S. V. University, 1992.

Naghdi, N. Folklore Medicinal Plants Used in Liver Disease: A Review. *Int. J. Green Pharm. (IJGP)* **2018,** *12*, 463–472.

Naziroglu, M.; Cay, M. Protective Role of Intraperitoneally Administered Vitamin E and Selenium on the Oxidative Defense Mechanisms in Rats with Diabetes Induced by Streptozotocin. *Biol. Stress Elem. Res.* **2001,** *47*, 475–488.

Ng, K. W.; Cao, Z. J.; Chen, H. B.; Zhao, Z. Z.; Zhu, L.; Yi, T. Oolong Tea: A Critical Review of Processing Methods, Chemical Composition, Health Effects, and Risk. *Crit. Rev. Food Sci. Nutr.* **2018,** *58* (17), 2957–2980. https://doi.org/10.1080/10408398.2017.1347556

Nishida, C.; Mucavele, P. Monitoring the Rapidly Emerging Public Health Problem of Overweight and Obesity: The WHO Global Database on Body Mass Index. *SCN News* **2005,** *29*, 5–12.

Nolan, P. B.; Carrick-Ranson, G.; Stinear, J. W.; Reading, S. A.; Dalleck, L. C. Prevalence of Metabolic Syndrome and Metabolic Syndrome Components in Young Adults: A Pooled Analysis. *Prev. Med. Rep.* **2017,** *7*, 211–215. https://doi.org/10.1016/j.pmedr.2017.07.004

Noor, H.; Ashcroft, S. J. Pharmacological Characterization of the Anti-Hyperglycemic Properties of Tinosporacrispa Extract. *J. Ethnopharmacol.* **1998,** *62* (1), 7–13. https://doi.org/10.1016/s0378-8741 (98)00008-7

Nyakudya, T. T.; Mukwevho, E.; Nkomozepi, P.; Erlwanger, K. H. Neonatal Intake of Oleanolic Acid Attenuates the Subsequent Development of High Fructose Diet-Induced Non-Alcoholic Fatty Liver Disease in Rats. *J. Dev. Origins Health Dis.* **2018,** *9* (5), 500–510. https://doi.org/10.1017/S2040174418000259

O'Neill, S.; O'Driscoll, L. Metabolic Syndrome: A Closer Look at the Growing Epidemic and Its Associated Pathologies. *Obes. Rev.* **2015,** *16* (1), 1–12. https://doi.org/10.1111/obr.12229

Oben, J. E.; Enyegue, D. M.; Fomekong, G. I.; Soukontoua, Y. B.; Agbor, G. A. The Effect of *Cissus quadrangularis* (CQR-300) and a Cissus Formulation (CORE) on Obesity and Obesity Induced Oxidative Stress. *Lipids Health Dis.* **2007,** *6*, 4.https://doi.org/10.1186/1476-511X-6-4; 6: 4

Oben, J. E.; Ngondi, J. L.; Momo, C. N.; Agbor, G. A.; Sobgui, C. S. The Use of a Cissus Quadrangularis/Irvingia gabonensis Combination in the Management of Weight Loss: A Double Blind Placebo-Controlled Study. *Lipids Health Dis.* **2008,** *7*, 12. https://doi.org/10.1186/1476-511X-7-12

Oberley, L. W. Free Radicals and Diabetes. *Free Radic. Biol. Med.* **1988,** *5* (2), 113–124. https://doi.org/10.1016/0891-5849 (88)90036-6

Ou-Yang, S. H.; Jiang, T.; Zhu, L.; Yi, T. Dioscorea nipponica Makino: A Systematic Review on Its Ethnobotany, Phytochemical and Pharmacological Profiles. *Chem. Central J.* **2018,** *12* (1), 57.https://doi.org/10.1186/s13065-018-0423-4

Oyedemi, S.; Koekemoer, T.; Bradley, G.; van de Venter, M.; Afolayan, A. In Vitro Anti-Hyperglycemia Properties of the Aqueous Stem Bark Extract from Strychnos henningsii (Gilg). *Int. J. Diab. Dev. Countries* **2013,** *33* (2), 120–127. https://doi.org/10.1007/s13410-013-0120-8

Pari, L.; Amarnath Satheesh, M. Antidiabetic Activity of Boerhavia diffusa L. Effect on Hepatic Key Enzymes in Experimental Diabetes. *J. Ethnopharmacol. 2004a*, *91* (1), 109–113. https://doi.org/10.1016/j.jep.2003.12.013

Pari, L.; Amarnath Satheesh, M. Antidiabetic Effect of Boerhavia diffusa: Effect on Serum and Tissue Lipids in Experimental Diabetes. *J. Med. Food 2004b*, *7* (4), 472–476. https://doi.org/10.1089/jmf.2004.7.472

Pari, L.; Latha, M. Antidiabetic Effect of *Scoparia dulcis*: Effect on Lipid Peroxidation in Streptozotocin Diabetes. *Gen. Physiol. Biophys.* **2005,** *24* (1), 13–26.

Pari, L.; Latha, M. Antihyperlipidemic Effect of *Scoparia dulcis* (Sweet Broom Weed) in Streptozotocin Diabetic Rats. *J. Med. Food* **2006,** *9* (1), 102–107. https://doi.org/10.1089/jmf.2006.9.102

Pari, L.; Maheswari, J. U. Hypoglycaemic Effect of *Musa sapientum* L. in Alloxan-Induced Diabetic Rats. *J. Ethnopharmacol.* **1999,** *68* (1–3), 321–325. https://doi.org/10.1016/s0378-8741 (99)00088-4

Pari, L.; Umamaheswari, J. Antihyperglycaemic Activity of *Musa sapientum* Flowers: Effect on Lipid Peroxidation in Alloxan Diabetic Rats. *Phytother. Res.* **2000,** *14* (2), 136–138. https://doi.org/10.1002/ (sici)1099-1573 (200003)14:2<136::aid-ptr607>3.0.co;2-k

Pari, L.; Venkateswaran, S. Protective Role of *Phaseolus vulgaris* on Changes in the Fatty Acid Composition in Experimental Diabetes. *J. Med. Food* **2004,** *7* (2), 204–209. https://doi.org/10.1089/1096620041224120

Park, Y. S.; Yoon, Y.; Ahn, H. S. *Platycodon grandiflorum* Extract Represses Up-Regulated Adipocyte Fatty Acid Binding Protein Triggered by a High Fat Feeding in Obese Rats. *World J. Gastroenterol.* **2007,** *13* (25), 3493–3499. https://doi.org/10.3748/wjg.v13.i25.3493

Peeters, A.; Barendregt, J. J.; Willekens, F.; Mackenbach, J. P.; Al Mamun, A. A.; Bonneux, L.; NEDCOM, the Netherlands Epidemiology and Demography Compression of Morbidity Research Group. Obesity in Adulthood and Its Consequences for Life Expectancy: A Life-Table Analysis. *Ann.f Intern. Med.* **2003,** *138* (1), 24–32. https://doi.org/10.7326/0003-4819-138-1-200301070-00008

Preuss, H. G.; Bagchi, D.; Bagchi, M.; Rao, C. V.; Dey, D. K.; Satyanarayana, S. Effects of a Natural Extract of (-)-Hydroxycitric Acid (HCA-SX) and a Combination of HCA-SX Plus Niacin-Bound Chromium and Gymnemasylvestre Extract on Weight Loss. *Diab., Obes. Metabol.* **2004,** *6* (3), 171–180. https://doi.org/10.1111/j.1462-8902.2004.00328.x

Preuss, H. G.; Jarrell, S. T.; Scheckenbach, R.; Lieberman, S.; Anderson, R. A. Comparative Effects of Chromium, Vanadium and Gymnemasylvestre on Sugar-Induced Blood Pressure Elevations in SHR. *J. Am. College Nutr.* **1998,** *17* (2), 116–123.https://doi.org/10.1080/07315724.1998.10718736

Ramachandran, A.; Snehalatha, C.; Viswanathan, V. Burden of Type 2 Diabetes and Its Complications—The Indian Scenario. *Curr. Sci.* **2002,** *83*, 1471–1476.

Ramesar, S.; Baijnath, H.; Govender, T.; Mackraj, I. Angiotensin I-Converting Enzyme Inhibitor Activity of Nutritive Plants in KwaZulu-Natal. *J. Med. Food* **2008,** *11* (2), 331–336. https://doi.org/10.1089/jmf.2007.569

Ramírez-Vélez, R.; Pérez-Sousa, M. Á, González-Ruíz, K.; Cano-Gutierrez, C. A.; Schmidt-RioValle, J.; Correa-Rodríguez, M.; Izquierdo, M.; Romero-García, J. A.; Campos-Rodríguez, A. Y.; Triana-Reina, H. R.; González-Jiménez, E. Obesity- and Lipid-Related Parameters in the Identification of Older Adults with a High Risk of Prediabetes According to the American Diabetes Association: An Analysis of the 2015 Health, Well-Being, and Aging Study. *Nutrients* **2019,** *11* (11), E2654.https://doi.org/10.3390/nu11112654

Rane, B. R.; Keservani, R. K. Introduction to Nutraceutical & Dietary Supplements. In: *Nutraceutical and Dietary Supplements*. In Keservani, R. K., Sharma, A. K., Kesharwani, R. K., Eds.; Apple Academic Press, CRC Press, Taylor & Francis Group, 2020a; Chapter 3; pp 35–55. ISBN: 9781771888738.

Rane, B. R.; Patil A. S.; Keservani, R. K.; Jain, A. S. Novel Approaches in Herbal Formulation. In: *Enhancing the Therapeutic Efficacy of Herbal Formulations through Novel Drug Delivery Systems*; Kesharwani, R. K.; Keservani, R. K.; Sharma, A. K., Eds.; IGI Global International Publisher: Pennsylvania, 2020b; Chapter 2; pp 43–68. ISBN13: 9781799844532.

Rane, B. R.; Tadavi, S. A.; Keservani, R. K. Naturopathy. In *Herbal Product Development*; Sharma, A. K.; Keservani, R. K.; Gautam, S. P., Eds.; Apple Academic Press, CRC Press, Taylor & Francis Group, 2020c; Chapter 12; pp 321–347. ISBN: 9781771888776

Rao, B. K.; Rao, C. H. Hypoglycemic and Anti-Hyperglycemic Activity of Syzygium alternifolium (Wt.) Walp. Seed Extracts in Normal and Diabetic Rats. *Phytomedicine* **2001,** *8* (2), 88–93. https://doi.org/10.1078/0944-7113-00015

Rao, B. K.; Kesavulu, M. M.; Giri, R.; Appa Rao, C. Antidiabetic and Hypolipidemic Effects of *Momordica cymbalaria* Hook. Fruit Powder in Alloxan-Diabetic Rats. *J. Ethnopharmacol.* **1999,** *67* (1), 103–109. https://doi.org/10.1016/s0378-8741 (99)00004-5

Rastogi, S.; Pandey, M. M.; Rawat, A. K. Traditional Herbs: A Remedy for Cardiovascular Disorders. *Phytomedicine* **2016,** *23* (11), 1082–1089. https://doi.org/10.1016/j.phymed.2015.10.012

Rodriguez-Casado, A. The Health Potential of Fruits and Vegetables Phytochemicals: Notable Examples. *Crit. Rev. Food Sci. Nutr.* **2016,** *56* (7),1097–1107. https://doi.org/10.1080/10408398.2012.755149

Rodriguez-Colon, S. M.; Mo, J.; Duan, Y.; Liu, J.; Caulfield, J. E.; Jin, X.; Liao, D. Metabolic Syndrome Clusters and the Risk of Incident Stroke: The Atherosclerosis Risk in Communities (ARIC) Study. *Stroke* **2009,** *40* (1), 200–205. https://doi.org/10.1161/STROKEAHA.108.523035

Roongpisuthipong, C.; Kantawan, R.; Roongpisuthipong, W. Reduction of Adipose Tissue and Body Weight: Effect of Water Soluble Calcium Hydroxycitrate in Garcinia atroviridis on the Short Term Treatment of Obese Women in Thailand. *Asia Pac. J. Clin. Nutr.* **2007,** *16* (1), 25–29.

Rouhi-Boroujeni, H.; Heidarian, E.; Rouhi-Boroujeni, H.; Deris, F.; Rafieian-Kopaei, M. Medicinal Plants with Multiple Effects on Cardiovascular Diseases: A Systematic Review. *Curr. Pharma. Design* **2017,** *23* (7), 999–1015.https://doi.org/10.2174/1381612822666161021160524

Sabu, M. C.; Kuttan, R. Anti-Diabetic Activity of Medicinal Plants and Its Relationship with Their Antioxidant Property. *J. Ethnopharmacol.* **2002,** *81* (2), 155–160. https://doi.org/10.1016/s0378-8741 (02)00034-x

Sachdewa, A.; Khemani, L. D. A Preliminary Investigation of the Possible Hypoglycemic Activity of *Hibiscus rosa-sinensis*. *Biomed. Environ. Sci.* **1999,** *12* (3), 222–226.

Saleem, R.; Ahmad, M.; Hussain, S. A.; Qazi, A. M.; Ahmad, S. I.; Qazi, M. H.; Ali, M.; Faizi, S.; Akhtar, S.; Husnain, S. N. Hypotensive, Hypoglycaemic and Toxicological Studies on the Flavonol C-Glycoside Shamimin from *Bombax ceiba*. *Planta Medica* **1999,** *65* (4), 331–334. https://doi.org/10.1055/s-1999-14060

Sasidharan, S.; Sumathi, V.; Jegathambigai, N. R.; Latha, L. Y. Anti-Hyperglycaemic Effects of Ethanol Extracts of *Carica papaya* and *Pandanus amaryfollius* Leaf in

Streptozotocin-Induced Diabetic Mice. *Nat. Product Res.* **2011,** *25* (20), 1982–1987. https://doi.org/10.1080/14786419.2010.523703

Satheesh, M. A.; Pari, L. Antioxidant Effect of Boerhavia diffusa L. in Tissues of Alloxan Induced Diabetic Rats. *Indian J. Exp. Biol.* **2004,** *42* (10), 989–992.

Saxena, A. M.; Bajpai, M. B.; Murthy, P. S.; Mukherjee, S. K. Mechanism of Blood Sugar Lowering by a Swerchirin-Containing Hexane Fraction (SWI) of *Swertia chirayita. Indian J. Exp. Biol.* **1993,** *31* (2), 178–181.

Scartezzini, P.; Speroni, E. Review on some plants of Indian traditional medicine with antioxidant activity. *Journal of Ethnopharmacology* **2000,** *71* (1–2), 23–43. https://doi.org/10.1016/s0378-8741 (00)00213-0

Seth, S. D.; Sharma, B. Medicinal Plants in India. *Indian J. Med. Res.* **2004,** *120* (1), 9–11.

Sharma, S. R.; Dwivedi, S. K.; Swarup, D. Hypoglycaemic, Anti-Hyperglycaemic and Hypolipidemic Activities of *Caesalpinia bonducella* Seeds in Rats. *J. Ethnopharmacol.* **1997,** *58* (1), 39–44. https://doi.org/10.1016/s0378-8741 (97)00079-2

Singh, P.; Kesharwani, R. K.; Keservani, R. K. Antioxidants and Vitamins: Roles in Cellular Functions and Metabolism. In: *Sustained Energy for Enhanced Human Function and Activity*; Bagchi, D., Ed.; Elsevier, 2017; Ch 24. ISBN: 9780128054130

Smith, C.; Krygsman, A. Hoodia gordonii Extract Targets Both Adipose and Muscle Tissue to Achieve Weight Loss in Rats. *J. Ethnopharmacol.* **2014,** *155* (2), 1284–1290. https://doi.org/10.1016/j.jep.2014.07.018

Sofowora, A. Research on Medicinal Plants and Traditional Medicine in Africa. *J. Altern. Complem. Med.* **1996,** *2* (3), 365–372. https://doi.org/10.1089/acm.1996.2.365

Somani, R.; Kasture, S.; Singhai, A. K. Antidiabetic Potential of *Butea monosperma* in Rats. *Fitoterapia* **2006,** *77* (2), 86–90. https://doi.org/10.1016/j.fitote.2005.11.003

Somova, L. I.; Shode, F. O.; Ramnanan, P.; Nadar, A. Antihypertensive, Anti-Atherosclerotic and Antioxidant Activity of Triterpenoids Isolated from Olea Europaea, Subspecies Africana Leaves. *J. Ethnopharmacol.* **2003,** *84* (2–3), 299–305. https://doi.org/10.1016/s0378-8741 (02)00332-x

SSA Statistics South Africa. South Africa Demographic and Health Survey 2016: Key Indicator Report. https://www.statssa.gov.za/publications/Report%2003-00-09/Report%20 03-00-092016.pdf. (accessed April 22, 2019).

Subramoniam, A.; Pushpangadan, P.; Rajasekharan, S.; Evans, D. A.; Latha, P. G.; Valsaraj, R. Effects of *Artemisia pallens* Wall. On Blood Glucose Levels in Normal and Alloxan-Induced Diabetic Rats. *J. Ethnopharmacol.* **1996,** *50* (1), 13–17. https://doi.org/10.1016/0378-8741 (95)01329-6

Tabasum, S.; Khare, S. Safety of Medicinal Plants: An Important Concern*Int. J. Pharm. Biol. Sci.* **2016,** *7*, 237–243.

Tang, Y. N.; He, X. C.; Ye, M.; Huang, H.; Chen, H. L.; Peng, W. L.; Zhao, Z. Z.; Yi, T.; Chen, H. B. Cardio Protective Effect of Total Saponins from Three Medicinal Species of Dioscorea Against Isoprenaline-Induced Myocardial Ischemia. *J. Ethnopharmacol.* **2015,** *175*, 451–455. https://doi.org/10.1016/j.jep.2015.10.004

Tasos, E.; Huang, A.; Timimi, L. J.; Wych, J.; Mander, A. P.; Chowienczyk, P. J.; Wilkinson, I. B.; Mukhtar, O. *Randomised Controlled Trials of Anti-Hypertensive Therapy in Sub-Saharan Africa-A Systematic Review*. Available SSRN3327351, 2019.

Tomlinson, T. R.; Akerele, O. *Medicinal Plants: Their Role in Health and Biodiversity*; University of Pennsylvania Press, 2015.

Tormo, M. A.; Gil-Exojo, I.; Romero de Tejada, A.; Campillo, J. E. Hypoglycaemic and Anorexigenic Activities of an Alpha-Amylase Inhibitor from White Kidney Beans (*Phaseolus vulgaris*) in Wistar Rats. *Br. J. Nutr.* **2004,** *92* (5), 785–790. https://doi.org/10.1079/ bjn 20041260

Tune, J. D.; Goodwill, A. G.; Sassoon, D. J.; Mather, K. J. Cardiovascular Consequences of Metabolic Syndrome. *Transl. Res.: J. Lab. Clin. Med.* **2017,** *183*, 57–70. https://doi.org/10.1016/j.trsl.2017.01.001

Udani, J.; Hardy, M.; Madsen, D. C. Blocking Carbohydrate Absorption and Weight Loss: A Clinical Trial Using Phase 2 Brand Proprietary Fractionated White Bean Extract. *Altern. Med. Rev.* **2004,** *9* (1), 63–69.

Udani, J.; Singh, B. B. Blocking Carbohydrate Absorption and Weight Loss: A Clinical Trial Using a Proprietary Fractionated White Bean Extract. *Altern. Ther. Health Med.* **2007,** *13* (4), 32–37.

Urías-Silvas, J. E.; Cani, P. D.; Delmée, E.; Neyrinck, A.; López, M. G.; Delzenne, N. M. Physiological Effects of Dietary Fructans Extracted from Agave Tequilana Gto. and Dasylirion spp. *Br. J. Nutr.* **2008,** *99* (2), 254–261. https://doi.org/10.1017/S0007114507795338

Van Der Merwe, M. T.; Pepper, M. S. Obesity in South Africa. *Obes. Rev.* **2006,** *7* (4), 315–322. https://doi.org/10.1111/j.1467-789X.2006.00237.x

Van Heerden, F. R. Hoodia Gordonii: A Natural Appetite Suppressant. *J. Ethnopharmacol.* **2008,** *119* (3), 434–437. https://doi.org/10.1016/j.jep.2008.08.023

Vijayvargia, R.; Kumar, M.; Gupta, S. Hypoglycemic Effect of Aqueous Extract of *Enicostemma littorale* Blume (Chhotachirayata) on Alloxan Induced Diabetes Mellitus in Rats. *Indian J. Exp. Biol.* **2000,** *38* (8), 781–784.

Wang, W.; Wang, W. X.; Sun, B. H.; Zhao, D. Z.; Gao, P. [Effect of Haidonghua Powder (HDHP) on Hypothalamic Obesity in Rats].*Zhongguo Zhongyao Zazhi* **2000,** *25* (8), 490–492.

Wei, H.; Liu, S.; Liao, Y.; Ma, C.; Wang, D.; Tong, J.; Feng, J.; Yi, T.; Zhu, L. A Systematic Review of the Medicinal Potential of Mulberry in Treating Diabetes Mellitus. *Am. J. Chin. Med.* **2018,** *46* (8), 1743–1770. https://doi.org/10.1142/S0192415X1850088X

WHO. World Health Organization: Obesity. https://www.who.int/news-room/fact-sheets/detail/obesity-and-overweight (accessed April 22, 2019).

Yanovski, S. Z.; Yanovski, J. A. Long-Term Drug Treatment for Obesity: A Systematic and Clinical Review. *JAMA* **2014,** *311* (1), 74–86. https://doi.org/10.1001/jama.2013.281361

Yoshikawa, M.; Murakami, T.; Kadoya, M.; Matsuda, H.; Muraoka, O.; Yamahara, J.; Murakami, N. Medicinal Foodstuff: III—Sugar Beet—(1): Hypoglycemic Oleanolic Acid Oligoglycosides, Beta Vulgarosides I, II, III, and IV, from the Root of *Beta vulgaris* L. (Chenopodiaceae). *Chem. Pharma. Bull.* **1996,** *44* (6), 1212–1217. https://doi.org/10.1248/cpb.44.1212

Yoshikawa, M.; Murakami, T.; Yashiro, K.; Matsuda, H. Kotalanol, a Potent Alpha-Glucosidase Inhibitor with Thiosugar Sulfonium Sulfate Structure, from Antidiabetic Ayurvedic Medicine Salacia Reticulata. *Chem. Pharma. Bull.* **1998,** *46* (8), 1339–1340. https://doi.org/10.1248/cpb.46.1339

Ziauddin, K. S.; Anwar, M.; Mannan, A.; Khan, A. B. Clinical Efficacy of Terminalia Arjuna Roxb.in the Management of Hyperlipidaemia. *Hamdard Medicus* **2004,** *47*, 15–18.

CHAPTER 10

ROLE OF FLAVONOIDS IN VASODILATION

MHATRE K. J., PATIL J. R., and NIKALJE G. C.

Seva Sadan's R. K. Talreja College of Arts, Science and Commerce, Affiliated to University of Mumbai, Ulhasnagar, India.

ABSTRACT

Flavonoids are the secondary plant metabolites which are synthesized by plants for floral coloration, attraction of pollination agents, different physiological pathways, and cell cycle regulation. Evidence from various studies supports the notion that intake of flavonoid enriched diet may reduce risk of certain chronic diseases like cardiovascular diseases (CVD), neurodegenerative and few types of cancers. Of these, CVD is the main cause of death globally, yet it is largely preventable. Flavonoids are known to increase plasma Nitric Oxide levels, which cause vasodilation, prevent platelet aggregation, anti-atherogenic effect, and also reduce hypertension. All such biological properties of flavonoids have been studied in several clinical and experimental studies. This chapter explores the effects of flavonoids on human endothelial functioning, specifically in cardiovascular health and vasodilation.

10.1 INTRODUCTION

Flavonoids are the major phenolic compounds synthesized by plants. It consists of more than 5000 hydroxylated polyphenols which are responsible for several important functions like floral pigmentation, attracting pollinating

The Flavonoids: Extraction and Applications. Deepika Saini, Rajesh K. Kesharwani, & Raj K. Keservani (Eds.)

agents such as insects, amelioration of biotic stresses like microbial infection and/or act as cell cycle inhibitors (Kumar et al., 2013). The basic structure of flavonoids consists of a 15-carbon skeleton having 2 benzene rings connected by a heterocyclic linking chain of 3-carbons. These structures form flavan nucleus which is a basic structure of flavonoids (Kumar and Pandey, 2013). Flavonoids in plants are synthesized from phenylalanine, by phenylpropanoid metabolic pathway, via transformation of phenyl propanol into 4- coumarate Co-A. This leads to the synthesis of chalcone, structurally similar to flavonoids and it acts as a precursor for multiple flavonoids (Heim et al., 2002; Mahomoodally et al., 2005). A large variety of flavonoids have been identified and are classified on the basis of degree of oxidation and chemical structure (Faggio et al., 2017). Among them the first group comprises anthocyanins, flavan-3-ols, flavanones, flavones and the second group consists of isoflavonoids (Cook and Samman, 1996; Middleton, 1998). Flavanoids are involved in diverse biological activities in animals, bacteria, and plants. In plants, flavonoids are responsible for the color and smell of flowers, in fruits to attract seed-dispersing agents, the defence against oxidative stress and as UV filters. In addition, they protect plants from environmental stresses and act as signaling agents, detoxifiers, and defensive molecules against pathogens. Though there is a great diversity of flavonoids, their low bioavailability and biological activities such as vasodilation, anti-atherogenic, anti-inflammation, antioxidant, hypo-cholesterolemic, modulation of cell signaling, gene expression-related disease development, reduction of hypertension, and prevention of platelet aggregation in humans are dependent on their chemical nature. The fruits, nuts, seeds, vegetables, drinks like coffee, wine, and tea are rich in flavonoids and have significant antioxidant activity (Keservani et al., 2016a, 2016b). It has great value in various conditions such as Alzheimer's disease, atherosclerosis, and cancer (Rees et al., 2018). Flavonoids are major antioxidantsin human diet, their intake is higher than vitamin C (10 times) and Vitamin E /carotenoids (20 times). Because of their antioxidant, anticarcinogenic, anti-inflammatory, and antimutagenic action, flavonoids are used in cosmetic, medicinal, nutritional, and pharmaceutical applications (Art et al., 2000). The flavonoids vary in plants both qualitatively and quantitatively. These variations are influenced by species, age, and geographical distribution, cultivation method, plant part used, and storage (Costa et al., 2015). Moreover, abiotic factors such as sunlight exposure, soil type, precipitation, and/or number of fruits per tree are also responsible for flavonoid content in plants (Kumar and Pandey, 2013). Since the last decade, flavonoids have been explored to a greater extent for their potential health benefits. Due to heterogeneous structure of flavonoids, their study is long term and complex. Several studies have recommended that dietary flavonoids can be advantageous as lateral treatment in chronic inflammatory disease (Quinones et al., 2013; Lapuente et al., 2019). Among the known flavonoid rich plants, vegetables and fruits are extensively included in daily diet. For example, apples, berries, grapes, kale, onion, and tomatoes are enriched in flavones. Also, celery, chamomile, mint, red peppers, parsley, peels of citrus fruits, and drinks like red wine, tea have abundant amount of flavones.

These flavones play a vital role in vasodilation process, that is, expanding blood vessels. It requires relaxation of smooth muscles present in walls of arteries or large veins. This allows blood vessels to become more open and increase the blood flow which helps in lowering of blood pressure. The decrease in systemic vascular resistance (SVR) and increase in blood flow are essential components of vasodilation. The role of flavonoids in vasodilation is thoroughly discussed in the following chapter.

10.2 SOURCES AND BIOAVAILABILITY OF FLAVONOIDS

Polyphenolic compounds including Flavonoids are crucial for plant flavor and coloration (Ross and Kasum, 2002). They are performing several important functions in plants such as auxin transport, growth, pollinator attraction, reproduction, seed dispersal, protection against abiotic and biotic stresses, and resistance to predators (Bravo, 1998; Shah and Smith, 2020). Flavonoids abundantly found in many vegetables (like onion, lettuce), fruits (like apple, limes, cherries), grains, legumes, and tea; even in processed foods (wine and dark chocolate) (Panche et al., 2016; Janabi et al., 2020). The flavonoid content depends on source, species, sunlight exposure, season of collection etc. (D'Archivio et al., 2007). Physiologically, the flavonoids are not synthesized by humans and they depend on the plant sources of flavonoids (Koes et al., 2005). In humans, small intestine absorbs flavonoids and gets conjugated with glucuronic acid or sulfate or O-methylation, which varies greatly among all the classes of flavonoids, but overall, the absorption rate stays high when flavonoids are consumed as an integrated food (Nijveldt et al., 2001).

10.3 CLASSIFICATION OF FLAVANOIDS

Based on the basic ring structure, degree of unsaturation, oxidation status, and active functional groups flavonoids are categorized in six groups viz.

flavanones, flavones, flavan-3-ols, flavanols, anthocyanidins, and isoflavones (Peluso, 2006; Nishiumi et al., 2011; Ciumărnean et al., 2020).

10.3.1 FLAVONES

It consists of closed pyran ring with ketone in position 4 and positions 2 and 3 are bound by double bond (Panche et al., 2016). They possess a ketone in position 4 of the closed pyran ring and a double bond between positions 2 and 3. Flavones occur in several foods viz., celery, mint, red peppers, garlic, parsley, and chamomile flowers. These sources of flavones are rich in apigenin, luteolin, and tangeritin (Clark et al., 2015; Panche et al., 2016). The fruit peel of Citrus also contains flavones (polymethoxylated) like nobiletin, tageretin, and sinensetin (Panche et al., 2016). Luteolin reduces the blood pressure in hypertensive rats by lowering the proliferation of Angiotensin II-induced Vascular Smooth Muscle Cells, improving vasodilation in aorta, and inhibiting cAMP-specific phosphodiesterase which leads to the accumulation of cAMP (Su et al., 2015; Ciumărnean et al., 2020). Luteolin activates cAMP which in turn activates Protein kinase A (PKA). Further, nitric oxide synthase gets activated and increases nitric oxide level in endothelial cells. The accumulated nitric oxide enhances vascular relaxation through potassium and calcium channels (Liu et al., 2004; Ciumărnean et al., 2020).

10.3.2 FLAVONOLS

Flavonols contain the ketone group, which is unique among all flavonoids and acts as precursors of proanthocyanins (Panche et al., 2016). Flavonols have a hydroxyl group at the position 3 of closed pyran ring, which can be glycosylated (Panche et al., 2016). The mostly found flavonols are quercetin, kaempferol, fisetin, and myricetin which abundantly occur in broccoli, beans, kale, onion, lettuce, tea, tomatoes, apple, berries, strawberries, grapes, and wine (Kumar et al., 2013; Clark et al., 2015; Panche et al., 2016; Ciumărnean et al., 2020). Quercetin is known to exhibit an antihypertensive effect. When consumed, quercetin improves endothelial function and modulates the renin–angiotensin–aldosterone system (Larson et al., 2012). It causes vasodilation in the kidneys and reduces blood pressure level in diabetic patients (Ciumărnean et al., 2020). It showed improvement in hypertension, hyperinsulinemia, dyslipidemia, as well as produced anti-inflammatory effects in Vascular Adipose Tissue along with reduction in body weight among obese Zucker Rats (Rivera et al., 2008). Quercetin also gives relief from the oxidative stress of certain organs such as the heart and kidneys (Oyagbemi et al., 2018). It possesses antihypertensive properties. Its mechanism of action is associated with release of nitric oxide in endothelial cells (Bravo, 1998; Leeya et al., 2010).

10.3.3 FLAVAN-3-OLS

In this, at position 3 of closed pyran ring, the hydroxyl group is bounded and at positions 2 and 3, double bond is missing (Panche et al., 2016). It includes catechin, epicatechin, gallctocatechin, and oligomers (Su et al., 2015). The catechins are monomeric form of flavanols and they are found in many plants like apple, apricots, cocoa, pears, tea, etc. in the aglycone form (Manach et al., 2004). These catechins have positive impact on vascular functions with cardioprotective effect. In addition, they can reduce both diastolic and systolic blood pressure (D'Archivio et al., 2007; Rolnik et al., 2020). The other type of catechin, Epigallocatechin-3-gallate, is abundantly found in green tea. It is basically an ester of gallic acid and epigallocatechin. It has multiple effects on different physiological and pathological processes in humans. It shows bioactivities like anti-antherogenic action, antioxidant, and anti-inflammatory (Ludwig et al., 2004). Epicatechin is one more example of flavan-3-ol flavonoid. It is known to have an antihypertensive action. Consumption of an epicatechin-supplemented diet helps to decrease both blood pressure—systolic and diastolic. In addition, it reduces rigidity of myocardial tissues in rats having hypertrophic cardiomyopathy (Mohammed and Juan, 2008).

10.3.4 FLAVANONES

Flavanones are also called dihydroflavones. They have a saturated closed pyran ring, which is a unique structural feature of this class of flavonoids. Naringenin and hesperetin are occurin citrus fruit peels that are the important flavanones. It blocks free radicals' activity and acts as an antioxidant compound (reviewed in Mladenka et al. (2010)). Hesperetin is again a dietary flavanone, the main flavonoid in lemons and sweet oranges (Clark et al.,

2015). It gets easily and rapidly absorbed from the intestine. The resulting metabolites exhibit antihypertensive effect which may be result of mechanism by which under hypertension, G-hesperidin lowers blood pressure. In addition, antioxidant nature of hesperetin enhances nitric acid content and reduces calcium ion content and thereby carries out relaxation of vascular smooth muscles (Yamamoto et al., 2013; Liu et al., 2014). Naringenin is also found in grapefruit, and some herbs (Sharma et al., 2020). Several studies have revealed that naringin supplement diet is beneficial in the treatment of hypertension, diabetes, antiviral, antibacterial, anti-inflammatory, antiadipogenic, and metabolic syndrome. It is known to reduce blood pressure, provide protection against endothelial dysfunction, and regulate the nitric oxide level (Ikemura et al., 2012; Yang et al., 2012).

10.3.5 ANTHOCYANIDINS

Anthocyanidins are water-soluble pigments and are the only flavonoids that give the colors in plants, particularly fruits, flowers, and vegetables (Ciumărnean et al., 2020). They are known to impart blue, red, or purple color to fruits (forest fruits, black currants) (Clark et al., 2015). The color of anthocyanins, glycosides of anthocyanidins, is influenced by pH and available substituting groups. They occur in the peels of various berries and currants. They help the cardiovascular system by exerting endothelium-dependent vasodilation, as well as in hypertensive patients they lower the blood pressure (McKay et al., 2009).

10.3.6 ISOFLAVONES

The numerous Isoflavones are abundantly found in leguminous plants, especially in soyabean, lentils, peas, and even in microbes (Panche et al., 2016). The Soy isoflavones show structural similarity with estrogens of mammals; hence, these isoflavones are said to be estrogen receptor agonists. Daidzein and genistein are the prime and well-studied examples of this class (Jiang et al., 2005). A higher consumption of isoflavones converts the endogenous estrogens into their protective estrogen derivatives. Genistein shows an antihypertensive effect (Mita et al., 2013), while Daidzein reduces the oxidative stress causing damage and LDL oxidation while increasing the prostaglandins and nitric oxide production (Ciumărnean et al., 2020).

10.4 FLAVONOIDS-INDUCED EFFECTS IN RELATION TO VASODILATION

10.4.1 ANTIPLATELET EFFECTS

Platelets are comparatively smaller blood corpuscles that play a crucial role in the coagulation process so as to stop bleeding, that is, hemostasis (Periayah et al., 2017). Intact blood vessels prevent clotting, but any injury to the blood vessels activates them. It is the endothelial lining of blood vessels which regulates this process by secreting coagulation and platelet aggregation inhibitors, when intact. The moment this endothelial lining ruptures, it secretes von Willebrand factor (vWF`), a key molecule responsible for the maintenance of hemostasis in humans (Peyvandi et al., 2011). The moment blood vessel gets injured, an immediate response through the reflex is initiated by local sympathetic pain receptors. The damaged vessels undergo constriction to narrow down the size of injury to reduce the amount of blood loss, resulting in vascular spasm. Collagen, present in the walls of blood vessels, gets exposed to the blood when blood vessels are damaged. Such collagen promotes platelets adhesion at the injury site, where damaged collages fibers are exposed. Platelets, when coming in contact with such collage, release cytoplasmic granules which contain serotonin, ADP, and thromboxane A2 (Sangkuhl et al., 2011). These platelet secretions ultimately result in vasoconstriction. To fix the remaining problem, platelets activate other platelets and simultaneously initiate coagulation cascade to form a blood clot. It is the platelet activation that plays an important central role in both protective hemostasis and pathological thrombosis through multiple physiological pathways. Excessive platelet aggregation and several chronic diseases are closely related, like diabetes, hypertension, and several other cardio-vascular diseases (Barale and Russo 2020). Adhesion proteins responsible for platelet aggregation when secreted in higher concentrations due to excessive platelet activation, thrombi get generated, which can be a cause of development of various thrombotic diseases (Mohammed and Juan, 2008; Rolnik et al., 2020). This is mainly due to the clogging of thrombi in narrow blood vessels, which creates blockages in and/or near the affected blood vessels. Thromboxane A2, which is released during platelet activation, is a very powerful vasoconstrictor as well as platelet activator. It is synthesized in platelets by cytoplasmic cyclooxygenase enzyme from the arachidonic acid obtained from the platelet membranes. Prostaglandin (PG) endoperoxide is formed as an intermediate, and is converted into

Thromboxane A2 either by TXA_2 or by PGI_2 pathways. This is important for platelet activation and takes by cyclooxygenase 1 (Faggio et al, 2017). Since Thromboxane A2 is involved in the recruitment and activation of platelets, molecules that block COX can result in reducing the platelet activation and aggregation at/near the injury site. Flavonoids show several beneficial effects like improvement in endothelial functioning, reduction of platelet adhesion, and interference in lipid metabolism (Ludwig et al., 2004; Faggio et al, 2017). Flavonoids act as antagonist of thromboxane A2 (TP) receptors situated on the platelet membrane. TP levels are known to regulate expression of COX. In this case, the presence of flavonoids indirectly inhibits cyclooxygenase 1 activity, resulting in the inactivation of platelets (Mladenka et al., 2010). Presence of double bonds at C2-C3, and a keto group in C4 positions of flavonoids, are responsible for tight stearic binding of them to thromboxane A2 receptor. Genistein and diadzene (non-glucuronidated) hinder the platelet aggregation. This is possible since these non-glucuronidated flavonoids bind to surface receptors of thromboxane A2 (Munoz et al., 2009). Collagen plays a vital role in the initiation of platelet aggregation. Flavonoids not only intervene in the arachidonic acid metabolism but also intervene the collagen metabolism, thus reducing the platelet aggregation. Platelets bind collagen directly or indirectly through binding of glycoprotein Ib/IX matrix to vWF. Oxidative stress caused due reactive oxygen species like Superoxide anion can activate collagen-induced platelet aggregation. This activation is achieved by increasing level of calcium in cells and the inositol pathway. NADPH oxidase is a family of enzymes that transfers electron to oxygen to generate reactive oxygen species like peroxide. Studies conducted in the last 2 decades have shown that catechin, kaempferol, or quercetin flavonoids can minimize the oxidative stress. These flavonoids act as inhibitors of NADPH-oxidase (Ludwig et al., 2004; Wang et al., 2015) and are thus responsible for the antiplatelet effect. Plant extracts exert antiplatelet effect by different mechanisms. By inhibiting the levels of intracellular calcium ions, thrombi formation is inhibited and reorganization of cytoskeletal elements is possible. This is the most important mechanism to show an antiplatelet effect. There are several other mechanisms by which it is achieved. These mechanisms involve inhibition pathways like inhibition of thrombocyte secretion, protein C breakdown, platelet activation factor, and phospholipase C. One more approach by which antiplatelet activity is shown is by blocking thromboxane formation and increasing intracellular cAMP and cGMP levels. Enzymes like phospholipase A and few tyrosine kinases are also inhibited by flavonoids (Bunting et al., 1983). Glycoprotein (GP) IIb/IIIa inhibitors bind to the platelet receptors and prevent binding of fibrinogen and vWF, which is a crucial step in platelet aggregation.

Flavonoids like quercetin or catechin work by increasing the nitric oxide levels, which decrease the expression of glycoprotein IIB/IIIa complex. This leads to blocking platelet aggregation, in vitro (Hamed et al., 2008).

10.4.2 HYPERTENSION

Hypertension is a clinical condition wherein blood vessels experience high diastolic and/or systolic pressure for prolonged time (Raghupathy et al., 2014). Hypertensive cardiovascular disease is the chief cause of illness and death due to high diastolic blood pressure, obesity, sedentary lifestyle, smoking, stress, age, etc. (Raghupathy et al., 2014). Nitric oxide (NO) synthesized by the endothelium in one of the most important molecules for the health of blood vessels. Being vasodilator, NO regulates vascular tone by relaxing blood vessels and thus the blood pressure. NO formed by endothelium of vessels, when released in blood, binds to hemoglobin or diffuses in smooth muscles (Rifkind et al., 2018). In smooth muscles it binds and triggers guanylyl cyclase. This leads to the activation of the cGMP-protein kinase G cascade pathway in the vascular smooth muscles. Being a cascade, a series of chemical reactions that occur within the effector cell till the final result is achieved. After the activation of cGMP-protein kinase G cascade, cGMP by several mechanisms enhances relaxation of smooth muscle (Francis et al., 2010). Elevated levels of cGMP inhibit the uptake of calcium ions within cells, activate $K+$ channels which results in hyperpolarization, and thus lead to vasodilation. Another mechanism is the activation of cGMP-dependent kinases. The light chains of myosin get phosphorylated by activated cGMP-dependent kinase, which reduces the vasoconstriction of vascular smooth muscles (Gao et al., 2016; Puzserova and Bernatova, 2016). Vasodilation increases lumen of blood vessels creating more space for blood to flow and reduces hypertension. Flavones show antihypertensive activity by targeting cGMP protein kinase G cascade. Flavones activate the cGMP protein kinase A cascade, which causes excessive synthesis and secretion of endothelial NO, which in turn achieves vasodilation by modulating levels of calcium and potassium ions so as to bring hyperpolarization of smooth muscles adjacent to endothelium (Liu et al, 2004; Calderone et al., 2004; Su et al, 2015). Flavonols follow a different mechanism to show antihypertensive action. Kaempferol and quercetin are the flavonols, which regulate the renin–angiotensin–aldosterone system (RAAS) to show an antihypertensive effect. RAAS is one of the major regulators of blood pressure, since it regulates blood volume, electrolyte balance, and vasodilation. RAAS interacts with the sympathetic nervous system, by positive feedback mechanism, improvement in endothelial dysfunction, and regulating vascular smooth muscles contraction (Leeya et al., 2010; Larson et al., 2012; Thamcharoen et al., 2015). Nitric Oxide Synthase 3 (NOS3) is a gene responsible for NO synthesis. The above-mentioned mechanisms activate NOS3 and thus influence the levels of plasma NO, thereby increasing vasodilation and reducing hypertension. Endothelin 1 is a vasoconstrictor secreted by endothelial cells and is used in counteracting the mechanism exhibited by NO. Its suppression can improvise endothelial functioning (Rivera et al., 2008; Larson et al., 2012). Vascular denudation changes the structural characteristics and thus the function. Flavonoids like Quercetin and kaempferol show vasodilation in intact blood vessels by elevation of the levels on NO locally, but they fail to show vasodilation, in vitro, in denuded vessels (Rivera et al., 2008). Similarly, a flavanone naringenin also fails to show vasodilation in denuded endothelium even though it possesses the ability to activate voltage-dependent calcium and potassium channels in nondenuded endothelium. This results in hyperpolarization of adjacent smooth muscles, which ultimately results in loosing of vascular tone, and thus vasodilation (Saponara et al., 2009) and hence, naringenin is used in the hypertension treatment (Ikemura et al., 2012; Yang et al., 2012). Hesperidin metabolite hesperetin-7-O-glucuronide enhances vasodilation (endothelium-dependent) and lowers blood pressure (Yamamoto et al., 2013) by reducing influx of intracellular calcium ions and blocking the voltage-gated calcium channels (Liu et al., 2014).

10.4.3 ANTIATHEROGENIC EFFECTS

Atherogenesis is the formation of plaque in the internal layer of arteries which progressively leads to atherosclerosis, which is characterized by inflammation and lipid accumulation in the arterial wall (ref et al., 2017). It ultimately causes acute myocardial infarction. The accumulation of cholesterol by macrophages in the subendothelial cells results in the formation of lesion (Tamminen et al., 1999). Macrophages and endothelial cells accumulate the oxidize LDL, and later this oxidized LDL acts as a key factor for atherosclerosis (Juan et al., 2001). Several studies reported the antiatherogenic effects of flavonoids. Flavonoids and other antioxidants protect the cells from oxidized LDL by increasing resistance of cells against the cytotoxic effect of oxidized LDL and by their radical scavenging activity which inhibits the oxidation of LDL (Juan et al., 2001). Lee et al. (2001) reported reduced hepatic acyl-CoA:cholesterol acyltransferase activity and downregulation of VCAM-1 (Vascular cell adhesion molecule-1) and MCP-1 (Monocyte Chemotactic Protein-1) gene expression in high cholesterol-fed rabbits supplemented with citrus flavonoids, 0.1% naringin, and 0.05% naringenin. The mice fed on pomegranate juice supplements rich in anthocyanidins and proanthocyanidins showed reduction of macrophage cholesteryl ester accumulation, lipid peroxides, and atherosclerosis by 44% (Aviram et al., 2008). But citrus flavonoid naringenin-treated mice exhibited more than 70% reduction in atherosclerosis (Mulvihill and Huff, 2010). Salvamani et al. (2014) studied the antiatherosclerotic effects of few flavonoids like catechin, kaempferol, fisetin, gossypetin, myricetin, rutin, quercetin, and naringenin. They found that myricetin and fisetin have more potential against atherosclerosis.

10.4.4 BLOOD VESSELS DILATION

Flavonoids can activate endothelial nitric oxide synthase (eNOS) phosphorylation which further induces release of nitric oxide (NO) in endothelial cells (Khoo et al., 2010). Another possible mechanism of eNOS activation is flavonoid-induced redox-sensitive activation of the PI3K/AKT pathway by reactive oxygen species (ROS) or interaction of flavonoids with membrane estrogen receptors (ER) (Si et al., 2013). The PI3K/AKT pathway can also be involved in the synthesis of endothelium-derived hyperpolarizing factor (EDHF) through the activation of potassium channels (IKca and SKca) or induced by gap junctions, which are communicated by the electrical interaction between the endothelial cells and the vascular smooth muscle layer (Bełtowski and Jamroz-Wiśniewska, 2014). Flavonoids can induce vasorelaxation via endothelium-independent pathways such as blockage of Ca^{2+} channels or inhibition of phosphodiesterase (PDEs) enzymes, preventing cyclic guanosine monophosphate (cGMP) from degradation and consequent increase in protein kinase G (PKG), activity leading to smooth muscle cells relaxation (Abdallah et al., 2020). (sGC) = soluble guanylyl-cyclase; (GTP) = guanosine triphosphate (Figure 10.1).

10.5 CONCLUSION

Flavonoids are involved in the vasodilation process which helps in the prevention of cardiovascular diseases. The role of flavonoids is not restricted to cardiovascular systems; it is extended to other vital systems by associating with cascade of molecular signaling and regulation of cellular functions. The working principle of flavonoids involves enhanced activity of nitrous oxide enzyme. It promotes the vascular relaxation which helps to reduce arterial blood pressure. However, still there is paucity of information about molecular regulators associated with flavonoids induced vasodilation in the prevention and treatment of CVDs. Further studies must be focused on omics approaches to understand the mechanism of flavonoids at the molecular and cellular level.

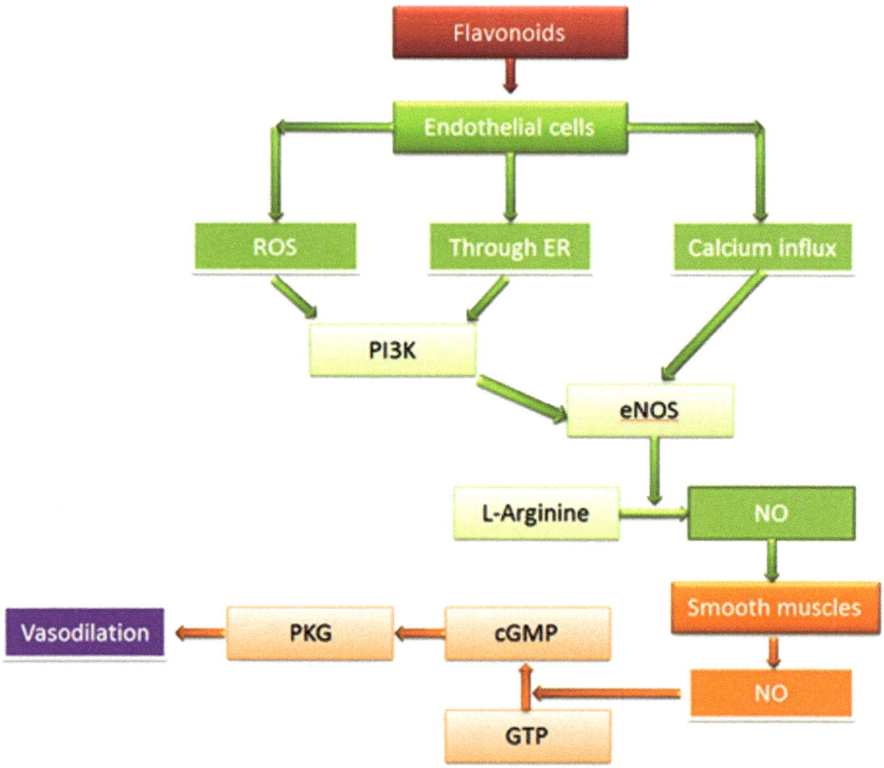

FIGURE 10.1 Mechanism of Flavonoids-induced blood vessels dilation. Flavonoids can induce release of nitric oxide (NO) through endothelial nitric oxide synthase (eNOS). This can be achieved through three mechanisms, namely, through calcium influx, estrogen receptors (ER), and reactive oxygen species (ROS). This NO, synthesized from L-arginine, increases protein kinase G levels in smooth muscles of blood vessels and results in vasodilation.

KEYWORDS

- **flavonoids**
- **human health**
- **nitrous oxide**
- **vasodilation**

REFERENCES

Abdallah, H. M.; Hassan, N. A.; El-Halawany, A. M.; Mohamed, G. A.; Safo, M. K.; El-Bassossy, H. M. Major Flavonoids from *Psiadia punctulata* Produce Vasodilation via Activation of Endothelial Dependent NO Signaling. *J. Adv. Res.* **2020**, *24*, 273–279.

Arts, I. C. W.; Van De Putte, B.; Hollman, P. C. H. Catechin Contents of Foods Commonly Consumed in The Netherlands. 1. Fruits, Vegetables, Staple Foods, and Processed Foods. *J. Agric. Food Chem.* **2000,** *48* (5), 1746–1751.

Athanasiou, L. S.; Fotiadis, D. I.; & Michalis, L. K. *Atherosclerotic Plaque Characterization Methods Based on Coronary Imaging*; Academic Press, 2017.

Aviram, M.; Volkova, N.; Coleman, R. et al. Pomegranate Phenolics from the Peels, Arils, and Flowers Are Antiatherogenic: Studies In Vivo in Atherosclerotic Apolipoprotein E-Deficient (E0) Mice and In Vitro in Cultured Macrophages and Lipoproteins. *J. Agric. Food Chem.* **2008,** *56*, 1148–1157.

Barale, C.; Russo, I. Influence of Cardiometabolic Risk Factors on Platelet Function. *Int. J. Mol. Sci.* **2020,** *21* (2), 623. https://doi.org/10.3390/ijms21020623

Bełtowski, J.; Jamroz-Wiśniewska, A. Hydrogen Sulfide and Endothelium-Dependent Vasorelaxation. *Molecules* **2014**, *19* (12), 21183–21199.

Bravo, L. Polyphenols: Chemistry, Dietary Sources, Metabolism, and Nutritional Significance. *Nutr. Rev.* **1998,** *56* (11), 317–333.

Bunting, S.; Moncada, S.; Vane, J. R. The Prostacyclin-Thromboxane A2 Balance: Pathophysiological and Therapeutic Implications. *Br. Med. Bull.* **1983,** *39* (3), 271–276.

Calderone, V.; Chericoni, S.; Martinelli, C.; Testai, L.; Nardi, A.; Morelli, I.; Breschi, M. C.; Martinotti, E. Vasorelaxing Effects of Flavonoids: Investigation on the Possible Involvement of Potassium Channels. *Naunyn Schmiedebergs Arch. Pharmacol.* **2004,** *370* (4), 290–298.

Ciumărnean, L.; Milaciu, M. V.; Runcan, O.; Vesa, Ș. C.; Răchișan, A. L.; Negrean, V.; Perné, M. G.; Donca, V. I.; Alexescu, T. G.; Para, I.; Dogaru, G. The Effects of Flavonoids in Cardiovascular Diseases. *Molecules* **2020,** *25* (18), 4320.

Clark, J. L.; Zahradka, P.; Taylor, C. G. Efficacy of Flavonoids in the Management of High Blood Pressure. *Nutr. Rev.* **2015,** *73* (12), 799–822.

Cook, N. C.; Samman, S. Flavonoids—Chemistry, Metabolism, Cardioprotective Effects, and Dietary Sources. *J. Nutr. Biochem.* **1996,** *7* (2), 66–76.

Costa, D. C.; Costa, H. S.; Albuquerque, T. G.; Ramos, F.; Castilho, M.; Silva, A. S. Advances in phenolic Compounds Analysis of aromatic plants and their potential applications. *Trends Food Sci. Technol.* **2015,** *45* (2), 336–354.

D'Archivio, M.; Filesi, C.; Di Benedetto, R.; Gargiulo, R.; Giovannini, C.; Masella, R. Polyphenols, Dietary Sources and Bioavailability. *Ann. Ist. Super. Sanita* **2007,** *43* (4), 348–361.

Faggio, C.; Sureda, A.; Morabito, S.; Silva, A. S.; Mocan, A.; Nabavi, S. F.; Nabavi, S. M. Flavonoids and Platelet Aggregation: A Brief Review. *Eur. J. Pharmacol.* **2017,** *807*, 91–101.

Gao, Y.; Chen, T.; Raj, J. U. Endothelial and Smooth Muscle Cell Interactions in the Pathobiology of Pulmonary Hypertension. *Am. J. Respir. Cell Mol. Boil.* **2016,** *54* (4), 451–460.

Hamed, M. S.; Gambert, S.; Bliden, K. P.; Bailon, O.; Anand, S.; Antonino, M. J.; Hamed, F.; Tantry, U.; Gurbel, P. A. Dark Chocolate Effect on Platelet Activity, C-Reactive Protein and Lipid Profile: A Pilot Study. *South. Med. J.* **2008,** *101* (12), 1203–1208.

Heim, K. E.; Tagliaferro, A. R.; Bobilya, D. J. Flavonoid Antioxidants: Chemistry, Metabolism and Structure-Activity Relationships. *J. Nutr. Biochem.* **2002,** *13* (10), 572–584.

Ikemura, M.; Sasaki, Y.; Giddings, J. C.; Yamamoto, J. Preventive Effects of Hesperidin, Glucosyl Hesperidin and Naringin on Hypertension and Cerebral Thrombosis in Stroke-Prone Spontaneously Hypertensive Rats. *Phytother. Res.* **2012,** *26* (9), 1272–1277.

Janabi, A. H. W.; Kamboh, A. A.; Saeed, M.; Xiaoyu, L.; BiBi, J.; Majeed, F. et al. Flavonoid-Rich Foods (FRF), A Promising Nutraceutical Approach Against Lifespan-Shortening Diseases. *Iran. J. Basic Med. Sci.* **2020,** *23* (2), 140.

Jiang, R. W.; Lau, K. M.; Lam, H. M.; Yam, W. S.; Leung, L. K.; Choi, K. L.; Waye, M. M. Y.; Mak, T. C.; Woo, K. S.; Fung, K. P. A Comparative Study on Aqueous Root Extracts of *Pueraria thomsonii* and *Pueraria lobata* by Antioxidant Assay and HPLC Fingerprint Analysis. *J. Ethnopharmacol.* **2005,** *96* (1–2), 133–138.

Juan, D.; Perez-Vizcaino, F.; Jimenez, J.; Tamargo, J.; Zarzuelo, A. Flavonoids and Cardiovascular Diseases. *Stud. Nat. Prod. Chem.* **2001,** *25*, 565–605.

Keservani, R. K.; Sharma, A. K.; Kesharwani, R. K. Nutraceutical and Functional Foods for Cardiovascular Health. In *Food Process Engineering*; Apple Academic Press, CRC Press, 2016a; pp 257–278; 291–312. ISBN: 9781771884020.

Keservani, R. K.; Sharma, A. K.; Kesharwani, R. K. Medicinal Effect of Nutraceutical Fruits for the Cognition and Brain Health. *Scientifica* **2016b,** Article ID 3109254, 10 pp. DOI: 10.1155/2016/3109254.

Khoo, N. K.; White, C. R.; Pozzo-Miller, L.; Zhou, F.; Constance, C. et al. Dietary Flavonoid Quercetin Stimulates Vasorelaxation in Aortic Vessels. *Free Radic. Biol. Med.* **2010,** *49* (3), 339–347.

Koes, R.; Verweij, W.; Quattrocchio, F. Flavonoids: A Colorful Model for the Regulation and Evolution of Biochemical Pathways. *Trends Plant Sci.* **2005,** *10* (5), 236–242.

Kumar, S.; Pandey, A. Chemistry and Biological Activities of Flavonoids: An Overview. *Sci. World J.* **2013,** *2013*, 1–16.

Kumar, S.; Mishra, A.; Pandey, A. Antioxidant Mediated Protective Effect of *Parthenium hysterophorus* Against Oxidative Damage Using In Vitro Models. *BMC Complement. Altern. Med.* **2013,** *13* (1), 1–9.

Lapuente, M.; Estruch, R.; Shahbaz, M.; Casas, R. Relation of Fruits and Vegetables with Major Cardiometabolic Risk Factors, Markers of Oxidation, and Inflammation. *Nutrients* **2019,** *11* (10), 2381.

Larson, A. J.; Symons, J. D.; Jalili, T. Therapeutic Potential of Quercetin to Decrease Blood Pressure: Review of Efficacy and Mechanisms. *Adv. Nutr.* **2012,** *3* (1), 39–46.

Lee, C. H.; Jeong, T. S.; Choi, Y. K.; Hyun, B. H.; Oh, G. T.; Kim, E. H.; Bok, S. H. Anti-Atherogenic Effect of Citrus Flavonoids, Naringin and Naringenin, Associated with Hepatic ACAT and Aortic VCAM-1 and MCP-1 in High Cholesterol-Fed Rabbits. *Biochem. Biophys. Res. Commun.* **2001,** *284* (3), 681–688.

Leeya, Y.; Mulvany, M. J.; Queiroz, E. F.; Marston, A.; Hostettmann, K.; Jansakul, C. Hypotensive Activity of an N-Butanol Extract and Their Purified Compounds from Leaves of *Phyllanthus acidus* (L.) Skeels in Rats. *Eur. J. Pharmacol.* **2010,** *649* (1–3), 301–313.

Liu, D.; Homan, L. L.; Dillon, J. S. Genistein Acutely Stimulates Nitric Oxide Synthesis in Vascular Endothelial Cells by a Cyclic Adenosine 5'-Monophosphate-Dependent Mechanism. *Endocrinology* **2004,** *145* (12), 5532–5539.

Liu, Y.; Niu, L.; Cui, L.; Hou, X.; Li, J.; Zhang, X.; Zhang, M. Hesperetin Inhibits Rat Coronary Constriction by Inhibiting Ca2+ Influx and Enhancing Voltage-Gated K+ Channel Currents of the Myocytes. *Eur. J. Pharmacol.* **2014,** *735*, 193–201.

Ludwig, A.; Lorenz, M.; Grimbo, N.; Steinle, F.; Meiners, S.; Bartsch, C.; Stangl, K.; Baumann, G.; Stangl, V. The Tea Flavonoid Epigallocatechin-3-Gallate Reduces Cytokine-Induced VCAM-1 Expression and Monocyte Adhesion to Endothelial Cells. *Biochem. Biophys. Res. Commun.* **2004,** *316* (3), 659–665.

Mahomoodally, M. F.; Gurib-Fakim, A.; Subratty, A. H. Antimicrobial Activities and Phytochemical Profiles of Endemic Medicinal Plants of Mauritius. *Pharm. Boil.* **2005,** *43* (3), 237–242.

Manach, C.; Scalbert, A.; Morand, C.; Rémésy, C.; & Jiménez, L. Polyphenols: Food Sources and Bioavailability. *Am. J. Clin. Nutr. 2004, 79* (5), 727–747.

McKay, D. L.; Chen, C. Y. O.; Saltzman, E.; Blumberg, J. B. *Hibiscus sabdariffa* L. Tea (Tisane) Lowers Blood Pressure in Prehypertensive and Mildly Hypertensive Adults. *J. Nutr.* **2009,** *140* (2), 298–303.

Middleton, E. Effect of Plant Flavonoids on Immune and Inflammatory Cell Function. *Adv. Exp. Med. Biol.* **1998,** *439*, 175–182.

Mita, M.; Tanaka, H.; Yanagihara, H.; Nakagawà, J. I.; Hishinuma, S.; Sutherland, C.; Walsh, M. P.; Shoji, M. Membrane Depolarization-Induced RhoA/Rho-Associated Kinase Activation and Sustained Contraction of Rat Caudal Arterial Smooth Muscle Involves Genistein-Sensitive Tyrosine Phosphorylation. *J. Smooth Muscle Res.* **2013,** *49*, 26–45.

Mladenka, P.; Zatloukalova, L.; Filipsky, T.; Hrdina, R. Cardiovascular Effects of Flavonoids Are Not Caused Only by Direct Antioxidant Activity. *Free Radic. Boil. Med.* **2010,** *49* (6), 963–975.

Mohammed, E. H.; Juan, A. R. Platelet Signalling Abnormalities in Patients with Type 2 Diabetes Mellitus: A Review. *Blood Cells Mol. Dis.***2008,** *41* (1), 119–123.

Mulvihill, E. E.; Huff, M. W. Antiatherogenic Properties of Flavonoids: Implications for Cardiovascular Health. *Can. J. Cardiol. 2010, 26*, 17A–21A.

Munoz, Y.; Garrido, A.; Valladares, L. Equol Is More Active Than Soy Isoflavone Itself to Compete for Binding to Thromboxane A2 Receptor in Human Platelets. *Thromb. Res.***2009,** *123* (5), 740–744.

Nijveldt, R. J.; Nood, E. V.; Van Hoorn D. E. C.; Boelens, P. G.; Van Norren, K.; Van Leeuwen, P. A.; Flavonoids: A Review of Probable Mechanisms of Action and Potential Applications. *Am. J. Clin. Nutr.* **2001,** *74* (4), 418–425, https://doi.org/10.1093/ajcn/74.4.418.

Nishiumi, S.; Miyamoto, S.; Kawabata, K.; Ohnishi, K.; Mukai, R.; Murakami, A.; Ashida, H.; Terao, J. Dietary Flavonoids as Cancer-Preventive and Therapeutic Biofactors. *Front. Biosci.* **2011,** *3* (3), 1332–1362.

Oyagbemi, A. A.; Omobowale, T. O.; Ola-Davies, O. E.; Asenuga, E. R.; Ajibade, T. O.; Adejumobi, O. A.; Arojojoye, O. A.; Afolabi, J. M.; Ogunpolu, B. S.; Falayi, O. O. Quercetin Attenuates Hypertension Induced by Sodium Fluoride via Reduction in Oxidative Stress and Modulation of HSP 70/ERK/PPARγ Signaling Pathways. *BioFactors* **2018,** *44* (5), 465–479.

Panche A. N.; Diwan A. D.; Chandra S. R. Flavonoids: An Overview. *J. Nutr. Sci.* **2016,** *5*, e47. DOI: 10.1017/jns.2016.41

Peluso, M. R. Flavonoids Attenuate Cardiovascular Disease, Inhibit Phosphodiesterase, and Modulate Lipid Homeostasis in Adipose Tissue and Liver. *Exp. Boil. Med.* **2006,** *231* (8), 1287–1299.

Periayah, M. H.; Halim, A. S.; Mat Saad, A. Z. Mechanism Action of Platelets and Crucial Blood Coagulation Pathways in Hemostasis. *Int. J. Hematol. Oncol. Stem Cell Res.* **2017,** *11* (4), 319–327.

Peyvandi, F.; Garagiola, I.; Baronciani, L. Role of von Willebrand Factor in the Haemostasis. *Blood Transf.* **2011,** *9* (Suppl 2), s3–s8. https://doi.org/10.2450/2011.002S.

Puzserova, A.; Bernatova, I. Blood Pressure Regulation in Stress: Focus on Nitric Oxide-Dependent Mechanisms. *Physiol. Res.* **2016,** *65*, S309–S342.

Quinones, M.; Miguel, M.; Aleixandre, A. Beneficial Effects of Polyphenols on Cardiovascular Disease. *Pharmacol. Res.* **2013,** *68* (1), 125–131.

Raghupathy, A.; Kannuri, N. K.; Pant, H.; Khan, H.; Franco, O. H.; Di, A. E.; Prabhakaran, D. Hypertension in India. *J. Hypertens.* **2014,** *32* (6), 170–1177.

Rees, A.; Dodd, G. F.; Spencer, J. P. The Effects of Flavonoids on Cardiovascular Health: A Review of Human Intervention Trials and Implications for Cerebrovascular Function. *Nutrients* **2018,** *10* (12), 1852.

Rifkind, J. M.; Mohanty, J. G.; Nagababu, E.; Salgado, M. T.; Cao, Z. Potential Modulation of Vascular Function by Nitric Oxide and Reactive Oxygen Species Released From Erythrocytes. *Front. Physiol.* **2018,** *9*, 690.

Rivera, L.; Morón, R.; Sanchez, M.; Zarzuelo, A.; Galisteo, M. Quercetin Ameliorates Metabolic Syndrome and Improves the Inflammatory Status in Obese Zucker Rats. *Obesity* **2008,** *16* (9), 2081–2087.

Rolnik, A.; Zuchowski, J.; Stochmal, A.; Olas, B. Quercetin and Kaempferol Derivatives Isolated from Aerial Parts of *Lens culinaris* Medik as Modulators of Blood Platelet Functions. *Ind. Crop. Prod.* **2020,** *152*, 112536.

Ross, J. A.; Kasum, C. M. Dietary Flavonoids: Bioavailability, Metabolic Effects, and Safety. *Annu. Rev. Nutr.* **2002,** *22* (1), 19–34.

Salvamani, S.; Gunasekaran, B.; Shaharuddin, N. A.; Ahmad, S. A.; Shukor, M. Y. Anti-Artherosclerotic Effects of Plant Flavonoids. *BioMed Res. Int.* **2014,** Article ID 480258. DOI: 10.1155/2014/480258.

Sangkuhl, K.; Shuldiner, A. R.; Klein, T. E.; Altman, R. B. Platelet Aggregation Pathway. *Pharmacogen. Genom.* **2011,** *21* (8), 516–521.

Saponara, S.; Testai, L.; Iozzi, D.; Martinotti, E.; Martelli, A.; Chericoni, S.; Sgaragli, G.; Fusi, F.; Calderone, V. (+/−)-Naringenin as Large Conductance Ca2+-Activated K+ (BKCa) Channel Opener in Vascular Smooth Muscle Cells. *Br. J. Pharmacol.* **2009,** *149* (8), 1013–1021.

Shah, A.; Smith, D. L. Flavonoids in Agriculture: Chemistry and Roles in, Biotic and Abiotic Stress Responses, and Microbial Associations. *Agronomy* **2020,** *10* (8), 1209.

Sharma, Anil K.; Keservani, Raj K.; Gautam Surya P. *Herbal Product Development*; Apple Academic Press, CRC Press, Taylor & Francis Group, 2020. ISBN: 9781771888776.

Su, J.; Xu, H. T.; Yu, J. J.; Gao, J. L.; Lei, J.; Yin, Q. S.; Li, B.; Pang, M. X.; Su, M. X.; Mi, W. J.; et al. Luteolin Ameliorates Hypertensive Vascular Remodeling Through Inhibiting the Proliferation and Migration of Vascular Smooth Muscle Cells. *Evid. Based Complement. Altern. Med.* **2015,** *2015*, 1–14.

Tamminen, M.; Mottino, G.; Qiao, J. H.; Breslow, J. L.; Frank, J. S. Ultrastructure of Early Lipid Accumulation in ApoE-Deficient Mice. *Arter. Thromb. Vasc. Boil.* **1999,** *19* (4), 847–853.

Thamcharoen, N.; Susantitaphong, P.; Wongrakpanich, S.; Chongsathidkiet, P.; Tantrachoti, P.; Pitukweerakul, S.; Avihingsanon, Y.; Praditpornsilpa, K.; Jaber, B. L.; Eiam-Ong, S. Effect of N- and T-Type Calcium Channel Blocker on Proteinuria, Blood Pressure and Kidney Function in Hypertensive Patients: A Meta-Analysis. *Hypertens. Res.* **2015,** *38* (12), 847–855.

Wang, S. B.; Jang, J. Y.; Chae, Y. H.; Min, J. H.; Baek, J. Y.; Kim, M.; Park, Y.; Hwang, G. S.; Ryu, J. S.; Chang, T. S. Kaempferol Suppresses Collagen-Induced Platelet Activation by Inhibiting NADPH Oxidase and Protecting SHP-2 from Oxidative Inactivation. *Free Radic. Boil. Med.* **2015,** *83*, 41–53.

Yamamoto, M.; Jokura, H.; Hashizume, K.; Ominami, H.; Shibuya, Y.; Suzuki, A.; Hase, T.; Shimotoyodome, A. Hesperidin Metabolite Hesperetin-7-O-Glucuronide, But Not Hesperetin-3'-O-Glucuronide, Exerts Hypotensive, Vasodilatory, and Anti-Inflammatory Activities. *Food Funct.* **2013,** *4* (9), 1346.

Yang, Y.; Li, P. Y.; Cheng, J.; Mao, L.; Wen, J.; Tan, X. Q.; Liu, Z. F.; Zeng, X. R. Function of BKCa Channels Is Reduced in Human Vascular Smooth Muscle Cells from Han Chinese Patients with Hypertension. *Hypertension* **2012,** *61* (2), 519–525.

CHAPTER 11

ANTIMICROBIAL AND ANTI-INFLAMMATORY ACTION OF FLAVONOIDS

KHEMCHAND R. SURANA[1], EKNATH D. AHIRE[2], SUNIL K. MAHAJAN[3], DHANANJAY M. PATIL[4], and KHANDERAO R. JADHAV[5]

[1]*Department of Pharmaceutical Chemistry, Shreeshakti Shaikshanik Sanstha, Divine College of Pharmacy, Satana, Nashik, India*

[2]*Department of Pharmaceutics, MET's Institute of Pharmacy, Adgaon Nashik, India*

[3]*Department of Pharmaceutical Chemistry, MGV's Pharmacy College, Panchvati, Nashik, India*

[4]*Department of Pharmaceutics, Shreeshakti Shaikshanik Sanstha, Divine College of Pharmacy, Satana, Nashik, India*

[5]*Department of Pharmaceutics, RGS College of Pharmacy, Anjaneri, Nashik, India*

ABSTRACT

Flavonoids are one of the most important categories of small molecular secondary metabolites created in several elements of the plant. They show a good variety of medical specialty and helpful health effects for humans, which include, among others, antioxidative activity, radical scavenging capacity, coronary cardiovascular disease hindrance and antiatherosclerotic, hepatoprotective, antimicrobial, anti-inflammatory, and antitumor activities.

The Flavonoids: Extraction and Applications. Deepika Saini, Rajesh K. Kesharwani, & Raj K. Keservani (Eds.)

Hence, flavonoids are gaining high attention from the pharmaceutical and care industries. Notably, plants synthesize flavonoids in response to microbic infection, and these compounds are found to be a potent antimicrobial agent against a wide range of infective microorganisms in vitro. Flavonoids are identified as potent antioxidants with the potential to attenuate tissue harm or fibrosis. Consequently, various studies in vitro and in animal models have found that flavonoids have the potential to inhibit the onset and development of inflammatory diseases. Antimicrobial and anti-inflammatory action of flavonoids results from their various biological activities, which cannot appear terribly specific at first. Moreover, some plant flavonoids manifest an ability to reverse the antibiotic resistance and enhance action of this antibiotic medication. Hence, the event and application of flavonoid-based drugs may well be a promising approach for antibiotic-resistant infections. This chapter aims to enhance our understanding of the biological and molecular roles of plant flavonoids as antimicrobial and anti-inflammatory.

11.1 INTRODUCTION

One of the biggest categories of naturally occurring polyphenolic compounds are the flavonoids (Bylka et al., 2004). This cluster of plant pigments is essentially answerable for the colors of the many fruits and flowers, and over 4000 flavonoid compounds are characterized and classified per chemical structure (Geissman, 1969). The word flavonoid comes from the Latin flavus which suggests yellow, but some flavonoids are red, blue, purple, or white (Murray, 1996). They were discovered, alongside aliment C, in 1928 by prince consort Szent-Gyorgyi, who is referred to as then bioflavinoid (Chandra et al., 2017).

The compounds with C6-C3-C6 compounds within which the two C6 teams are substituted benzol rings, and also the C3 is associated acyclic chain which contains a pyran ring (Mills et al., 2000). Flavonoids occur as O- or C-glycosides or within the Free State as aglycones with chemical group or methoxyl groups present on the aglycone. The flavonoids could also be divided into seven types: flavones, flavonols, flavonones, chalcones, xanthones, isoflavones, and biflavones (Bylka et al., 2004). A number of the simplest food sources of flavonoids are red wine, apples, blueberries, bilberries, onions, soy merchandise, and tea (Adamczak et al., 2019). The common daily intake of flavonoids within us is between 150 and 200 μg. Varied medicative plants contain therapeutic amounts of flavonoids that are accustomed to treating disorders of the peripheral circulation to lower blood pressure, to enhance aquaresis, and act as anti-inflammatory, antispasmodic, and anti-allergic agents (Robinson, 1980; Wc, 2002). The various medical specialty effects of flavonoids are joined to their ability to act as sturdy antioxidants and free radical scavengers, to chelate metals, and to move with enzymes, nucleoside receptors, and biomembranes (Balch and Balch, 2000). Some flavonoids additionally possess antimicrobial and anti-inflammatory activity (Alka et al., 2012).

11.2 ANTIMICROBIAL ACTIVITY OF FLAVONOIDS

A number of flavones, flavonols, flavanones, and isoflavones, also as a number of their methoxy, isoprenyl, and acylated derivatives, show antibacterial drug activity (Bartels and Miller, 2003). Most microorganisms may be divided into two groups based on a stain developed by the Danish Dr. Hans Christian Gram in 1884. This procedure is predicated on the actual fact that Gram-positive bacteria retain a crystal violet-iodine complicated through decolorization with alcohol or solvent and gram-negative bacteria don't (Özçelik et al., 2011). Gram-positive bacteria possess a thick peptidoglycan layer in their cell membrane whereas Gram-negative bacteria possess a skinny peptidoglycan layer plus a lipopolysaccharide outer membrane. Gram staining works by differentiating bacteria by the chemical and physical properties of their cell walls.

11.2.1 FLAVONES

It is reportable that chrysin (5,7-dihydroxyflavone), within the quantity of five microgram, reserved the expansion of the Gram-negative bacilli escherichia (*E. coli*) and genus *Pseudomonas aeruginosa* (*Ps. aeruginosa*) at a rate akin to that of antibiotic drug (Harborne and Williams, 2000). Baicalein, on the other hand, that has an extra chemical group at C-6, did not inhibit the growth of gram-negative bacteria and showed a weak impact against the gram-positive, spore-forming bacilli (*B. subtilis*), and therefore the gram-positive eubacterium cocci aureus (*S. aureus*) (Khameneh et al., 2019). Basile et al. reported that apigenin, vitexin, saponarin, apigenin, and lucenin 2-O-glycoside, and luteolin 7-O-glycoside acted by selection solely toward bound Gram-negative bacilli: *Proteus vulgaris* (*P. vulgaris*), *Proteus caryophylloid* dicot genus (*P. mirabilis*), notation aeruginosa, *E. coli*, enterobacteria pneumonia (*K. pneumoniae*), and *Enterobacter cloacae*

(*E. cloacae*); however, they showed no activity against the Gram-positive cocci *S. aureus* and *Enterococcus faecalis* (*E. faecalis*). The best activity and the broadest spectrum of activity was noted for saponarin with a minimum repressive concentration (MIC) of 4–2048 μg/mL; weaker activity was noted for apigenin, conjointly the remaining flavones were inactive or active solely against single strains. MIC or negligible repressive concentration is the lowest concentration of associate degree antimicrobial compound which will inhibit the expansion of the organism being tested (Bylka et al., 2004). It is found that the MIC for apigenin against *P. vulgaris, Notation aeruginosa, E. coli,* and *K. pneumoniae* ranged from 54 to 219 μg/mL which apigenin is also proven to move against the spore-former *B. subtilis* (Bashir et al., 2013).

11.2.2 FLAVONOLS

It is said that the varied quercetin mono- and diglycosides examined, quercetin-3-O-rhamnoside exhibited the most powerful activity toward *Pseudomonas maltophilia* (*Ps. maltophilia*) and *E. cloacae*, while the other quercetin-O-glycosides (3-O-glucoside, 3-O-galactoside, 3'-O-galactoside, 3-O-rutinoside, 3-O-galactoglucoside, and 7-O-glucoside) exhibited prone activity or no activity toward those small organisms (Strohl et al., 2001).

11.2.3 ACYLATED FLAVONOLS

It has been examined a number of acylated kaempferol-3-O-glucosides. The best toward Gram-positive microorganism are compounds having one or cis-pcoumaroyl groups. The MICs for kaempferol-3-O-(3",6"- di-O-Z-p-coumaroyl)-β-glucopyranoside, kaempferol-3-O- (3"-Z-p-coumaroyl)-(6"-O-E-feruloyl)-β-glucopyranoside, and chloromycetin toward Bacilli caryophylloid dicot genus (*B. subtilis*) have been 4, 16, and 2 μg/mL, severally, and toward enzymepositive *S. aureus* are 8, 32, and 64 μg/mL, respectively, but toward the coagulase-bad staph epidermidis (*S. epidermidis*) they need to be 2, 4, and 4 μg/mL, respectively (Ali et al., 1998). The presence of the protein coagulase may be a sign that a particular species of staphylococci is a lot of virulent (Jungbauer and Medjakovic 2012). Iwagawa et al. aforesaid that the varied acetylated derivatives of quercetin, quercetin 3-arabinopyranoside-2"-gallate suppressed the rise of *E. coli* (Basile et al., 1999). Mitrokosta et al. examined tiliroside [kaempferol-3-O-β-D- (6"-E-p-coumaroyl)-glucopyranoside] and plantanoside [kaempferol-3-O-a-L-(2",

3"-di-E-p-coumaroyl) rhamnoside] against the Gram-positive cocci *S. aureus*, *S. epidermidis*, and B and cluster *F. streptococci*. They need to be boot tested toward the Gram-negative bacilli *E. coli*, *notation aeruginosa*, and *K. pneumoniae*. Plantanoside showed medicine activity stronger than that of tiliroside, which established to be infirm active to inactive (Narayana et al., 2001).

11.2.4 FLAVONONES

Six flavonones have been examined for antibacterial activity (which included naringenin, taxifolin, and dihydrokaempferide); most effective naringenin showed activity toward *E. coli, S. aureus,* and *E. faecalis* (Barbieri et al., 2017). The flavanones pinocembrin and pinocembrin chalcone, 0.1 and 1 μg, respectively, inhibited the increase of S. aureus. Wachter et al. said that the isomeric compounds 5,7,4'-trihydroxy-6-methyl-8-isoprenylflavonone and 5,7,4'- trihydroxy 8-methyl-6-isoprenylflavonone, by using the agar diffusion method, had been said to inhibit increase of *S. aureus* in an awareness of 0.1 mg/mL (Wang et al., 2019). Rahman and Gray in 2002 said that five 3'-dihydroxy-4'-methoxy-6", 6"-dimethylchromeno-(7, eight, 2", 3") flavanone and five 7, 4'-trihydroxy-6,eight-di-(3-methylbut 2- enyl)-flavanone had been said to be energetic toward each Gram-fine and Gram-poor micro-organism (Lambert and Hammond, 1973).

11.2.5 ISOFLAVONONES

It is confirmed that the isoflavonones containing prenyl organizations had the very best interest toward Gram-fine micro-organism which includes *S. aureus* and *B subtilis*. This interest becomes finest while the prenyl organizations had been positioned at positions C-6 or C-8 in ring A and C-3' or C-5' in ring B (Kuete et al., 2007).

11.2.6 LIPOPHILIC FLAVONOIDS

It is said that the lipophilic compound 7, 8- dihydroxyflavone exhibited susceptible interest toward *S. aureus*, and no interest toward *P. vulgaris*. These effects can be associated with the excessive lipid content material of the mobileular wall of *P. vulgaris*, which may also have trapped the 7,

8-dihydroxyflavone. The cell wall of *S. aureus*, missing a lipid layer, can be penetrated. The flavonoids exhibited a more potent impact on DNA synthesis in *P. vulgaris*, even as showing a more potent impact on RNA synthesis in *S. aureus* (Karlíčková et al., 2015).Wang et al measured the interest of some of lipophilic flavonoids toward *B. cereus*. The effects indicated that the presence of hydroxyl groups at positions C-5 and C-7 become very crucial for interest, while the presence of an extra methoxyl organization at C7 or dihydroxyl organizations at C-3' and C-4' extensively decreased interest. The maximum energetic compound becomes discovered to be 3, 3'-dimethoxy-5, 7, 4'-trihydroxyflavone. It inhibited the increase of *B. cereus* with the use of TLC plates at 0.25 mg. As a factor of reference, quercetin is energetic while it is found with the quantity of 2.5 μg (Ohemeng et al., 1993). Van Puyvelde et al. said that 3-O-methylquercetin confirmed interest toward Gram-fine micro-organism, mainly *S. aureus* and *B. subtilis* (Rauter et al., 2013).

11.2.7 AGLYCONES OF FLAVONOIDS

The activities of the aglycones of flavones, flavonols, flavanones, and isoflavones are toward *P. vulgaris* and *S. aureus*. The aglycones maximum active toward *P. vulgaris* have been myricetin, and morin, quercetagetin, datiscetin, and robinetin. Myricetin and robinetin confirmed the best interest toward S aureus. The different flavonoids studied, chrysin, luteolin, apigenin, apigenin 7,4'-dimethylether, kaempferol, quercetagetin, eriodictyol, hesperidin, and taxifolin did now no longer inhibit the increase of the micro-organism mentioned (Russo et al., 2012).

11.2.8 STRUCTURE-ACTIVITY RELATIONSHIPS

It is mentioned that the huge variety of myricetin interest, each toward Gram-advantageous and Gram-negative micro-organism, became associated with its inhibition of protein synthesis. They suggested that most effective polyhydroxylated derivatives of flavonoids, besides for flavone, which incorporates no hydroxyl agencies, have been lively toward MRSA. The presence of as a minimum one hydroxyl institution in jewelry A or B at C-3, 5, 7 became vital for activity. Compounds without hydroxyl agencies in ring B (pinocembrin, chrysin, galangin) or compounds wherein the hydroxyl became changed with a methoxy institution (kaempferide, tamarixetin) grew to become out to be inactive. Substitution of aglycones with glycones, (myricetin glucoside, quercetin glucoside) ended in a lack of activity toward the strains of micro-organism examined (Soto, 2013).

11.3 ANTI-INFLAMMATORY EFFECTS OF FLAVONOIDS

Flavonoids have anti-inflammatory drug homes through absolutely distinct mechanisms similar to inhibition of restrictive enzymes and transcription factors which have a totally critical position within the control of mediators involved in inflammation. Flavonoids are also mighty antioxidants capable of scavenge free radicals and of decreasing their formation. Consequently, flavonoids have a deep effect on many immune cells and immune mechanisms which can be critical in the inflammatory processes (Lee et al., 1997; Kim et al., 1999) (Table 11.1).

TABLE 11.1 Anti-Inflammatory Properties and Mechanisms of Flavonoids.

Flavonoids properties	Mechanism	Consequences
Inhibition of regulatory enzymes	Inhibition of protein kinases, phosphodiesterases	Decreased signal transduction, decreased cell activation
Antioxidant properties	Scavenger activity, inhibition of free radicals production	Decreased in oxidative stress
Impact on arachidonic acid metabolism	Inhibition of PLA2, COX, LOX	Decreased release of proinflammatory mediators such as prostaglandins, thromboxanes, leukotrienes, etc.
Modulation of gene expression	Modulation of transcription factors: NF-κB, GATA-3, STAT-6	Decreased transcription of proinflammatory genes
Modulation of immune cells	Inhibition of cell activation, maturation, signaling transduction, secretory processes	Immature cell stage (i.e., DCs), decreased cell proliferation, decreased release of proinflammatory cytokines

Source: Adapted from Lee et al., 1997; Kim et al., 1999

11.3.1 INHIBITION OF PROTEIN KINASES AND TRANSCRIPTION FACTORS

Protein kinases are involved in signal transduction throughout cell activation in inflammation. In that respect, certain flavonoids can target more than one imperative kinases which can be involved in more than one signaling pathways (Gil et al., 1997). The inhibition of kinases which include phosphoinositol kinase, protein kinase C, phosphatidylinositol kinase, tyrosine kinase or cyclin-established kinase-4 via way of exclusive styles of flavonoids was pronounced (Escrig et al., 1997). Flavonoids also can modulate protein kinases via way of the inhibition of transcription factors, which include NF-Kb (Kimata et al., 2000). This transcription factor regulates numerous cytokines, chemokines, and cell adhesion molecules concerned in infection. IκB acts as an inhibitory molecule of NF-κB; however, throughout the route of inflammation, IκB is phosphorylated and degraded. Consequently, NF-κB is translocated from the cytoplasm to the nucleus in which it induces the expression of various pro-inflammatory genes (Alarcon De La Lastra and Villegas, 2005). Flavonoids were mentioned to regulate the activity of IκB and NF-κB, with a right away effect on cell activation (Benavente-Garcia and Castillo, 2008). Flavonoids also can adjust master regulatory transcription factors for CD4 + T helper 2 (Th2) cytokines which includes GATA-3, and signal transducer and activator of transcription 6 (STAT-6) (Elliott and Chithan, 2017).

11.3.2 INHIBITION OF PHOSPHODIESTERASES

Flavonoids can inhibit phosphodiesterases which includes cAMP phosphodiesterase. cAMP is a 2D messenger molecule key withinside the regulation of various cell functions throughout inflammation. High levels of cAMP were related to anti-inflammatory features. The enzymes phosphodiesterases can hydrolyze cAMP to maintain regular levels. The inhibitory effects of flavonoids on phosphodiesterases have the capacity to dam cAMP degradation and prolong cAMP signaling (Maleki et al., 2019; Spagnuolo et al., 2018).

Antioxidant activity tissue damage throughout inflammation produces free radicals which includes oxygen-derived radicals (or reactive oxygen species, ROS), nitrogen-derived radicals (or reactive nitrogen species, RNS), which have adverse effects in cell characteristic (Kris-Etherton et al., 2004). Free radicals have unpaired electrons that lead them to noticeably reactive and adverse for lipids, proteins, and DNA. Free radicals affect cellular membranes via way of lipid peroxidation and proteins and nucleic acids via way of oxidative damage. High manufacturing of free radicals collectively with low sequestration of transition metallic ions and occasional free radical scavenging interest bring about oxidative stress (Hämäläinen et al., 2007). Flavonoids have antioxidant interest because of their inhibitory effects at the manufacturing of free radicals and their scavenger activity for ROS, RNS, and different reactive species. The antioxidant houses of flavonoids are derived from their chemical structure, the particular substitution patterns in the structures and the phenolic hydrogens that allow them to behave as hydrogen-donating molecules (Joo et al., 2015).

11.3.3 IMPACT ON ARACHIDONIC ACID METABOLISM

Arachidonic acid is released throughout inflammation from phospholipids contained withinside the plasma membranes via way of the enzyme phospholipase A2 (PLA2). Arachidonic acid is then metabolized via way of exclusive oxygenases, which includes cyclooxygenase (COX) and lipoxygenase (LOX), for you to produce prostaglandins, thromboxanes, leukotrienes, and different inflammatory mediators (Kang et al., 2011). Flavonoids have the capacity to inhibit the enzymes involved withinside the metabolism of arachidonic acid lowering the release of the inflammatory mediators derived from this pathway. For example, flavonoids can inhibit the biosynthesis of prostaglandins, thromboxanes, leukotrienes via way of inhibition of the enzymes PLA2, COX (Read, 1995).

11.3.4 EFFECTS IN IMMUNE CELLS

The specific properties and activities of flavonoids have an effect on cell activation, maturation, signaling transduction, cytokine manufacturing, or secretory techniques in numerous immune cells. For instance, flavonoids were shown to inhibit maturation of dendritic cells (DCs) via way of suppression of maturation markers which includes CD80, CD86, that are molecules important for CD4 + T cell activation and they may be upregulated throughout the maturation of DCs. This might translate into an inhibitory impact withinside the secretion of cytokines and proliferative reaction of CD4 + T cells (Serafini et al., 2010). Flavonoids additionally have a power at the inflammatory reaction of DCs through modulation of iron metabolism (Ma et al., 2019). Several research have determined that sure flavonoids can decrease the release of histamine or prostaglandin from mast cells or inhibit the manufacturing of pro-inflammatory cytokines or chemokines in mast cells, neutrophils, and different immune cells. The effect of flavonoids in cell signaling is supported via way of research that display that they could bind to cytokine receptors which includes IL-17RA subunit of the IL-17 receptor, inflicting attenuation in its signaling. Flavonoids also can inhibit downstream signaling from receptors such the excessive affinity receptor for

IgE (FcεRI) and different receptors on the site of inflammation (Kim et al., 2004). Flavonoids could have inhibitory outcomes on monocyte adhesion function or lower the activation and proliferative reaction or certain immune cells implicated in continual inflammation (Ciumărnean et al., 2020).

11.4 CONCLUSION

The antibacterial activity of flavonoids against each Gram-positive and Gram-negative micro-organism has been pronounced. Activity toward Gram-positive micro-organism (*S. aureus*) changed into verified, especially via the way of compounds that contained hydroxyl groups in ring B. Only those 3-methoxy flavones having additional hydroxyl groups at C-5 and C-7 showed activity toward *B. cereus*. Flavones such as asapigenin and its C- and O- glycosides had been now no longer active. Activity toward Gram-negative bacilli (*E. coli, P. aeruginosa, P. vulgaris,* and *K. pneumoniae*) changed into verified via the way of the flavones apigenin, vitexin, and saponarin, while flavonoid compounds having or 3 hydroxyl groups in rings A or B had been active toward Gram-positive micro-organism. Glycosides in popular had been much less energetic; however, acetyl derivatives of kaempferol-3-glucoside confirmed a few activities toward *S. aureus* and *B. cereus*. Their activity, however, changed into directly associated with the presence of an acyl substitution withinside the cis form.

Due to the particular action of mechanisms and enormous in vivo interest, flavonoids are taken into consideration to be realistic applicants for brand new anti-inflammatory drugs. There is growing proof to signify that inflammatory techniques make a contribution to the gush of occasions that cause the innovative neuronal harm found in neurodegenerative problems which include Parkinson's ailment and Alzheimer's ailment. Therefore, remedy regimes aimed toward modulating neuroinflammatory techniques might also additionally act step by step development of those devastating brain issues. The organic actions expressed via way of flavonoids replicate their various modes of action in inflammation. The inhibitory effect on inflammatory cells, in particular mast cells, seems to be more potent than that of any other clinically available drug. The fact that flavonoids are capable of modulating the activities of various mediators of inflammation suggests their capacity in influencing infection and ought to allow the improvement of latest pharmacological anti-inflammatory agents which might also additionally offer novel insights into the regulation of the inflammatory process. Although the numerous research posted with in vitro processes allow identifying molecular mechanisms of flavonoids effects, data generated have to be confirmed in humans. It is likewise important to be careful while studying capacity useful outcomes of flavonoids administration, due to the fact those molecules might also additionally act as pro-oxidants at high doses.

KEYWORDS

- **flavonoids**
- **antimicrobial**
- **anti-inflammatory**
- **flavones**
- **flavonols**

REFERENCES

Adamczak, A.; Ożarowski, M.; Karpiński, T. M. Antibacterial Activity of Some Flavonoids and Organic Acids Widely Distributed in Plants. *J. Clin. Med.* **2019,** *9* (1), 109.

Alarcon De La Lastra, C.; Villegas, I. Resveratrol as an Anti-inflammatory and Anti-aging Agent: Mechanisms and Clinical Implications. *Mol. Nutr. Food Res.* **2005,** *49* (5), 405–430.

Ali, R. M.; Houghton, P. J.; Raman, A.; Hoult, J. R. S. Antimicrobial and Antiinflammatory Activities of Extracts and Constituents of Oroxylum Indicum (L.) Vent. *Phytomedicine* **1998,** *5* (5), 375–381.

Alka, J.; Padma, K.; Chitra, J. Antifungal Activity of Flavonoids of Sida Acuta Burm f. against Candida Albicans. *Int. J. Drug Dev. Res.* **2012,** *4* (3), 92–96.

Balch, J. F.; Balch, P. A. In *Prescription for Nutritional Healing*; New York: A very. *Penguin Putnam Inc. Vemonia amygdalina used by wild chimpanzees. Tetrahedron* **2000,** *48*, 625–630.

Barbieri, R.; Coppo, E.; Marchese, A.; Daglia, M.; Sobarzo-Sánchez, E.; Nabavi, S. F.; Nabavi, S. M. Phytochemicals for Human Disease: An Update on Plant-Derived Compounds Antibacterial Activity. *Microbiol. Res.* **2017,** *196*, 44–68.

Bartels, C. L.; Miller, S. J. Dietary Supplements Marketed for Weight Loss. *Nutr. Clin. Pract.* **2003,** *18* (2), 156–169.

Bashir, H. S.; Mohammed, A. M.; Magsoud, A. S.; Shaoub, A. M. Isolation of Three Flavonoids from Withania somnifera Leaves (Solanaceae) and their Antimicrobial Activities. *J. Forest Prod. Ind.* **2013,** *2* (5), 39.

Basile, A.; Giordano, S.; López-Sáez, J. A.; Cobianchi, R. C. Antibacterial Activity of Pure Flavonoids Isolated from Mosses. *Phytochemistry* **1999,** *52* (8), 1479–1482.

Benavente-Garcia, O.; Castillo, J. Update on uses and Properties of Citrus Flavonoids: New Findings in Anticancer, Cardiovascular, and Anti-inflammatory Activity. *J. Agric. Food Chem.* **2008,** *56* (15), 6185–6205.

Blumenthal M. In *The ABC Clinical Guide to Herbs*; American Botanical Council: Austin, 2003; p 239.

Bylka, W.; Matlawska, I.; Pilewski, N. A. Natural Flavonoids as Antimicrobial Agents. *JANA* **2004,** *7* (2), 24–31.

Chandra, H.; Bishnoi, P.; Yadav, A.; Patni, B.; Mishra, A. P.; Nautiyal, A. R. Antimicrobial Resistance and the Alternative Resources with Special Emphasis on Plant-Based Antimicrobials—A Review. *Plants* **2017,** *6* (2), 16.

Ciumărnean, L.; Milaciu, M. V.; Runcan, O.; Vesa, Ș. C.; Răchișan, A. L.; Negrean, V.; Perné, M. G.; Donca, V. I.; Alexescu, T. G.; Para, I.; Dogaru, G. The Effects of Flavonoids in Cardiovascular Diseases. *Molecules* **2020,** *25* (18), 4320.

Elliott, M.; Chithan, K. The Impact of Plant Flavonoids on Mammalian Biology: Implications for Immunity, Inflammation and Cancer. In *The Flavonoids*; Routledge, 2017; pp 619–652.

Escrig, V.; Ubeda, A.; Ferrandiz, M. L.; Darias, J.; Sanchez, J. M.; Alcaraz, M. J.; Paya, M. Variabilin: A Dual Inhibitor of Human Secretory and Cytosolic Phospholipase A2 with Anti-inflammatory Activity. *J. Pharmacol. Exp. Ther.* **1997,** *282* (1), 123–131.

Geissman, T. A. In *Organic Chemistry of Secondary Plant Metabolism* (No. 581.13 G4), 1969.

Gil, B.; Sanz, M. J.; Terencio, M. C.; Gunasegaran, R.; Payá, M.; Alcaraz, M. J. Morelloflavone, A Novel Biflavonoid Inhibitor of Human Secretory Phospholipase A2 with Anti-inflammatory Activity. *Biochem. Pharmacol.* **1997,** *53* (5), 733–740.

Hämäläinen, M.; Nieminen, R.; Vuorela, P.; Heinonen, M.; Moilanen, E. Anti-Inflammatory Effects of Flavonoids: Genistein, Kaempferol, Quercetin, and Daidzein Inhibit STAT-1 and NF-κB Activations, Whereas Flavone, Isorhamnetin, Naringenin, and Pelargonidin Inhibit only NF-κB Activation along with their Inhibitory Effect on iNOS Expression and NO Production in Activated Macrophages. *Mediators Inflamm.* **2007,** *2007*.

Harborne, J. B.; Williams, C. A. Advances in Flavonoid Research since 1992. *Phytochemistry* **2000,** *55* (6), 481–504.

Joo, M.; Kim, H. S.; Kwon, T. H.; Palikhe, A.; Zaw, T. S.; Jeong, J. H.; Sohn, U. D. Anti-inflammatory Effects of Flavonoids on TNBS-Induced Colitis of Rats. *Korean J. Physiol. Pharmacol.* 2015, *19* (1), 43–50.

Jungbauer, A.; Medjakovic, S. Anti-inflammatory Properties of Culinary Herbs and Spices that Ameliorate the Effects of Metabolic Syndrome. *Maturitas* **2012,** *71* (3), 227–239.

Kang, S. R.; Park, K. I.; Park, H. S.; Lee, D. H.; Kim, J. A.; Nagappan, A.; Kim, E. H.; Lee, W. S.; Shin, S. C.; Park, M. K.; Han, D. Y. Anti-inflammatory Effect of Flavonoids Isolated from Korea Citrus Aurantium L. on Lipopolysaccharide-Induced Mouse Macrophage RAW 264.7 Cells by Blocking of Nuclear Factor-kappa B (NF-κB) and Mitogen-Activated Protein Kinase (MAPK) Signalling Pathways. *Food Chem.* **2011,** *129* (4), 1721–1728.

Karlíčková, J.; Macáková, K.; Říha, M.; Pinheiro, L. M. T.; Filipský, T.; Hořňasová, V.; Hrdina, R.; Mladěnka, P. Isoflavones Reduce Copper with Minimal Impact on Iron In Vitro. *Oxid. Med. Cell. Longev.* **2015,** *2015*.

Kim, H. P.; Son, K. H.; Chang, H. W.; Kang, S. S. Anti-inflammatory Plant Flavonoids and Cellular Action Mechanisms. *J. Pharmacol. Sci.* **2004,** 0411110005–0411110005.

Kim, H. K.; Son, K. H.; Chang, H. W.; Kang, S. S.; Kim, H. P. Inhibition of Rat Adjuvant-Induced Arthritis by Ginkgetin, A Biflavone from Ginkgo biloba Leaves. *Planta medica* **1999,** *65* (05), 465–467.

Kimata, M.; Shichijo, M.; Miura, T.; Serizawa, I.; Inagaki, N.; Nagai, H. Effects of Luteolin, Quercetin and Baicalein on Immunoglobulin E-mediated Mediator Release from Human Cultured Mast Cells. *Clin. Exp. Allergy* **2000,** *30* (4), 501–508.

Khameneh, B.; Iranshahy, M.; Soheili, V.; Bazzaz, B. S. F. Review on Plant Antimicrobials: A Mechanistic Viewpoint. *Antimicrob. Resist. Infect. Control* **2019,** *8* (1), 1–28.

Kris-Etherton, P. M.; Lefevre, M.; Beecher, G. R.; Gross, M. D.; Keen, C. L.; Etherton, T. D. Bioactive Compounds in Nutrition and Health-Research Methodologies for Establishing Biological Function: The Antioxidant and Anti-inflammatory Effects of Flavonoids on Atherosclerosis. *Annu. Rev. Nutr.* **2004,** *24*, 511–538.

Kuete, V.; Simo, I. K.; Ngameni, B.; Bigoga, J. D.; Watchueng, J.; Kapguep, R. N.; Etoa, F. X.; Tchaleu, B. N.; Beng, V. P. Antimicrobial Activity of the Methanolic Extract, Fractions and Four Dlavonoids from the Twigs of Dorstenia Angusticornis Engl.(Moraceae). *J. Ethnopharmacol.* **2007,** *112* (2), 271–277.

Lambert, P. A.; Hammond, S. M. Potassium Fluxes, First Indications of Membrane Damage in Micro-Organisms. *Biochem. Biophys. Res. Commun.* **1973,** *54* (2), 796–799.

Lee, H. S.; Ko, H. R.; Ryu, S. Y.; Oh, W. K.; Kim, B. Y.; Ahn, S. C.; Mheen, T. I.; Ahn, J. S. Inhibition of Phospholipase Cγ1 by the Prenylated Flavonoids from Sophora Flavescens. *Planta medica* **1997,** *63* (03), 266–268.

Löf, D.; Schillén, K.; Nilsson, L. Flavonoids: Precipitation Kinetics and Interaction with Surfactant Micelles. *J. Food Sci.* **2011,** *76* (3), N35–N39.

Ma, Q.; Jiang, J. G.; Yuan, X.; Qiu, K.; Zhu, W. Comparative Antitumor and Anti-inflammatory Effects of Flavonoids, Saponins, Polysaccharides, Essential Oil, Coumarin and Alkaloids from Cirsium japonicum DC. *Food Chem. Toxicol.* **2019,** *125*, 422–429.

Maleki, S. J., Crespo, J. F.; Cabanillas, B. Anti-inflammatory Effects of Flavonoids. *Food Chem.* **2019,** *299*, 125124.

Mills, S.; Mills, S. Y.; Bone, K. In *Principles and Practice of Phytotherapy: Modern Herbal Medicine*; Churchill Livingstone, 2000.

Murray, M. T. In *Encyclopedia of Nutritional Supplements*; Prima Pub., 1996.

Narayana, K. R.; Reddy, M. S.; Chaluvadi, M. R.; Krishna, D. R. Bioflavonoids Classification, Pharmacological, Biochemical Effects and Therapeutic Potential. *Indian J. Pharmacol.* **2001,** *33* (1), 2–16.

Ohemeng, K. A.; Schwender, C. F.; Fu, K. P.; Barrett, J. F. DNA Gyrase Inhibitory and Antibacterial Activity of Some Flavones (1). *Bioorg. Med. Chem. Lett.* **1993,** *3* (2), 225–230.

Özçelik, B.; Kartal, M.; Orhan, I. Cytotoxicity, Antiviral and Antimicrobial Activities of Alkaloids, Flavonoids, and Phenolic Acids. *Pharm. Biol.* **2011,** *49* (4), 396–402.

Rauter, A. P.; Herold, B. J.; Horton, D.; Moss, G.; Schomburg, I.; Hellwich, K. H.; Ennis, M. Nomenclature of Flavonoids. *Int. Union Pure Appl. Chem.* 2013.

Read, M. A. Flavonoids: Naturally Occurring Anti-inflammatory Agents. *Am. J. Pathol.* **1995,** *147* (2), 235.

Robinson, T. In *The Organic Constituents of Higher Plants: Their Chemistry and Interrelationships* (No. 581.192 R6 1980), 1980.

Russo, P.; Del Bufalo, A.; Cesario, A. Flavonoids Acting on DNA Topoisomerases: Recent Advances and Future Perspectives in Cancer Therapy. *Curr. Med. Chem.* **2012,** *19* (31), 5287–5293.

Serafini, M.; Peluso, I.; Raguzzini, A. In *Flavonoids as Anti-inflammatory Agents*, Proceedings of the Nutrition Society, 2010; vol *69* (3), pp 273–278.

Soto, S. M. Role of Efflux Pumps in the Antibiotic Resistance of Bacteria Embedded in a Biofilm. *Virulence* **2013,** *4* (3), 223–229.

Spagnuolo, C.; Moccia, S.; Russo, G. L. Anti-inflammatory Effects of Flavonoids in Neurodegenerative Disorders. *Eur. J. Med. Chem.* **2018,** *153*, 105–115.

Strohl, W. A.; Rouse, H.; Fisher, B. D. In *Lippincott's Illustrated Reviews: Microbiology*, 2001; pp 26–27.

Wang, M.; Firrman, J.; Liu, L.; Yam, K. A Review on Flavonoid Apigenin: Dietary Intake, ADME, Antimicrobial Effects, and Interactions with Human Gut Microbiota. *BioMed. Res. Int.* **2019,** *2019*.

Wc, E. Trease and Evans. In *Pharmacognosy*, 15th Ed.; New York, 2002.

CHAPTER 12

ROLE OF FLAVONOIDS IN DERMATOLOGY AND COSMETIC PREPARATIONS

EKNATH D. AHIRE[1], SHUBHAM J. KHAIRNAR[1], KHEMCHAND R. SURANA[2], and SANJAY J. KSHIRSAGAR[1]

[1]*Department of Pharmaceutics, MET's Institute of Pharmacy, Affiliated to Savitribai Phule Pune University, BKC, Adgaon, Nashik, Maharashtra, India*

[2]*Department of Pharmaceutical Chemistry, Divine College of Pharmacy, Satana, Nashik, Maharashtra, India*

ABSTRACT

Flavonoids are dietary substances that belong to the phytochemicals, or plant chemicals, class of compounds. There are about 5000 different varieties of flavonoids, and a single food item can contain hundreds of them. Flavonoids are formed up of a basic polyphenolic ring structure (see the chapter on Flavonoids) with numerous side chains connected, giving the compound a variety of properties (see the chapter on Flavonoids). Despite their antioxidant properties, flavonoids' ability to absorb ultraviolet (UV) radiation and change signaling pathways that regulate cellular function appears to support their beneficial effects on skin health. Flavonoids administered topically are not vulnerable to first-pass metabolism. Topical application can deliver pharmacological doses; however there are concerns about solubility, stability, and penetration. In vitro testing has been done to determine whether specific flavonoids formulations have the ability to permeate excised human skin samples.

The Flavonoids: Extraction and Applications. Deepika Saini, Rajesh K. Kesharwani, & Raj K. Keservani (Eds.)

12.1 INTRODUCTION

Due to first-pass metabolism in the GI tract and liver, dietary flavonoids are dramatically altered before reaching the skin. Topical flavonoids administration or instillation is an effective technique to acquire pharmacological quantities of parent chemicals in the skin; nevertheless, the chemical composition of topical formulations has a significant impact on stability and bioavailability. Green tea polyphenols have photoprotective properties when consumed or applied topically. Animal and in vitro studies reveal that genistein administered topically has photoprotective properties. By absorbing UVB and so protecting the skin, certain flavonoids may act as sunscreens (Panche et al., 2016). Various flavonoids suppress inflammatory enzymes and may help to reduce skin inflammation caused by UV exposure. By regulating endogenous defense mechanisms in the skin, flavonoids have the ability to regulate the response to environmental agents such as UVR and procarcinogens. Flavonoids may have an effect on wound healing and blood vessel health, but additional human research is needed to determine their therapeutic value (Yao et al., 2004; Harborne et al., 2013).

Flavonoids are polyphenols that are divided into six groups: anthocyanins, flavanones, flavones, flavonols, isoflavones, and flavanols. Many fruits, vegetables, flowers, cereals, and other plant-derived foods contain anthocyanins, which are responsible for their characteristic red, blue, and purple hue. Anthocyanidins are sugar-free aglycones, whereas anthocyanins are anthocyanin glycosides (Tapas et al., 2008). Cyanidin, delphinidin, malvidin, peonidin, pelargonidin, and petunidin are the six most frequent anthocyanidins. Flavanols, which are found in tea, fruits, cocoa, and chocolate, are another type of flavonoid. They come as monomers (catechins) and polymers (catechinols) (Di Carlo et al., 2005). Epicatechin, gallocatechin, and epigallocatechin are free catechins, while epicatechin gallate and epigallocatechin gallate are bound catechins. The most frequent flavanones include hesperidin (HES) and its metabolite hesperetin, as well as naringenin. Another flavonoid subclass is flavones, which include isosinensetin, sinensetin, tangeretin, luteolin, nobiletin, apigenin, chrysin, and galangin (Perez-Vizcaino and Fraga, 2018). The different concentration of flavones are found in foods and among that Parsley and oregano are the best source of flavones. Flavonols, which are present throughout the plant kingdom, are the most common flavonoids. The most prevalent dietary flavonols include kaempferol, quercetin, fisetin, isorhamnetin, and myricetin (Wang et al., 2000). The most common sources include berries, cowpeas, dock, kale, dark chocolate, parsley, oregano, capers, and green and black tea. It's important to keep in mind that the presence of certain flavonoids in vegetables and fruits is based on the crop variety, region, and type of production, as well as the plant's morphological component (Middleton, 1996; Dajas et al., 2003). Flavonoid content differences between plant species are normally minimal, although in a few occasions, extremely high quantities have been found, such as in particular berries and tea made from Quingmao tree leaves. Flavonoids are frequently consumed at a far higher rate in a plant-based diet than other phytochemicals and vitamins, such as ascorbic acid (vitamin C), tocopherol (vitamin E), or carotenoids. There are various databases of flavonoid content in foods that can be used to estimate flavonoid consumption in humans. Databases from the United States Department of Agriculture (USDA) and the online Phenol-Explorer database are the most often utilized (Agrawal, 2011; Serafini et al., 2010). As a result, comparing the results of nutritional assessments received from diverse data sources may yield different results. Because the database of flavonoid levels in foods at the time was poor, many of the surveys were conducted in the 1990s and may be underestimates. The year of release of specific databases should be taken into account for correct interpretation of the results when assessing dietary flavonoids intake (Dajas et al., 2003; Cazarolli et al., 2008; Xiao et al., 2011; Ullah et al., 2020). Figure 12.1 indicates the different types of flavonoid molecules.

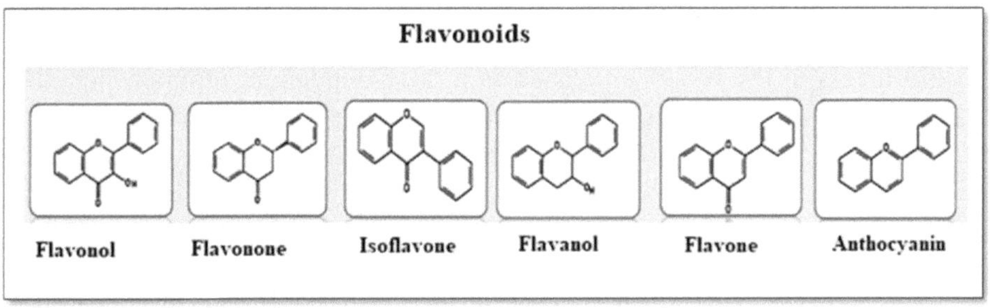

FIGURE 12.1 Different types of flavonoids.

12.2 ROLE OF FLAVONOIDS IN DERMATOLOGY

12.2.1 PHOTOPROTECTION

UVR damages the skin by causing erythema, edema, burnt cells, hyperplasia, inflammation, immunosuppression, photoaging, and photocarcinogenesis to mention a few. According to a study, certain flavonoids can diminish UVR-induced skin reactions in cell culture, animals, and humans (Sampaio et al., 2016; Chen et al., 2012; González et al., 2016).

12.2.2 GREEN TEA POLYPHENOLS

Heinrich et al., in a 12-week placebo-controlled trial, looked at the effects of dietary green tea polyphenol consumption on skin photoprotection, structure, and function in 60 healthy women aged 40–65 (Mansuri et al., 2021; Skotarczak et al., 2015; Verschooten et al., 2006).

Heinrich et al. conducted a short dose study to evaluate how green tea extract in capsule form affected participants, recognizing the difficulties of taking 1 L of beverage over the course of a day. A total of 15 healthy women were given either a placebo or 0.5, 1, or 2 g of encapsulated green tea extract (Boothby-Shoemaker et al., 2022). The immediate effects of one dose of encapsulated green tea extract were assessed by measuring serum flavonoid concentration and capillary blood flow to the dermis over the course of 4 h. Increased cutaneous blood flow may contribute to increased carrying capacity of oxygen and nutrients to the skin, a potential effect of flavonoids on skin health; hence dermal microcirculation was examined. All doses of green tea extract produced an equally rapid and temporary increase in cutaneous microcirculation (15–30 min) when compared to placebo (Verschooten 2012; Arruda et al., 2021; Jacob and Goldenberg, 2014). Epicatechin levels in the blood increased in a dose-dependent manner, peaking 2 h after eating. Photoprotection and skin structure were improved in healthy females after 12 weeks of administration of high-flavanol cocoa powder rich in epicatechin and catechin. A total of 24 female volunteers (18–25 years old) were randomly assigned to drink a high (326 mg) or low (27 mg) flavanol-containing cocoa beverage daily for 12 weeks in this double-blind study (Prasanth et al., 2020; Stahl et al., 2010; Afaq and Katiyar, 2011). High-flavanol cocoa powder reduced UV-induced erythema development, increased microcirculation, improved skin structure (roughness, scaling, volume, and wrinkles), and reduced TEWL, similar to green tea beverage; none of these parameters altered in the low-flavanol group. Furthermore, a single dose of high-flavanol (329 mg) cocoa beverage enhanced plasma epicatechin levels and cutaneous microcirculation quickly and transiently when compared to low-flavanol cocoa beverage. The ability of catechins to enhance cutaneous blood flow could explain their benefits on skin structure, texture, and water balance. The vascular advantages linked with flavanols may be due to vasodilatory mechanisms, which have also been proven for cocoa polyphenols in nonskin arteries (Stevanato et al., 2014; Wei et al., 2003; Saewan and Jimtaisong, 2015).

Katiyar et al. examined the effects of a topically applied green tea extract on a variety of UV-mediated skin responses in humans. Four to six volunteers (male and female, ages 25–55) were given a topical application of purified green tea extract (1 mg/cm^2 dissolved in acetone to sun-protected skin) containing a mixture of the four major polyphenols found in green tea, epigallocatechin-3-gallate, epicatechin, epigallocatechin, and epicatechin-3-gallate, in each study. Following topical therapy, skin regions were exposed to UVB for 20 or 30 min (4X MED). Punch samples were taken (both epidermis and dermis), and several endpoints were measured 24 or 48 h later. Pretreatment with pure green tea polyphenols reduced UV-induced inflammation, DNA damage, and the production of reactive nitrogen/oxygen species and free radicals when compared to vehicle-treated areas on the same subject. In a similar study design, Elmets et al. found that topically applied pure green tea extract had a protective and preventive effect on UV-induced photodamage, as measured by erythema development, the presence of sunburn cells, DNA damage, and the number of Langerhans cells, an immunosuppressive marker. A 5% solution comprising a mixture of the four principal green tea polyphenols (0.5 g of pure extract combined in ethanol/water) was the most effective at preventing UV-damage when compared to vehicle-treated locations in the same subjects. EGCG and ECG were also protective, though to a lesser level than the combined extract (Mierziak et al., 2014; Joachim et al., 2004).

In a double-blind trial, Camouse et al. used 10 volunteers to compare topical green and white tea extracts as photoprotective agents. Pretreatment with either tea extract (2.5 mg/cm^2 in organic solvent) protected against depletion of Langerhans cells and oxidative DNA damage caused by UVB radiation as compared to vehicle-treated regions on the same individual (2X MED). When UVR is absorbed by DNA, it causes strand breaking and the formation of cyclobutane pyrimidine dimers (CPDs), a photoproduct that forms an unfavorable covalent bond between pyrimidine bases. A cell can either heal the damage or sacrifice itself to protect the organism from mutations and malignant transformations (apoptosis). The key methods by which green tea polyphenols prevent UV-induced cellular damage appear to be stimulation of DNA repair pathways in the skin and influence on certain immune mediators known as cytokines. Because DNA damage induces immunosuppression, which is a risk factor for skin carcinogenesis, green tea polyphenols appear to function early in the UV-damage response in the skin (Aqil et al., 2020; Arct and Pytkowska, 2008).

12.2.3 GENISTEIN

Genistein's photoprotective properties have been studied in animals and in vitro human skin models. In hairless mice exposed to regular doses of acute and chronic UVB irradiation, pretreatment with topical genistein (60 min before UV exposure, 5 M) reduced skin roughness and wrinkles, as well as epidermal hyperproliferation. Topical genistein reduced CPD production and restored PCNA expression, a marker of proliferation and DNA repair, as found with EGCG, suggesting that its photoprotective advantages are attributable to its impact on UV-induced DNA damage. To test their findings in people, the authors conducted short research in six men, finding that topical genistein (5 M/cm^2) administered 30 min before UVB exposure (1X MED) prevented erythema development as measured by photography 24 h later. Furthermore, pretreatment with topical genistein reduced CPD production and elevated PCNA expression in human reconstituted skin samples in a dose-dependent manner (Singh and Agarwal, 2002; Benayad et al., 2014).

12.2.4 OTHER FLAVONOIDS

Silymarin is a flavonolignan, a flavonoid that contains both flavonoids and lignans. Milk thistle (*Silybum marianum*) seeds contain silymarin, which contains the bioactive flavonoid silibinin. Topical silymarin, like green tea polyphenols, has been shown in animal studies to prevent UV-induced photodamage and photocarcinogenesis. According to a research using primary cultures of normal human epidermal keratinocytes (NHEKs) and transgenic mice, topical silymarin reduces UV-induced apoptosis and lowers CPD synthesis in the epidermis. Topical silymarin also contributes to photoprotection by upregulating DNA repair processes in cells and animals weak in nucleotide excision repair, according to the research (Ruiz-Cruz et al., 2017).

12.2.5 SUNSCREEN EFFECT

Topically applied flavonoids may provide skin protection by absorbing UVR before it interacts with and destroys cellular components, resulting in a sunscreen effect. Melanin, urocanic acid, amino acids, and nucleic acids are all major epidermal chromophores (molecules that absorb UV radiation). Flavonoids and polyphenols applied topically may also protect skin by absorbing UV radiation and inhibiting UV penetration. Honeybush extract (which contains the flavonoids hesperidin and xanthone mangiferin) and Pycnogenol® (a registered blend of naturally occurring mono- and oligomeric procyanidins) absorb UVB rays. Thus, when applied topically prior to UV exposure, these flavonoids would act as sunscreens (Dias et al., 2021; Arct and Pytkowska, 2008).

12.2.6 SUPPRESSION VERSUS PREVENTION

Whether flavonoid administration is used as a prophylactic or therapeutic measure is determined by the timing of the intervention. The majority of research focuses on the use of flavonoids to reduce UV-induced photodamage prior to UV exposure. Several flavonoids have been studied in the context of flavonoid administration following UV exposure. One or four hours after irradiation, hairless mouse skin was given genistein dissolved in acetone, and epidermal slices were taken and analyzed 24 h later. Both flavonoids reduced the number of burnt cells, epidermal hyperplasia, and immune suppression even when given after UV exposure. Widyarini et al. used isoflavone extracts (20 M) from red clover (*Trifolium pretence*) to protect hairless mice from acute UVR effects following UV exposure. Genistein and its isoflavone metabolites equol, isoequol, and dehydroequol significantly reduced UV-induced inflammation, edema, and immunosuppression. In a randomized, double-blind, placebo-controlled trial, Casetti et al. compared the efficacy of a luteolin-rich reseda extract (RE) to hydrocortisone, a common anti-inflammatory medicine, following UV exposure. A total of 40 healthy volunteers (18 years and older, both sexes) were exposed to UVB (1.5X MED) before being treated with a topical nanoparticle formulation of RE (2.5%), hydrocortisone (0.1%), or vehicle (glycerol). When compared to vehicle, both RE and hydrocortisone effectively reduced UVB-induced erythema (Dias et al., 2021).

The value of inhibiting the sunburn reaction to reduce skin damage is a point of contention. Primary defense against the harmful effects of UVR is to avoid sun damage in the first place (Patel and Patel, 2019).

12.2.7 PHOTOAGING

12.2.7.1 GREEN TEA POLYPHENOLS

The long-term benefits of green tea polyphenol supplementation on clinical and histological signs of photoaging were investigated in a 2-year, double-blind, placebo-controlled experiment. A total of 56 healthy female volunteers (ages 25–75) were given 250 mg green tea polyphenols twice daily or a placebo for 2 years. A dermatologist examined photodamaged facial skin at 0, 6, 12, and 24 months for wrinkles, hyperpigmentation, depigmentation, lentigines (liver spots), pore size, roughness, erythema, telangiectasias (permanent dilatation of superficial blood vessels), and total sun damage. After 12 months, green tea outperformed placebo in some skin measures, but there were no significant differences between the groups after 24 months, with both groups showing improvements in overall sun damage and elastosis (abnormal elastin accumulation) (Dias et al., 2021; Moghrovyan et al., 2019).

A tiny, double-blind, placebo-controlled pilot trial looked at the effects of a green tea extract combined oral and topical treatment on skin appearance and histology in female patients with mild photoaging.

For 8 weeks, 40 healthy women were given either a green tea treatment (10% green tea extract cream applied to the face and arms plus 300 mg green tea extract as an oral supplement) or a placebo cream and supplement. The placebo and treatment groups both graded wrinkles and roughness the same way, although several patients in the green tea group complained of irritation, dryness, and sun sensitivity at the application site. Physicians discovered no significant variations in skin appearance between the treatment and placebo groups. Only the elastic tissue content improved in the green tea group compared to the placebo group on histological evaluation (Hong et al., 2014; Takahashi et al., 2004).

12.2.7.2 GENISTEIN

The isoflavone genistein has been studied for its ability to reverse indications of photoaging in postmenopausal women, as estrogen has a substantial effect on skin aging. A total of 30 postmenopausal women who took a concentrated soy extract (100 mg daily for 6 months) showed significant improvements in skin thickness, elastic fiber content, collagen fiber content, and vasculature in a gluteal skin biopsy after 6 months of treatment compared to baseline. In a randomized, double-blind, estrogen-controlled study, Moraes et al. investigated the effects of topical isoflavones on morphological parameters in postmenopausal facial skin. A total of 40 participants were given estrogen (0.01% 17-estradiol) or isoflavone (4% genistein) gel to apply to their facial skin for 24 weeks. Topical estrogen significantly improved all parameters when compared to baseline and isoflavone therapy. Isoflavones significantly increased epidermal thickness and blood vessel number after 24 weeks of treatment, though to a smaller extent than estrogen treatment (Yakoub et al., 2018; Mierziak et al., 2014).

12.2.7.3 XENOBIOTIC METABOLISM

Skin serves as a physical as well as a biological barrier. The skin's xenobiotic metabolism inactivates potentially hazardous molecules, providing a second line of defense against toxins that penetrate the skin's surface. The process of turning a foreign chemical component into an inert substance that can be safely removed from the body through a series of enzyme reactions is known as xenobiotic metabolism. Members of the cytochrome P450 (CYP) family of enzymes contribute in phase I metabolism, also known as activation, in which oxygen is used to build a reactive site on the xenobiotic molecule. The attachment of a water-soluble functional group to the reactive site of the phase I metabolite is known as conjugation. Finally, the solubilized compound is ejected from the cell in phase III (Abdel-Daim et al., 2020; Korać and Khambholja, 2011).

Monoinduction of phase I enzymes without simultaneous induction of phase II enzymes can result in the creation of "activated" chemicals that can harm cells. UVB radiation causes epidermal CYP 1A1 and 1B1 to be activated in a time and dose-dependent manner. Many chemicals from foreign substrates are activated by CYP 1A1 and 1B1, including polycyclic aromatic hydrocarbons (PAH), a well-known family of procarcinogens. UVB could thereby accelerate the activation of environmental pollutants by inducing phase I enzymes, increasing the mutagenesis load in the epidermis. Another strategy for reducing UV-induced photodamage could be flavonoid-mediated enzyme regulation in xenobiotic metabolism (Del et al., 2017).

Flavonoids that target phase I or phase II components of the cellular detoxification pathway have varying effects on xenobiotic metabolism in the skin. When applied topically to mouse skin, the flavonols myricetin and quercetin decrease the activity of aryl hydrocarbon hydrolase (a phase I enzyme), potentially inhibiting the metabolic activation of procarcinogens and the generation of DNA adducts. Flavonoids that activate phase II enzymes, on the other hand, may aid in the inactivation of CYP-generated metabolites. In comparison to vehicle-treated control mice, oral administration of silibinin, the active component of silymarin, for 15 days dramatically increased phase II enzyme activity (glutathione S-transferase and quinone reductase) in mouse skin (Ruiz-Cruz et al., 2017).

12.2.7.4 WOUND HEALING

Scars, particularly keloid scars, have been treated with onion extract, which is high in the flavonoids quercetin and kaempferol. Onion extract and quercetin promote matrix metalloproteinase-1 (MMP-1) production in human skin fibroblasts and hairless mice skin, according to Cho and colleagues. MMPs are enzymes released by epidermal keratinocytes and dermal fibroblasts in response to UV radiation, oxidative stress, and inflammatory cytokines. UVR stimulates three MMPs: collagenase, stromelysin, and gelatinase, which all cleave and destroy collagen in the skin, contributing to photoaging. The balance of MMP-1 and tissue inhibitor matrix metalloproteinase-1 (TIMP-1) enzymatic activity influences the quantity of extracellular matrix (including collagen) formed at the wound site. Quercetin may influence extracellular matrix deposition during wound healing to reduce hypertrophic scarring (Aslam et al., 2018; Chirumbolo, 2012; Carvalho et al., 2021).

12.3 ROLE OF FLAVONOIDS IN COSMETOLOGY

12.3.1 ANTI-INFLAMMATORY ACTIVITY

The anti-inflammatory activities of flavonoids demonstrate their multiactivity. This action is often used in the field of cosmetology. Most plant extracts contain flavonoids, which are utilized to calm or lessen the irritability of many raw ingredients used in cosmetics. These chemicals' activity originates from their complex interactions with proinflammatory factors and enzymes that are either directly or indirectly engaged in the initiation or progression of inflammatory stages. Flavonoids' ability to scavenge free radicals limits the oxidative processes of membrane lipids, which contribute to arachidonic acid synthesis. Due to their affinity for proteins and metals, chelation flavonoids (e.g., apigenin glycosides found in chamomile) inactivate 5-lipoxygenase and cyclooxygenase, which play a vital role in the translation of arachidonic acid into proinflammatory leukotrienes (LTs) and prostaglandins. Another mode of action is the suppression of enzymes involved in cell activation and production of regulatory chemicals such as tumor necrosis factor and other cytokines, which propagate the inflammatory stage in a tissue. The creation and release of nitric oxide, a molecule that regulates a variety of physiologic functions, including instigating and propagating inflammation, have been demonstrated to be altered by a number of flavonoids. The anti-inflammatory and antiirritant actions of flavonoids are mostly owing to their blood vessel effects. Flavonoids diminish tissue congestion and have a strong antiedematous action. Flavonoids, as a result, aid in the reduction of inflammatory symptoms (Rathee et al., 2009; Pan et al., 2010; Serafini et al., 2010).

12.3.2 ANTIMICROBIAL ACTIVITY

The activity of flavonoids against microbes is the second most important aspect of their biological activity. Antibacterial and antifungal properties are found in several flavonoids. Protection against diseases produced by viruses, bacteria, and fungi is one of the key activities of these chemicals in plants. The activity isn't just restricted to phytopathogenic bacteria; it could also be used to protect the skin. Antiviral activity is one of the most essential aspects of the action. Flavonoids belong to the sparse group of compounds that selectively inhibit the proliferation of viruses such as herpes simplex, polio, and pseudorabies. Antiviral action is also demonstrated by a number of synthetic compounds (e.g., quercetin dimethyl ether). Despite multiple tests, there is still a scarcity of evidence in the literature on the structure–activity link. Biocidal effect against higher organisms has been demonstrated in various plant extracts (Górniak et al., 2019; Cushnie and Lamb, 2005; Osonga et al., 2019).

12.3.3 PROTECTIVE EFFECT ON BLOOD VESSEL WALLS

Many flavonoids are known to protect against telangiectasias and petechias, which are caused by blood vessel rupture. As a result, the idea that these chemicals have a "strengthening" impact on blood vessel walls is widely held. Blood vascular resistance is referred to as "resistance" in medicine. Reduced blood vessel resistance is characterized by their extreme fragility, tendency to rupture, and sensitivity to blood extravasation via the skin. In practice, the decrease in blood vessel resistance is evaluated by the drop in so-called critical petechial pressure, which causes blood vessel rupturing when applied to a specific skin surface for a specific amount of time. Rutin and its derivatives, in particular, have a protective impact on blood vessel walls. Capillary fragility increases dramatically when hesperidin deficiency is present, but the alterations are reversed when flavonoids are added. Epicatechin, catechin, and hesperetin show similar activity (Arct and Pytkowska, 2008; Asgary et al., 2005; Duarte et al., 2002).

12.4 CONCLUSION

The photoprotective properties of polyphenols, catechins, and genistein in green tea extract are the subject of the majority of studies on flavonoids and skin health. The flavanol subclass has been demonstrated to provide photoprotective advantages in humans through both oral intake and topical application. Despite their photoprotective potential, distribution is a problem that can affect how commercially available formulations penetrate and function in human skin. Flavonoids have a wide spectrum of effects due to their specific and nonspecific affinity for a variety of proteins throughout the cell. Although there is evidence that flavonoids physically limit UV penetration, impact DNA repair, minimize oxidative damage, attenuate the inflammatory response, maintain immunological function, and trigger cytoprotective pathways, the exact mechanisms by which flavonoids protect skin from UVR damage are unknown. Flavonoids are multiactive chemicals that have antioxidant and calming properties and are frequently used in cosmetics. Antiepileptic medications contain them as well. Despite their multifunctional properties, flavonoids are underutilized. The purpose of this study was to show how flavonoids can be employed as active ingredients in cosmetics. The influence of cosmeceuticals on skin–blood microcirculation to improve the factor-limiting tissue growth and renewal in all skin types is one of the most important of these possibilities.

ACKNOWLEDGMENTS

We, thank you to the MET's, Institute of Pharmacy, BKC, affiliated under Savitribai Phule Pune University, Nashik. EDA wants to thank Ministry of Tribal Affairs Govt. of India for providing financial assistance in the form of fellowship (Award No-202021-NFST-MAH-01235).

KEYWORDS

- **flavonoids**
- **dermatology**
- **cosmetic**
- **formulation**
- **cosmeceuticals**

REFERENCES

Abdel-Daim, M. M.; Mahmoud, O. M.; Al Badawi, M. H.; Alghamdi, J.; Alkahtani, S.; Salem, N. A. Protective Effects of Citrus Limonia Oil Against Cisplatin-Induced Nephrotoxicity. *Environ. Sci. Poll. Res.* **2020,** *27* (33), 41540–41550.

Afaq, F.; K Katiyar, S. Polyphenols: Skin Photoprotection and Inhibition of Photocarcinogenesis. *Mini Rev. Med. Chem.* **2011,** *11* (14), 1200–1215.

Agrawal, A. D. Pharmacological Activities of Flavonoids: A Review. *Int. J. Pharm. Sci. Nanotechnol.* **2011,** *4* (2), 1394–1398.

Andersen, O. M.; Markham, K. R. In *Flavonoids: Chemistry, Biochemistry and Applications*; CRC Press, 2005.

Aqil, M.; Chaudhuri, A.; Qadir, A. Herbal Cosmeceuticals: New Opportunities in Cosmetology. *Trends Phytochem. Res.* **2020,** *4* (3), 117–142.

Arct, J.; Pytkowska, K. Flavonoids as Components of Biologically Active Cosmeceuticals. *Clin. Dermatol.* **2008,** *26* (4), 347–357.

Arruda, R. L.; de Sousa Garcia, N. O.; Souza, N. F.; da Silva, F. M.; Arruda, E. L.; da Conceição, E. C. Natural Photoprotectors: A Literature Review. *Res. Soc. Develop.* **2021,** *10* (5), e0810514603–e0810514603.

Asgary, S.; Naderi, G. H.; Askari, N. Protective Effect of Flavonoids Against Red Blood Cell Hemolysis by Free Radicals. *Exp. Clin. Cardiol.* **2005,** *10* (2), 88.

Aslam, M. S.; Ahmad, M. S.; Riaz, H.; Raza, S. A.; Hussain, S.; Qureshi, O. S.; Maria, P.; Hamzah, Z.; Javed, O. Role of Flavonoids as Wound Healing Agent. In *Phytochemicals-Source of Antioxidants and Role in Disease Prevention*, 2018; pp 95–102.

Boothby-Shoemaker, W. T.; Mohammad, T. F.; Ozog, D. M.; Lim, H. W. Photoprotection by Clothing: A Review. *Photodermatol. Photoimmunol. Photomed.* **2022,** *38* (5), 478–488.

Carvalho, M. T.; Araújo-Filho, H. G.; Barreto, A. S.; Quintans-Júnior, L. J.; Quintans, J. S.; Barreto, R. S. Wound Healing Properties of Flavonoids: A Systematic Review Highlighting the Mechanisms of Action. *Phytomedicine* **2021,** *90*, 153636.

Cazarolli, L. H.; Zanatta, L.; Alberton, E. H.; Bonorino Figueiredo, M. S. R.; Folador, P.; Damazio, R. G.; Pizzolatti, M. G.; Barreto Silva, F. R. M. Flavonoids: Prospective Drug Candidates. *Mini Rev. Med. Chem.* **2008,** *8* (13), 1429–1440.

Chen, L.; Hu, J. Y.; Wang, S. Q. The Role of Antioxidants in Photoprotection: A Critical Review. *J. Am. Acad. Dermatol.* **2012,** *67* (5), 1013–1024.

Chirumbolo, S. Flavonoids in Propolis Acting on Mast Cell-Mediated Wound Healing. *Inflammopharmacology* **2012,** *20* (2), 99–101.

Cushnie, T. T.; Lamb, A. J. Antimicrobial Activity of Flavonoids. *Int. J. Antimicrob. Agents* **2005,** *26* (5), 343–356.

Dajas, F.; Rivera-Megret, F.; Blasina, F.; Arredondo, F.; Abin-Carriquiry, J. A.; Costa, G.; Echeverry, C.; Lafon, L.; Heizen, H.; Ferreira, M.; Morquio, A. Neuroprotection by Flavonoids. *Braz. J. Med. Biol. Res.* **2003,** *36* (12), 1613–1620.

Del Río, J. A.; Díaz, L.; Ortuño, A. Properties and Use of Citrus Flavonoids. In *Phytochemicals in Citrus: Applications in Functional Foods*, 2017; vol 5, pp 323–331.

Di Carlo, G.; Mascolo, N.; Izzo, A. A.; Capasso, F. Flavonoids: Old and New Aspects of a Class of Natural Therapeutic Drugs. *Life Sci.* **1999,** *65* (4), 337–353.

Dias, R.; Oliveira, H.; Fernandes, I.; Simal-Gandara, J.; Perez-Gregorio, R. Recent Advances in Extracting Phenolic Compounds from Food and their use in Disease Prevention and as Cosmetics. *Crit. Rev. Food Sci. Nutr.* **2021,** *61* (7), 1130–1151.

Dias, M. C.; Pinto, D. C.; Silva, A. Plant Flavonoids: Chemical Characteristics and Biological Activity. *Molecules* **2021,** *26* (17), 5377.

Duarte, J.; Jiménez, R.; O'Valle, F.; Galisteo, M.; Pérez-Palencia, R.; Vargas, F.; Pérez-Vizcaíno, F.; Zarzuelo, A.; Tamargo, J. Protective Effects of the Flavonoid Quercetin in Chronic Nitric Oxide Deficient Rats. *J. Hypertens.* **2002,** *20* (9), 1843–1854.

González, S.; Gilaberte, Y.; Juarranz, A. Oral and Systemic Photoprotection. In *Principles and Practice of Photoprotection*; Adis: Cham, 2016; pp 387–403.

Górniak, I.; Bartoszewski, R.; Króliczewski, J. Comprehensive Review of Antimicrobial Activities of Plant Flavonoids. *Phytochem. Rev.* **2019,** *18* (1), 241–272.

Harborne, J. B.; Mabry, T. J. In *The Flavonoids: Advances in Research*; Springer, 2013.

Harborne, J. B.; Marby, H.; Marby, T. J. In *The Flavonoids*; Springer, 2013.

Hong, J.; Hu, J. Y.; Liu, J. H.; Zhou, Z.; Zhao, A. F. In Vitro Antioxidant and Antimicrobial Activities of Flavonoids from Panax Notoginseng Flowers. *Nat. Prod. Res.* **2014,** *28* (16), 1260–1266.

Korać, R. R.; Khambholja, K. M. Potential of Herbs in Skin Protection from Ultraviolet Radiation. *Pharmacogn. Rev.* **2011,** *5* (10), 164.

Jacob, S.; Goldenberg, A. Photoprotection: Fundamental in Organ Transplant Patients. *Facial Plast. Surg.* **2014,** *30* (06), 688–689.

Joachim, J.; Abdelbary, G.; Prinderre, P.; Reynier, J. P.; Kister, J.; Piccerelle, P. Anti-free Radicals: Source, Role and Application in Dermo-Cosmetology. *Bull. Pharm. Sci. Assiut* **2004,** *27* (1), 63–68.

Mansuri, R.; Diwan, A.; Kumar, H.; Dangwal, K.; Yadav, D. Potential of Natural Compounds as Sunscreen Agents. *Pharmacogn. Rev.* **2021,** *15* (29).

Middleton, Jr. E. Biological Properties of Plant Flavonoids: An Overview. *Int. J. Pharmacogn.* 1996, *34* (5), 344–348.

Mierziak, J.; Kostyn, K.; Kulma, A. Flavonoids as Important Molecules of Plant Interactions with the Environment. *Molecules* **2014,** *19* (10), 16240–16265.

Moghrovyan, A.; Sahakyan, N.; Babayan, A.; Chichoyan, N.; Petrosyan, M.; Trchounian, A. Essential Oil and Ethanol Extract of Oregano (Origanum vulgare L.) from Armenian Flora as a Natural Source of Terpenes, Flavonoids and Other Phytochemicals with Antiradical, Antioxidant, Metal Chelating, Tyrosinase Inhibitory and Antibacterial Activity. *Curr. Pharm. Design* **2019,** *25* (16), 1809–1816.

Morganti, P. The Photoprotective Activity of Nutraceuticals. *Clin. Dermatol.* **2009,** *27* (2), 166–174.

Osonga, F. J.; Akgul, A.; Miller, R. M.; Eshun, G. B.; Yazgan, I.; Akgul, A.; Sadik, O. A. Antimicrobial Activity of a New Class of Phosphorylated and Modified Flavonoids. *ACS Omega* **2019,** *4* (7), 12865–12871.

Pan, M. H.; Lai, C. S.; Ho, C. T. Anti-inflammatory Activity of Natural Dietary Flavonoids. *Food Funct.* **2010,** *1* (1), 15–31.

Panche, A. N.; Diwan, A. D.; Chandra, S. R. Flavonoids: An Overview. *J. Nutr. Sci.* **2016,** *5*.

Patel, K.; Patel, D. K. The Beneficial Role of Rutin, A Naturally Occurring Flavonoid in Health Promotion and Disease Prevention: A Systematic Review and Update. In *Bioactive Food as Dietary Interventions for Arthritis and Related Inflammatory Diseases*, 2019; pp 457–479.

Perez-Vizcaino, F.; Fraga, C. G. Research Trends in Flavonoids and Health. *Arch. Biochem. Biophys.* **2018,** *646*, 107–112.

Prasanth, B.; Soman, A.; Jobin, J.; Narayanan, P. S.; John, A. P. Plants and Phytoconstituents having Sunscreen Activity. *World J. Curr. Med. Pharm. Res.* **2020,** 14–20.

Rathee, P.; Chaudhary, H.; Rathee, S.; Rathee, D.; Kumar, V.; Kohli, K. Mechanism of Action of Flavonoids as Anti-inflammatory Agents: A Review. In *Inflammation & Allergy-Drug Targets (Formerly Current Drug Targets-Inflammation & Allergy)(Discontinued)*, 2009; vol *8* (3), pp 229–235.

Ruiz-Cruz, S.; Chaparro-Hernández, S.; Hernández-Ruiz, K. L.; Cira-Chávez, L. A.; Estrada-Alvarado, M. I.; Ortega, L. E. G.; Mata, M. A. L. Flavonoids: Important Biocompounds in Food. In *Flavonoids: From Biosynthesis to Human Health*; Justino, J. G., Ed.; IntechOpen: London, UK, 2017; pp 353–369.

Saewan, N.; Jimtaisong, A. Natural Products as Photoprotection. *J. Cosmet. Dermatol.* **2015,** *14* (1), 47–63.

Sampaio, P. A.; ValenÃ, E. C.; Silva, F. S.; da Silva Almeida, J. R. G.; Rolim, L. A.; Nunes, X. P.; da Cruz AraÃºjo, E. C. Flavonoids as Photoprotective Agents: A Systematic Review. *J. Med. Plants Res.* **2016,** *10* (47), 848–864.

Serafini, M.; Peluso, I.; Raguzzini, A. Flavonoids as Anti-inflammatory Agents. *Proc. Nutr. Soc.* **2010,** *69* (3), 273–278.

Singh, R. P.; Agarwal, R. Flavonoid Antioxidant Silymarin and Skin Cancer. *Antioxid. Redox Signal.* **2002,** *4* (4), 655–663.

Skotarczak, K.; Osmola-Mankowska, A.; Lodyga, M.; Polanska, A.; Mazur, M.; Adamski, Z. Photoprotection: Facts and Controversies. *Eur. Rev. Med. Pharmacol. Sci.* **2015,** *19* (1), 98–112.

Stahl, W. Flavonoid-Rich Nutrients for the Skin. In *Nutrition for Healthy Skin*; Springer: Berlin, Heidelberg, 2010; pp 85–90.

Stevanato, R.; Bertelle, M.; Fabris, S. Photoprotective Characteristics of Natural Antioxidant Polyphenols. *Regul. Toxicol. Pharmacol.* **2014,** *69* (1), 71–77.

Takahashi, T.; Kokubo, R.; Sakaino, M. Antimicrobial Activities of Eucalyptus Leaf Extracts and Flavonoids from Eucalyptus Maculata. *Lett. Appl. Microbiol.* **2004,** *39* (1), 60–64.

Tapas, A. R.; Sakarkar, D. M.; Kakde, R. B. Flavonoids as Nutraceuticals: A Review. *Trop. J. Pharm. Res.* **2008,** *7* (3), 1089–1099.

Ullah, A.; Munir, S.; Badshah, S. L.; Khan, N.; Ghani, L.; Poulson, B. G.; Emwas, A. H.; Jaremko, M. Important Flavonoids and their Role as a Therapeutic Agent. *Molecules* **2020,** *25* (22), 5243.

Verschooten, L. Photoprotective and Therapeutic Effects of the Flavonoid Luteolin in Human Keratinocytes, 2012.

Verschooten, L.; Claerhout, S.; Van Laethem, A.; Agostinis, P.; Garmyn, M. New Strategies of Photoprotection. *Photochem. Photobiol.* **2006,** *82* (4), 1016–1023.

Wang, H. K. The Therapeutic Potential of Flavonoids. *Expert Opin. Investig. Drugs* **2000,** *9* (9), 2103–2119.

Wei, H.; Saladi, R.; Lu, Y.; Wang, Y.; Palep, S. R.; Moore, J.; Phelps, R.; Shyong, E.; Lebwohl, M. G. Isoflavone Genistein: Photoprotection and Clinical Implications in Dermatology. *J. Nutr.* **2003,** *133* (11), 3811S–3819S.

Xiao, Z. P.; Peng, Z. Y.; Peng, M. J.; Yan, W. B.; Ouyang, Y. Z.; Zhu, H. L. Flavonoids Health Benefits and their Molecular Mechanism. *Mini Rev. Med. Chem.* **2011,** *11* (2), 169–177.

Yakoub, A. R. B.; Abdehedi, O.; Jridi, M.; Elfalleh, W.; Nasri, M.; Ferchichi, A. Flavonoids, Phenols, Antioxidant, and Antimicrobial Activities in Various Extracts from Tossa Jute Leave (Corchorus olitorus L.). *Ind. Crops Prod.* **2018,** *118*, 206–213.

Yao, L. H.; Jiang, Y. M.; Shi, J.; Tomas-Barberan, F. A.; Datta, N.; Singanusong, R.; Chen, S. S. Flavonoids in Food and their Health Benefits. *Plant Foods Human Nutr.* **2004,** *59* (3), 113–122.

PART IV

Flavonoids in Agriculture

CHAPTER 13

FLAVONOIDS: A POTENTIAL FOR SUSTAINABLE AGRICULTURE

CHANDRA BISWAS[1] and CHINMOYEE BHATTACHARYYA[2]

[1]*Laboratory of Cancer Biology, National Centre for Cell Sciences, Pune, Maharashtra, India*

[2]*Department of Life Sciences, Dibrugarh University, Dibrugarh, Assam, India*

ABSTRACT

Flavonoids are a large group of polyhydroxy flavonoid derivatives synthesized widely in plants. Their health benefits have been documented for more than 150 years, with the first published article on the subject appearing in 1870. Throughout this time, there has been an ongoing debate on how they work biologically. There are over 4000 flavonoids discovered to date, with the vast majority of these being dietary products found in fruits, vegetables, and herbs. Only a small fraction of them has been shown to be pharmacologic. In plants, flavonoids are capable of preventing oxidative reactions in plant tissues and providing protection against UV radiation, drought, and salt stress. They also play a significant role in plant–microbe interactions, pollination, and in prevention of pests and other microorganisms. Majority of Indian population lives on agriculture as the country is blessed with fertile land and suitable climate for growing crops. Though this field gains a generous amount of income, but every year it faces a huge loss due to drought, attack of pests, weeds, and diseases, and to mitigate these, synthetic chemicals are used that can have adverse side effects in the long run. Finding a natural alternative strategy for preventing biotic and abiotic stresses in

The Flavonoids: Extraction and Applications. Deepika Saini, Rajesh K. Kesharwani, & Raj K. Keservani (Eds.)

paddy fields is a demand of time. Flavonoids can be a very effective potential candidate for finding a natural way. Using this versatile low molecular weight chemical for environment-friendly and sustainable agriculture is of utmost concern as the synthetic chemical fertilizers/pesticides are not only expensive but also detrimental for the environment and human health. This chapter aims to discuss the role played by flavonoids in plants and its potential use in agriculture.

13.1 INTRODUCTION

Flavonoids are a phenylpropanoid-derived product of the secondary metabolism of plants, which are capable of performing many protective functions, such as, they act as allelochemicals, antiherbivory, and antimicrobial factors, helping in symbiosis and pollination, and so on (Dixon and Pasinetti, 2020). There are diverse groups of flavonoids based on where the aromatic ring is attached; among which flavonoids and flavones are most commonly found in plants, while distribution of others such as flavanols, flavanones, dihydrochalcones, and dihydroflavones is limited. In plant tissues, flavonoids are mostly found in free form as aglycones or in conjugation with sugar as glycosides (Di Carlo et al., 1999). These compounds are found in almost all vascular as well as nonvascular plants and are also referred to as nutraceuticals, as they possess nutritional and medicinal values (Palma-Tenango et al., 2021). These versatile low molecular weight compounds play a crucial role in mitigating biotic and abiotic stresses in plants, thereby providing protection against UV radiation, drought, salt stress, and reactive oxygen species (Shah and Smith, 2020). They are also known to take part in mycorrhizal symbiosis and plant–microbe interactions enhancing nutrients availability and supply to the plant (Cesco et al., 2012). Additionally, some flavonoids also help in combating attack of pests and other pathogens (Mierziak et al., 2014). As reported, the activity of flavonoids, whether it will be stimulatory or inhibitory, depends on the concentration of these compounds in soil or water solution in the rhizosphere (Weston and Mathesius, 2013).

Agriculture is one of the major contributors to the economy of a developing country like India, where almost 70% of the population depends on it for their livelihood. The geographic position and other physical factors such as climate and soil fertility have attributed toward the development of agriculture in the country. Though this field has a generous amount of income, as per the reports, India faces a loss of almost 50,000 crores due to diseases and attack of pests. Agrochemicals, mostly chemical fertilizers and pesticides, are being used to mitigate the pest invasions and increase productivity in this field at present time. The "green revolution" of the 1960s was the result of increasing food production from the same surface of land due to intensive use of chemical fertilizers containing nitrogen, potassium, and phosphorus and intensive irrigation (Borlaugh and Dowswell, 1993). Developing countries in the tropical belt of the world like India who has the industrial capacity of producing chemicals, invest on cheaper products like DDT and HCHs (Chernyak et al., 1996). But, use of these chemical pesticides and fertilizers has adverse effects on both the components of the environment, biotic as well as abiotic components. The residues of these chemical compounds pollute soil surface/water bodies and finally enter the human bodies with food and water causing severe health issues (Barcelo and Hennion, 1997).

The concepts of allelopathy and use of metabolomics are a few relatively new tools that can be used to develop an alternative to the conventional chemical fertilizers and pesticides used in farming. Allelopathy is defined as the effect of secondary metabolites produced by a plant on another plant species, which can be either direct or indirect (Weston and Mathesius, 2013). The concept of allelopathy can be used to enhance crop rotation strategies, as a natural remedy for weeds and pests and also as a natural agent for reducing stress and diseases. Scientists have identified a significant number of flavonoids from many bioactive root exudates and also have discussed the function of flavonoids in allelopathy and its involvement in soil interaction (Cho and Lee, 2015; Farooq et al., 2013). The use of metabolomics to study the contents of flavonoids can also prove beneficial for agriculture, as flavonoids play a great role in eliminating stress factors and protecting against the harmful effects of oxidative processes on biological plant physiology (Palma-Tenango et al., 2017). This chapter aims to discuss the biological roles of flavonoids in plants and their application to improve the agricultural sector using natural products while reducing the health hazards and environmental stresses due to the use of conventional synthetic chemicals.

13.2 CLASSIFICATION OF FLAVONOIDS WITH STRUCTURE

The basic structure of flavonoids is three phenolic rings, six carbons containing A and B rings attached to a 3-carbon containing C-ring, from which many subclasses and derivatives are formed by substituting the basic structure (Aherne and O'Brien, 2002). Based on where B-ring attached to C-ring, flavonoids can be subgrouped into three major categories: (1) 2-phenylbenzopyrans commonly known as flavonoids, in which B is linked to second position of C, (2) 3-benzopyrans commonly called isoflavonoids, where B is linked to third position of C, and (3) 4-benzopyrans aka neoflavonoids, where B is linked to fourth position of C (Panche et al., 2016; Samanta et al., 2011).

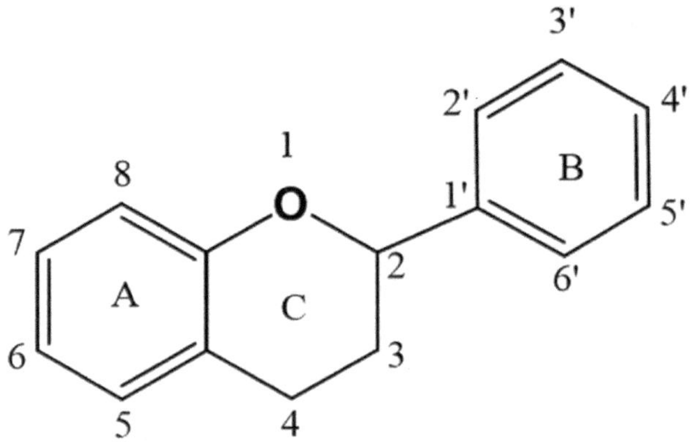

FIGURE 13.1 General structure of flavonoids.

Source: Adapted from Panche et al., 2016 and Samanta et al., 2011.

13.3 BIOLOGICAL FUNCTIONS OF FLAVONOIDS IN PLANTS

Flavonoids are the most disparate group of natural products, with over 9000 different phenolic compounds discovered in plants (Wang et al., 2011). Flavonols, flavanones, flavanonols, flavones, flavanols, isoflavonoids, chalcones, and anthocyanins are the major subgroups of flavonoids (Panche et al., 2016).

Their biological roles are vast in plants starting from crop yield to senescence.

- ***Crop yield and plant growth***

Agricultural yield of crops is directly associated with the ability of the soil to generate a good yield. We determine the soil quality by analyzing physiochemical factors like pH, total nitrogen content, WHC%, iron, organic matter, available P, exchangeable K, etc. Additionally, there are millions of microbes which may be beneficial or disadvantageous to the growth of the plants. Based on these parameters, crop yield is estimated. Plant growth and

TABLE 13.1 Classification of Flavonoids with Structure.

Name	Structure	Source	Sub classes with structure	
Flavonols	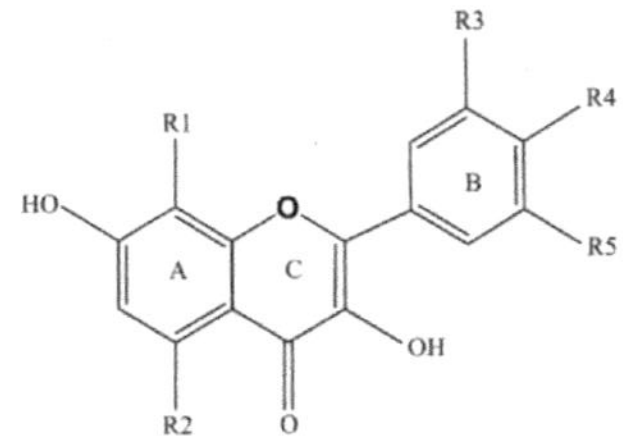	Fruits like strawberry and apple Vegetables such as onion and other green leafy vegetables Green and black tea and red wine	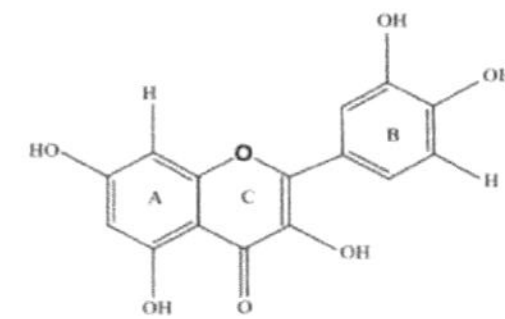 Quercetins	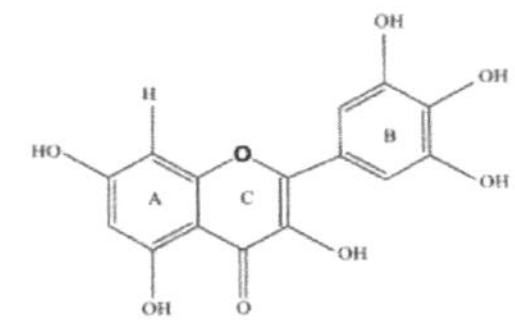 Myricetins
			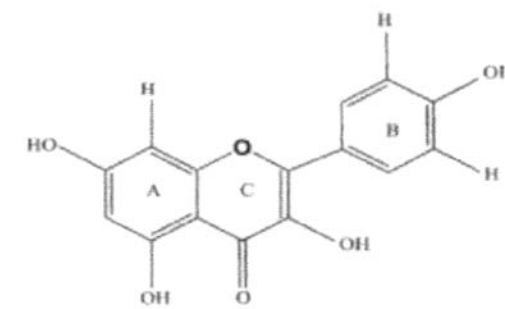 Kaempferols	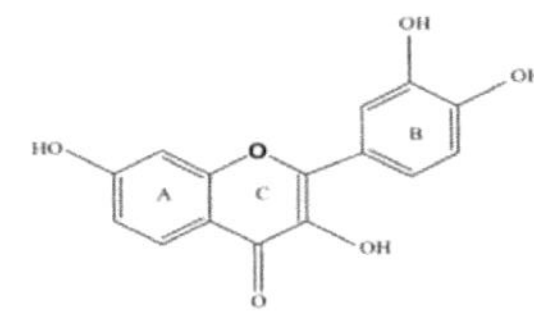 Fisetins
Flavones	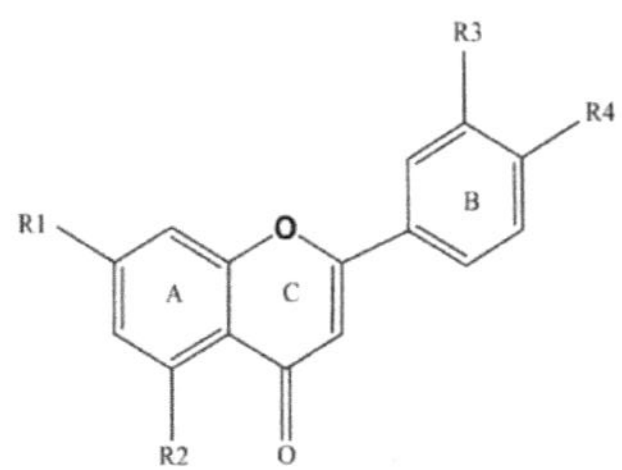	Fruits and vegetables like carrot and celery Herbs like parsley, peppermint Olive oil and pepper	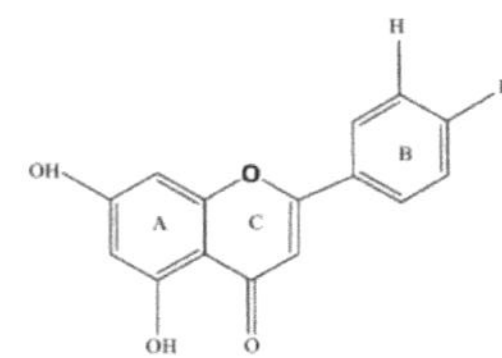 Chrysin	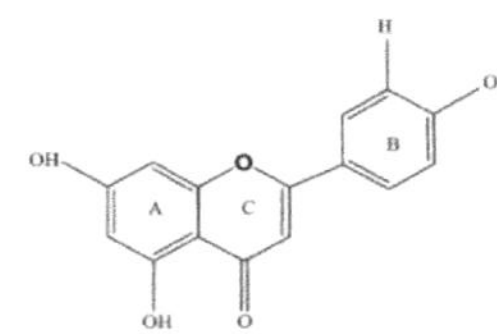 Apigenin
			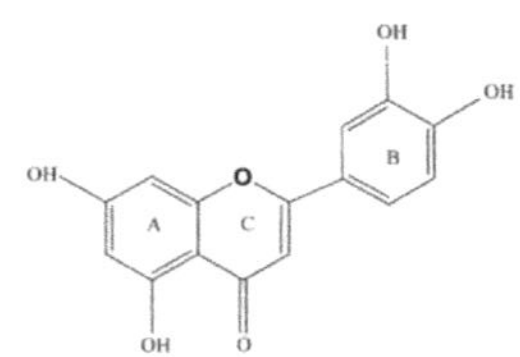 Luteolin	

TABLE 13.1 *(Continued)*

Name	Structure	Source	Sub classes with structure	
Anthocyanidins	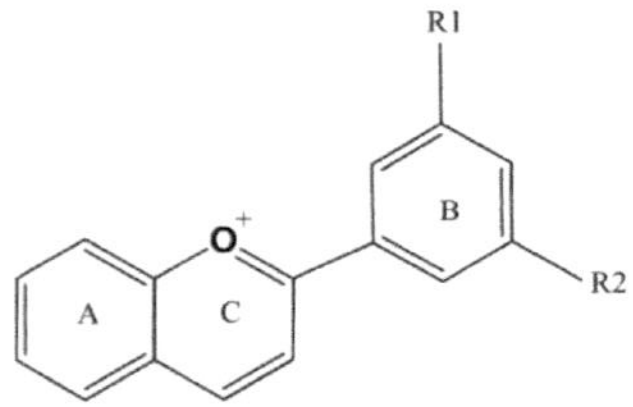	Colored part of many fruits such as grapes, berries, and vegetables	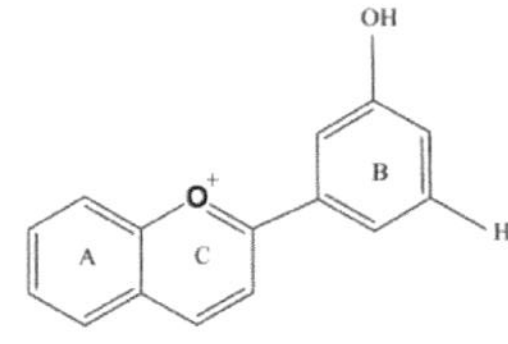 Cyanidin	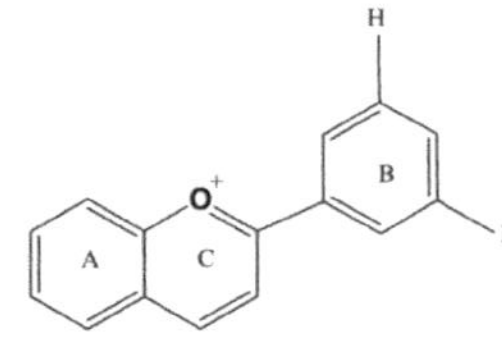 Pelargonidin
			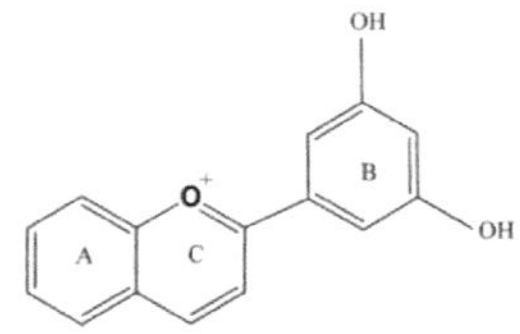 Delphinidin	
Flavanols	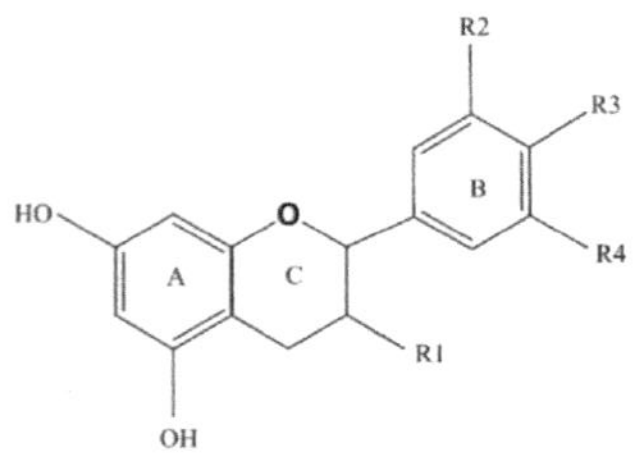	Grapes, apple, peach, wine and tea	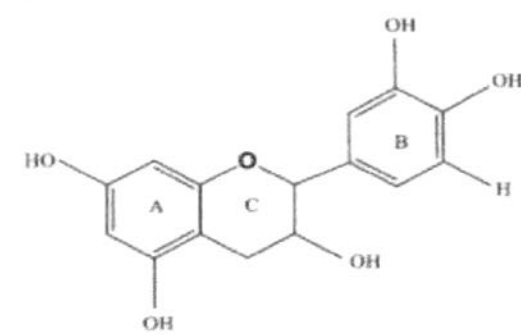 (+) Catechin	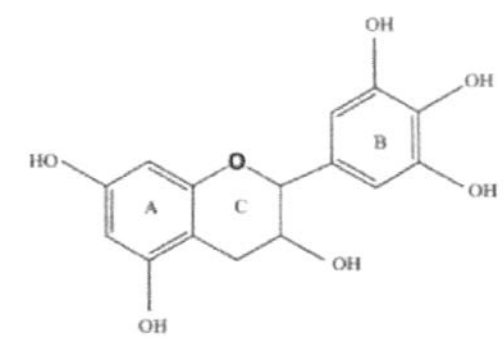 (−) Epigallocatechin
Chalcones	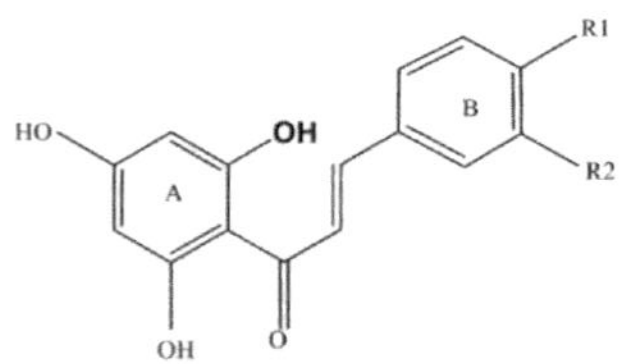	Berries, tomato, apple, wheat, and beer	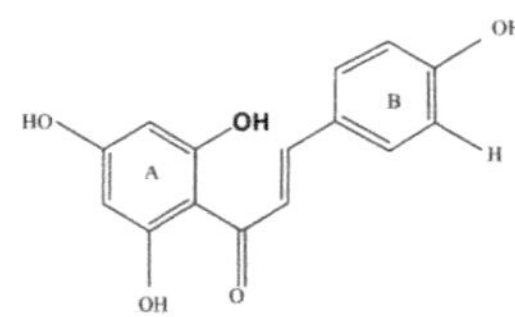 Naringenin chalcone	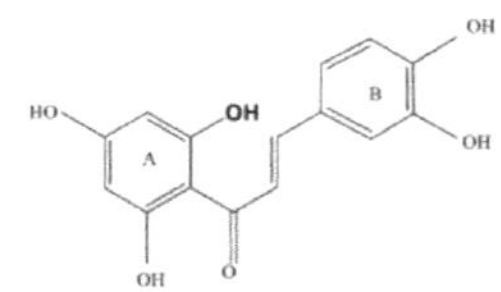 Eriodictyol chalcone

TABLE 13.1 *(Continued)*

Name	Structure	Source	Sub classes with structure
Flavanones	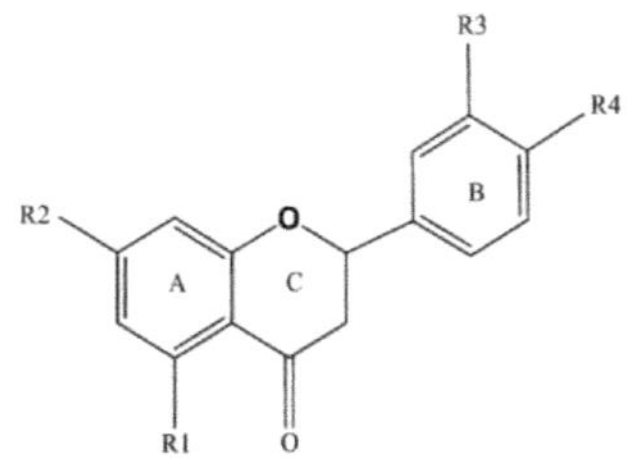	Hard tissues of citrus fruits like lemon, orange, and grape	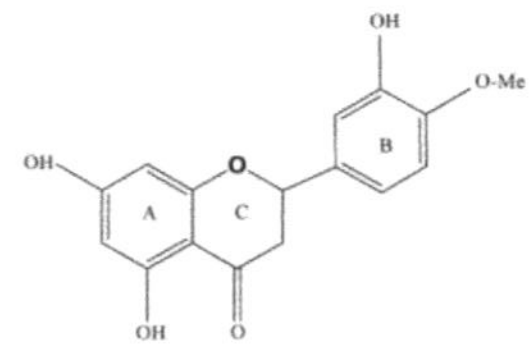 Hesperidin 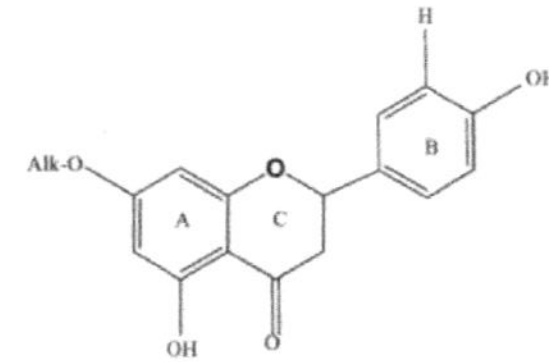 Naringin 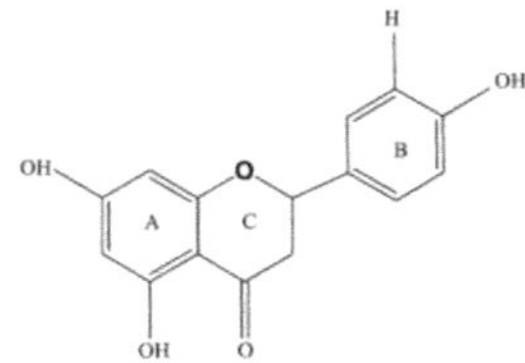 Naringenin
Isoflavonoids	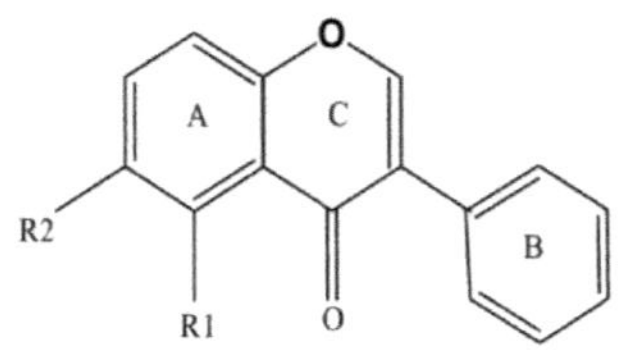	Legumes like soybean	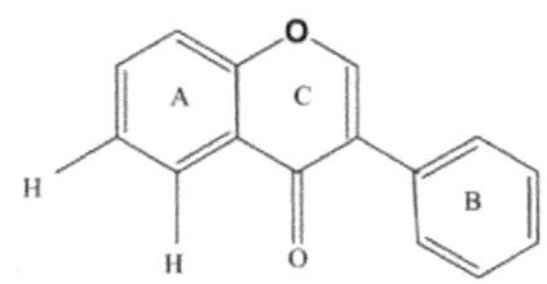 Daidzein 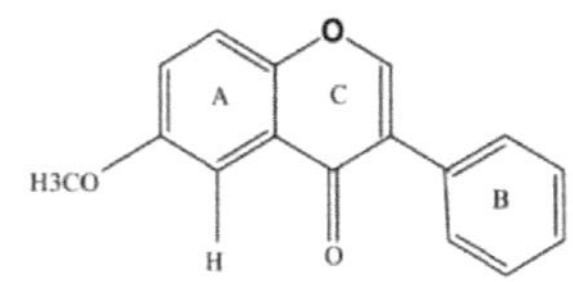 Glycitein 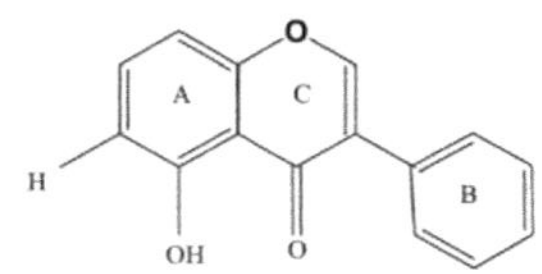 Genistein

Source: Adapted from Sultana et al., 2008; LópezLázaro, 2009; Heim et al., 2002; Erlund, 2004; Andersen and Markham, 2005; Tsao, 2010; Shah and Smith, 2020.

development are aided by flavonoids in both direct and indirect ways. The generation of certain flavonoids and their accumulation in specific plant tissues as a result of external or internal stimulus remain largely unknown. Flavonoids, on the other hand, are particularly efficient in specific microbe–plant interactions in the rhizosphere and can help with biotic and abiotic stress (Shah and Smith, 2020).

- ***Legume–rhizobium interaction***

Soil microorganisms, like rhizobia, interact with leguminous plants forming root nodules, which provide suitable condition for bacterial nitrogen fixation. This symbiotic relationship results in the conversion of atmospheric nitrogen (N_2) to ammonia that may be used by the plant. In order to initiate the symbiotic process, plants release chemoattractants in the form of flavonoids, which are particularly effective in the root tips, where attachment occurs (Shah and Smith, 2020; Hasan et al., 2012). Activation of nod genes in rhizobia is stimulated by these metabolites. In plant–microbe interactions, isoflavones are thought to be highly active and effective, and they're especially good at signaling and promoting nodulation through triggering nod gene systems. The flavonoids quercetin and luteolin, which induce the nod gene, were extracted from the roots and seeds of *Medicago sativa* L. Many flavonoids were identified to trigger nod genes in *Rhizobium melioti*, but not all of them, demonstrating that rhizobia is sensitive to specific flavonoids signals (Hatwig et al., 1991).

- ***UV scavengers***

Plants are resistant to UV radiation because flavonoids (anthocyanin) filter UVB by absorbing it and detoxify the reactive oxygen species (ROS) processed by photochemical reactions (Chen et al., 2015). When UVB radiation was used to investigate the impact of UV light on flavonoid content in the leaves of barley, it was demonstrated in the results that there were considerable increases in flavonoid (saponarin and lutonarin) concentrations when compared to control circumstances (absence of UVB) (Schmitz-Hoerner and Weissenbock, 2003). Furthermore, by absorbing specific wavelengths of UVB radiation, enhanced flavonoid assembling in response to UVB radiation may minimize impairment in exposed leaves. Plants grown in open-air circumstances, where they are open to full sunshine, have higher flavonoid content than those grown in greenhouses (Shah and Smith, 2020).

- ***Seed development and growth***

Seed development is a multifaceted operation that includes biochemical, morphological, and physiological changes. Flavonoids called anthocyanins have been found in the roots, blooms, shoots, stems, and leaves of tartary buckwheat (Peng et al., 2015; Park et al., 2011). Anthocyanins' antioxidant activity is proportional to their concentration (Watanabe, 2007; Gabrielska et al., 1999). The different concentrations of anthocyanin portions in the pericarp at different developmental stages of seed were determined in an experiment. Flavonoids including kaempferol and quercetin were also plentiful in flour. These findings suggest that immature seeds could be a valuable source of nutrient-dense plant material with therapeutic benefit (Song et al., 2016).

- ***Pollen tube germination***

Flavonols were studied in vitro to see how they affect tobacco pollen formation. The existence of flavonols in pollen that has matured appears to be a major factor in the influence of pollen diffusates on pollen germination and tube growth. Low molecular weight flavonoids have also been shown to advance pollen germination and tube growth. Small molecule elimination by dialysis, on the other hand, only slightly diminished the diffusate's germination-promoting impact. This could mean that high molecular weight molecules, like lipids and proteins, are also playing a role. It's possible that they work by binding the flavonols (YIstra et al., 1992).

13.3.1 USES OF FLAVONOIDS IN AGRICULTURE

- ***Flavonoids and biofertilizers***

Biofertilizers are living microorganisms, which live in the soil and interact with the roots of the plant to increase the nutrients availability and promote growth (Riaz et al., 2020). Growth and production of a particular crop directly depend on the nutrient's availability, the soil's fertility, and the phytomicrobiome of that particular land. The symbiotic microbes present in the phytomicrobiome aid the plant in disease resistance, prevention of stress, and nutrients availability. Secondary metabolites, particularly flavonoids, play a crucial role in plant–microbe interactions by several complex signal cascades in the rhizosphere (Shah and Smith, 2020).

Nitrogen fixation, a process to fix the atmospheric nitrogen as ammonia to make available for the plant, is done by the nitrogen-fixing microorganisms and the process is initiated by specific signaling molecules such as flavonoids and isoflavonoids exuded from the root tips of the legumes (Subramanian et al., 2007). Scientists have reported that flavonoids aid in plant–mycorrhizal symbiosis by stimulating the hyphal growth, germination of fungal spores, and colonization of roots, which enhance the availability of phosphorus, the second most essential molecule for growth and productivity of plant (Vierheilig et al., 1998; Shah and Smith, 2020). Becard et al., in their paper, reported that though flavonoids are not a compulsory signaling compound in arbuscular mycorrhizal symbiotic association, but if present they influence AM development to a great extent, which could be beneficial for proper growth and development of the crops (Becard et al., 1995).

Isoflavonoids are known to induce nodule formation in crops. Genistein, an isoflavonoid, together with *Bradyrhizobium*, when tested under salt stress, resulted in increased nitrogen fixation, photosynthesis, and nodulation both in saline as well as nonsaline environment (Subramanian et al., 2006). Further studies revealed that treatment of *Bradyrhizobium japonicum* preincubated with genistein increased the yield by 16% and the protein content by 70% in soybean (Belkheir et al., 2001). Formononetin, an AM-stimulating flavonoid, when tested under low phosphorus level resulted in increased phosphorus use efficiency, tuber development, plant dry matter, and also soil sporulation (Davies et al., 2005). Thus, exploiting flavonoids and their relationship with the phytomicrobiome can be proved beneficial for the development in the field of agriculture.

- ***Flavonoids and pest/disease resistance***

One of the major limiting factors for agriculture is invasion by insects and other pests, such as parasitic nematodes, fungi, and bacteria, which cause diseases in the field leading to major economic losses. Though use of chemical pesticides, fungicides, and insect repellents can reduce the amount of infestation, but these synthetic chemicals are expensive, not environment friendly, and also cause several health hazards to human when come in contact (Dakora, 1995). Flavonoids can serve as a good environmentally friendly alternative to these synthetic chemicals, which will not only help mitigate diseases but will also not be harmful to humans.

Flavonoids act as a protective chemical when plants are under the attack of parasitic nematodes and are synthesized in the roots. Lima bean roots secrete an isoflavonoid, coumestrol that acts as a phytoalexin when *Pratylenchus penetrans* invades the plant. Studies showed that soybean root secretes a protective, resistance-inducing isoflavonoid, glyceollin, in response to the attack of *Meloidogyne incognita* (Kaplan et al., 1980) and *Heterodera glycines* (Huang and Barker, 1991), resulting in decrease in the mobility of the nematodes and O_2 uptake. Oats, when infected with *H. avenae* and *P. neglectus*, release a glycoside, namely O-methyl-apigenin-C-deoxyhexoside-O-hexoside that reduces the attack by parasitic nematodes (Soriano et al., 2004).

Again, flavonoids exhibit antifungal and antibacterial activities, and provide resistance against many destructive fungal and bacterial pathogens. *Phytophthora citrophthora* is a fungal pathogen of citrus plant against which accumulation of several derivatives of flavonoids such as nobiletin, heptamethoxyflavone, sinensetin, tangeretin, hesperetin, and naringenin was confirmed, among which the last two were most effective (del Río et al., 2004). Studies have revealed growth inhibition of *Pseudomonas mors-prunorum* and *Pseudomonas phaseolicola* in infected bean leaves by accumulation of antibacterial phytoalexin isoflavonoid coumestrol in response to the infection (Lyon and Wood, 1975). Several other studies also showed that derivatives of flavonoids can be effective against a variety of bacterial pathogens of soybean. Further, flavonoids are also considered as effective insect repellents. Flavonoids hinder the process of molting as well as reproduction in many insects by inhibiting ecdysone (the juvenile hormone), thereby preventing the pest infestation. Rotenone is one of the most effective isoflavonoids that can be used as a basis for commercial insect repellents (Shah and Smith, 2020). Most of the legumes use these defense mechanisms to control pathogens including bacteria, fungus, as well as a few leaf-eating insects (Dakora, 1995). Thus, exploiting the disease resistance and insect repellent capacities of different groups of flavonoids would definitely be helpful to implement environmentally friendly preventive measures in agriculture.

- ***Flavonoids and weed control***

Controlling weeds is another most challenging task in agriculture, as they grow rapidly hindering the growth of the crop. Weeds are unwanted plant species that compete for food, water, and light with the crops, and for controlling them, chemically synthesized weedicides are used. Though the chemical herbicides are very effective, they are not environmentally friendly as they leave residues on the surface, which can adversely affect the ecosystem and also have bad impact on human health. Further, weeds might develop resistance against the commercial herbicides in the long run. So, to devise a natural way to control the weeds, the concept of allelopathy had been studied (Shah and Smith, 2020). Several plants like sunflower, rice, wheat, barley, rye, sugarcane, and sorghum possess allelopathic nature and suppress weeds through release of allelochemicals, including flavonoids. Studies have shown that sunflower cultivars have suppressive effect on weeds of wheat crops when cultivated together or applied over the weed and the crops (Alsaadawi et al., 2012). Recently, a study showed that among the five phenolic compounds found in the extract of Tunisian plants, three were flavonoids which have showed significant herbicidal activity against *Trifolium incarnatum* by preventing seed germination and growth of the saplings (Kaab et al., 2020). Several other findings have also revealed that flavonoids have the potential to be a natural, eco-friendly alternative to synthetic chemical herbicides with no or very little amount of side effects.

- ***Flavonoids and mitigation of abiotic stress***

Plants are exposed to many unfavorable conditions due to changes in their environment and hence to sustain these harmful situations, plants possess defense mechanisms, amongst them flavonoids are one of the important groups of chemicals, which help in reducing the effects of abiotic stresses. Flavonoids help to sustain many adverse conditions like UV radiation, salt stress, drought, etc. (Shah and Smith, 2020). Flavonoids act as detoxifying agent (antioxidant) in plants by destroying the ROS and H_2O_2 in chloroplasts, thereby protecting the tissues from oxidative damage (Yamasaki et al., 1997). It has been reported that in cell culture and callus cultures of *Ginkgo biloba* and *Ononis arvensis*, respectively, after treatment with $CuSO_4$, there was a significant increase in the flavonoid content in response to heavy metal stress (Keilig and Muller, 2009). In an experiment with two different varieties of apples, Granny Smith variety which is anthocyanin free and Braeburn, it was found that when they were exposed to UV radiation, anthocyanin-free variety showed decrease in carotenoid and chlorophyll contents, whereas Braeburn showed significant increase in these two contents. The compound responsible for absorbing the UV radiation was found to be quercetin glycoside (Solovchenko et al., 2003). Barley leaves, when tested against UV radiation showed increase in accumulation of flavonoid content, which reduced the damage in the leaves by absorbing the UVB radiation (Tossi et al., 2012). Plants are also equipped with defense mechanisms against high salt concentration in its surrounding environment as well as against adverse temperature changes by activating antioxidants like flavonoids, carotenoids, etc. (Shah and Smith, 2020).

Other than these, there are many beneficial effects of flavonoids in plant, which could be used for developing environment-friendly mitigation strategies in the field of agriculture. Scientific studies along with commercial use of flavonoids are much needed for sustainable agriculture in the near future.

13.4 CONCLUSION

Despite current understanding, flavonoids are used in agriculture in a fairly limited way. Flavonoids are important not just for their roles in plants but also for their applications in medicine and nutraceuticals. They are a family found in various concentrations in a wide range of plant species. Flavonoids could be used to reduce stress in an environmentally friendly and long-term way. Flavonoids' phytotoxic and pesticidal properties provide apprehension about how successful these biochemicals could be in the field if used properly. Metabolomics is a useful method for studying flavonoids' metabolic activities in diverse agricultural crops. In integrated weed management, the use of bioflavonoids as natural herbicides is gaining popularity. To fully comprehend the whole extent of activity of flavonoids produced naturally and/or applied artificially, more research and investigations are required.

KEYWORDS

- **flavonoids**
- **sustainable agriculture**
- **antioxidants**
- **metabolomics**
- **biotic and abiotic stress**
- **mitigation**
- **natural alternative**
- **phenolic compounds**

REFERENCES

Aherne, S.; O'Brien, N. Dietary Flavonols: Chemistry, Food Content, and Metabolism. *Nutrition* **2002,** *18* (1), 75–81.

Alsaadawi, I. S.; Sarbout, A. K.; Al-Shamma, L. M. Differential Allelopathic Potential of Sunflower (Helianthus annuus L.) Genotypes on Weeds and Wheat (Triticum aestivum L.) Crop. *Arch. Agron. Soil Sci.* **2012,** *58* (10), 1139–1148.

Andersen, O. M.; Markham, K. R. In *Flavonoids: Chemistry, Biochemistry and Applications*; CRC Press: Boca Raton, FL, USA, 2005.

Barcelo, D.; Hennion, M. C. Trace Determination of Pesticides and their Degradation Products in Water. In *Techniques and Instrumentation in Analytical Chemistry*; Elsevier: Amsterdam, 1997; vol 19, p 542.

Becard, G.; Taylor, L. P.; Douds, D. D.; Pfeffer, P. E.; Doner, L. W. Flavonoids are not Necessary Plant Signal Compounds in Arbuscular Mycorrhizal Symbioses. *MPMI Mol. Plant Microbe Interact.* **1995,** *8* (2), 252–258.

Belkheir, A. M.; Zhou, X.; Smith, D. L. Variability in Yield and Yield Component Responses to Genistein Pre-incubated Bradyrhizobium Japonicum by Soybean [Glycine max (L.) Merr] Cultivars. *Plant Soil* **2001,** *229* (1), 41–46.

Borlaugh, N.; Dowswell, C. R. In *Fertilizer: To Nourish Infertile Soil that Feeds a Fertile Population that Crowds a Fragile World*, Proceeding of the 61st International Fertilizer Association Annual Conference, 1993.

Cesco, S.; Mimmo, T.; Tonon, G.; Tomasi, N.; Pinton, R.;Terzano, R.; et al. Plant-Borne Flavonoids Released into the Rhizosphere: Impact on Soil Bio-activities Related to Plant Nutrition: A Review. *Biol. Fertil. Soils* **2012,** *48* (2), 123–149.

Chen, H.; Gao, H.; Fang, X.; Ye, L.; Zhou, Y.; Yang, H. Effects of Allyl Isothiocyanate Treatment on Postharvest Quality and the Activities of Antioxidant Enzymes of Mulberry Fruit. *Postharvest Biol. Technol.* **2015,** *108*, 61–67.

Chernyak, S. M.; Rice, C. P.; McConnel, L. L. Evidence of Currently-Used Pesticides in Air, Ice, Fog, Seawater and Surface Microlayer in the Bering and Chuckchi Seas. *Mar. Pollut. Bull.* **1996,** *32*, 410–419.

Cho, M.; Lee, S. Phenolic Phytoalexins in Rice: Biological Functions and Biosynthesis. *Int. J. Mol. Sci.* **2015,** *16* (12), 29120–29133.

Dakora, F. D. Plant Flavonoids: Biological Molecules for Useful Exploitation. *Funct. Plant Biol.* **1995,** *22* (1), 87–99.

Davies, F. T.; Calderón, C. M.; Huaman, Z. Influence of Arbuscular Mycorrhizae Indigenous to Peru and a Flavonoid on Growth, Yield, and Leaf Elemental Concentration of Yungay' Potatoes. *HortScience* **2005,** *40* (2), 381–385.

Del Río, J. A.; Gómez, P.; Baidez, A. G.; Arcas, M. C.; Botía, J. M.; Ortuño, A. Changes in the Levels of Polymethoxyflavones and Flavanones as Part of the Defense Mechanism of Citrus sinensis (cv. Valencia Late) Fruits against Phytophthora citrophthora. *J. Agric. Food Chem.* **2004,** *52* (7), 1913–1917.

Di Carlo, G.; Mascolo, N.; Izzo, A.; Capasso, F. Flavonoids: Old and New Aspects of a Class of Natural Therapeutic Drugs. *Life Sci.* **1999,** *65* (4), 337–353.

Dixon, R.; Pasinetti, G. Flavonoids and Isoflavonoids: From Plant Biology to Agriculture and Neuroscience. *Plant Physiol.* **2010,** *154* (2), 453–457.

Erlund, I. Review of the Flavonoids Quercetin, Hesperetin, and Naringenin. Dietary Sources, Bioactivities, Bioavailability, and Epidemiology. *Nutr. Res.* **2004,** *24* (10), 851–874.

Farooq, M.; Bajwa, A. A.; Cheema, S. A.; Cheema, Z. A. Application of Allelopathy in Crop Production. *Int. J. Agric. Biol.* **2013,** *15*, 1367–1378.

Gabrielska, J.; Oszmiański, J.; Komorowska, M.; Langner, M. Anthocyanin Extracts with Antioxidant and Radical Scavenging Effect. *ZeitschriftfürNaturforschung C* **1999,** *54*, 319–324.

Hartwig, U. A.; Joseph, C. M.; Phillips, D. A. Flavonoids Released Naturally from Alfalfa Seeds Enhance Growth Rate of Rhizobium Meliloti. *Plant Physiol.* **1991,** *95*, 797–803.

Hassan, S.; Mathesius, U. The Role of Flavonoids in Root-Rhizosphere Signalling: Opportunities and Challenges for Improving Plant-Microbe Interactions. *J. Exp. Bot.* **2012,** *63*, 3429–3444.

Heim, K.; Tagliaferro, A.; Bobilya, D. Flavonoid Antioxidants: Chemistry, Metabolism and Structure-Activity Relationships. *J. Nutr. Biochem.* **2002,** *13* (10), 572–584.

Huang, J. S.; Barker, K. R. Glyceollin I in Soybean-Cyst Nematode Interactions: Spatial and Temporal Distribution in Roots of Resistant and Susceptible Soybeans. *Plant Physiol.* **1991,** *96* (4), 1302–1307.

Kaab, S. B.; Rebey, I. B.; Hanafi, M.; Hammi, K. M.; Smaoui, A.; Fauconnier, M. L.; ... Ksouri, R. Screening of Tunisian Plant Extracts for Herbicidal Activity and Formulation of a Bioherbicide Based on Cynara cardunculus. *S. Afr. J. Bot.* **2020,** *128*, 67–76.

Kaplan, D. T.; Keen, N. T.; Thomason, I. J. Studies on the Mode of Action of Glyceollin in Soybean Incompatibility to the Root Knot Nematode, Meloidogyne incognita. *Physiol. Plant Pathol.* **1980,** *16* (3), 319–325.

Keilig, K.; Ludwig-Mueller, J. Effect of Flavonoids on Heavy Metal Tolerance in Arabidopsis thaliana Seedlings. *Bot. Stud.* **2009,** *50* (3), 311–318.

Lopez-Lazaro, M. Distribution and Biological Activities of the Flavonoid Luteolin. *Mini Rev. Med. Chem.* **2009,** *9* (1), 31–59.

Lyon, F. M.; Wood, R. K. S. Production of Phaseollin, Coumestrol and Related Compounds in Bean Leaves Inoculated with Pseudomonas spp. *Physiol. Plant Pathol.* **1975,** *6* (2), 117–124.

Mierziak, J.; Kostyn, K.; Kulma, A. Flavonoids as Important Molecules of Plant Interactions with the Environment. *Molecules* **2014,** *19* (10), 16240–16265.

Palma-Tenango, M.; Soto-Hernández, M.; Aguirre-Hernández, E. In *Flavonoids in Agriculture*, 2021.

Panche, A.; Diwan, A.; Chandra, S. Flavonoids: An Overview. *J. Nutr. Sci.* **2016,** *5* (e47).

Park, N. I.; Li, X.; Suzuki, T.; Kim, S. J.; Woo, S. H.; Park, C. H.; Park, S. U. Differential Expression of Anthocyanin Biosynthetic Genes and Anthocyanin Accumulation in Tartary Buckwheat Cultivars 'Hokkai T8'and 'Hokkai T10'. *J. Agric. Food Chem.* **2011,** *59*, 2356–2361.

Peng, L. X.; Zou, L.; Su, Y. M.; Fan, Y.; Zhao, G. Effects of Light on Growth, Levels of Anthocyanin, Concentration of Metabolites in *Fagopyrum tataricum* Sprout Cultures. *Int. J. Food Sci. Technol.* **2015,** *50*, 1382–1389.

Riaz, U.; Mehdi, S. M.; Iqbal, S.; Khalid, H. I.; Qadir, A. A.; Anum, W.; et al. Bio-Fertilizers: Eco-friendly Approach for Plant and Soil Environment. In *Bioremediation and Biotechnology: Sustainable Approaches to Pollution Degradation*; Hakeem, K. R., Bhat, R. A., Qadri, H., Eds.; Springer: Cham, 2020; 188–214.

Samanta, A.; Das, G.; Das, S. K. Roles of Flavonoids in Plants. *Int. J. Pharm. Sci. Technol.* **2011,** *6*, 12–35.

Schmitz-Hoerner, R.; Weissenböck, G. Contribution of Phenolic Compounds to the UV-B Screening Capacity of Developing Barley Primary Leaves in Relation to DNA Damage and Repair under Elevated UV-B levels. *Phytochemistry* **2003,** *64*, 243–255.

Shah, A.; Smith, D. Flavonoids in Agriculture: Chemistry and Roles in, Biotic and Abiotic Stress Responses, and Microbial Associations. *Agronomy* **2020,** *10* (8), 1209.

Solovchenko, A.; Schmitz-Eiberger, M. Significance of Skin Flavonoids for UV-B-Protection in Apple Fruits. *J. Exp. Bot.* **2003,** *54* (389), 1977–1984.

Song, C.; Xiang, D. B.; Yan, L.; Song, Y.; Zhao, G.; Wang, Y. H.; Zhang, B. L. Changes in Seed Growth, Levels and Distribution of Flavonoids During Tartary Buckwheat Seed Development. *Plant Prod. Sci.* **2016,** *19* (4), 518–527.

Soriano, I. R.; Asenstorfer, R. E.; Schmidt, O.; Riley, I. T. Inducible Flavone in Oats (Avena sativa) is a Novel Defense against Plant-Parasitic Nematodes. *Phytopathology* **2004,** *94* (11), 1207–1214.

Subramanian, S.; Stacey, G.; Yu, O. Endogenous Isoflavones are Essential for the Establishment of Symbiosis between Soybean and Bradyrhizobium japonicum. *Plant J.* **2006,** *48* (2), 261–273.

Subramanian, S.; Stacey, G.; Yu, O. Distinct, Crucial Roles of Flavonoids during Legume Nodulation. *Trends Plant Sci.* **2007,** *12* (7), 282–285.

Sultana, B.; Anwar, F. Flavonols (Kaempeferol, Quercetin, Myricetin) Contents of Selected Fruits, Vegetables and Medicinal Plants. *Food Chem.* **2008,** *108* (3), 879–884.

Tossi, V.; Lombardo, C.; Cassia, R.; Lamattina, L. RETRACTED: Nitric oxide and flavonoids are systemically induced by UV-B in maize leaves. *Plant Sci.* **2012,** *193–194*, 103–109.

Tsao, R. Chemistry and Biochemistry of Dietary Polyphenols. *Nutrients* **2010,** *2* (12), 1231–1246.

Vierheilig, H.; Bago, B.; Albrecht, C.; Poulin, M. J.; Piché, Y. Flavonoids and Arbuscular-Mycorrhizal Fungi. In *Flavonoids in the Living System*, 1998; pp 9–33.

Wang, Y.; Chen, S.; Yu, O. Metabolic Engineering of Flavonoids in Plants and Microorganisms. *Appl. Microbiol. Biotechnol.* **2011,** *91*, 949–956.

Watanabe, M. An Anthocyanin Compound in Buckwheat Sprouts and its Contribution to Antioxidant Capacity. *Biosci. Biotechnol. Biochem.* **2007,** *71*, 579–582.

Weston, L.; Mathesius, U. Flavonoids: Their Structure, Biosynthesis and Role in the Rhizosphere, Including Allelopathy. *J. Chem. Ecol.* **2013,** *39* (2), 283–297.

Yamasaki, H.; Sakihama, Y.; Ikehara, N. Flavonoid-Peroxidase Reaction as a Detoxification Mechanism of Plant Cells against H2O2. *Plant Physiol.* **1997,** *115* (4), 1405–1412.

Ylstra, B.; Touraev, A.; Moreno, R. M. B.; Stöger, E.; Van Tunen, A. J.; Vicente, O.;... Heberle-Bors, E. Flavonols Stimulate Development, Germination, and Tube Growth of Tobacco Pollen. *Plant Physiol.* **1992,** *100* (2), 902–907.

INDEX

F

N

O

P

T

U

V

W

X